Y0-ABS-365

Phage and the Origins of Molecular Biology

The Cold Spring Harbor Symposia

I. (1933) Surface Phenomena
II. (1934) Aspects of Growth
III. (1935) Photochemical Reactions
IV. (1936) Excitation Phenomena
V. (1937) Internal Secretions
VI. (1938) Protein Chemistry
VII. (1939) Biological Oxidations
VIII. (1940) Permeability and the Nature of Cell Membranes
IX. (1941) Genes and Chromosomes: Structure and Organization
X. (1942) The Relation of Hormones to Development
XI. (1946) Heredity and Variation in Microorganisms
XII. (1947) Nucleic Acids and Nucleoproteins
XIII. (1948) Biological Applications of Tracer Elements
XIV. (1949) Amino Acids and Proteins
XV. (1950) Origin and Evolution of Man
XVI. (1951) Genes and Mutations
XVII. (1952) The Neuron
XVIII. (1953) Viruses
XIX. (1954) The Mammalian Fetus: Physiological Aspects of Development
XX. (1955) Population Genetics: The Nature and Causes of Genetic Variability in Population
XXI. (1956) Genetic Mechanisms: Structure and Function
XXII. (1957) Population Studies: Animal Ecology and Demography
XXIII. (1958) Exchange of Genetic Material: Mechanism and Consequences
XXIV. (1959) Genetics and Twentieth Century Darwinism
XXV. (1960) Biological Clocks
XXVI. (1961) Cellular Regulatory Mechanisms
XXVII. (1962) Basic Mechanisms in Animal Virus Biology
XXVIII. (1963) Synthesis and Structure of Macromolecules
XXIX. (1964) Human Genetics
XXX. (1965) Sensory Receptors
XXXI. (1966) The Genetic Code

Published by the Cold Spring Harbor Laboratory of Quantitative Biology, these volumes may be purchased by writing directly to:

Cold Spring Harbor Laboratory
Cold Spring Harbor, New York 11724

MAX DELBRÜCK, 1961

Phage and The Origins of Molecular Biology

Edited by

JOHN CAIRNS
Cold Spring Harbor Laboratory

GUNTHER S. STENT
University of California

JAMES D. WATSON
Harvard University

Cold Spring Harbor Laboratory of Quantitative Biology

1966

Library of Congress Catalog Card Number: 66-26455

Manufactured in the United States of America

All orders should be addressed to:
Cold Spring Harbor Laboratory of Quantitative Biology
Cold Spring Harbor, Long Island, New York 11724

DEDICATED BY THE AUTHORS TO

Max Delbrück

ON THE OCCASION OF HIS

SIXTIETH BIRTHDAY

Preface

This collection of stories, told by some biologists whose work and careers have been strongly influenced by Max Delbrück (or, in the case of a few, had influenced him), is dedicated to him on his 6oth birthday.

The reader will probably wonder why there should be a *Festschrift* in honor of a man still in his prime. Why not wait until he has joined the ranks of senior citizens whom custom favors as subjects for such commemoration? The answer lies in the exponentially increasing growth rate of science which has now reached such a pitch that one lifetime encompasses many scientific eras. Thus the singular contribution that Delbrück made some 25 years ago already seems to lie in a past so distant that the circumstances attending it are no longer easy to reconstruct. In another ten or twenty years, they might have been beyond recall. Besides paying homage to Delbrück as a prime mover and arbiter of nascent molecular biology, this book is an attempt therefore to write a history of a bygone age and put on record the network of interactions, folklore, and method of operation of the Phage Group that had Delbrück as its focal point.

Only one of the articles was not expressly written for the book, namely (and understandably) Delbrück's own 1949 essay, "A Physicist Looks at Biology." In this essay, a lecture given at the thousandth meeting of the Connecticut Academy of Arts and Sciences, Delbrück described the spirit of the 1930's and 1940's that moved many of the creators of molecular biology before the *dénouement* brought by the discovery of the structure of DNA in 1953. We are very grateful to the Connecticut Academy for permission to reprint it.

The reader will notice that not only the names of a few particular people but also the names of two scientific institutions keep recurring in the following pages—the California Institute of Technology and the Cold Spring Harbor Laboratory. These two places, at opposite ends of the American continent, were the Mecca and Medina of the Phage Group to which the faithful made their periodic *hadj* (though later the Institut Pasteur and the Cambridge Molecular Biology Laboratory were to cut into the pilgrim trade). Caltech, though not large by the standards of American universities, is a powerful and world-famous institution and was already renowned for its galaxy of outstanding physicists, chemists and biologists when Delbrück came there in the late 1930's. It was therefore a natural breeding ground for the ideas that were later to give the keys to the physical basis of heredity. But why Cold Spring Harbor, a small station of meager resources on the shores of Long Island Sound, should have played such an important role may not be as immediately obvious. It is fitting, therefore, to point out that this became possible through the imagination and enterprise of Milislav Demerec, who was Director at Cold Spring Harbor from 1941 until his retirement in 1960. Demerec was a prominent *Drosophila* geneticist when, in the early 1940's, he realized that bacteria and their viruses were likely to become the materials of choice for basic studies in genetics. He then not only moved his own work from classical to molecular genetics but also managed, at the same time, to conjure up the wherewithal for others to do likewise. For this the many contributors to this book whose scientific evolution took a decisive turn on a visit to Cold Spring Harbor are greatly in Dr. Demerec's debt. It seems singularly appropriate therefore that this book is being published by the laboratory he brought into prominence.

Although the essays presented here cover a period of some thirty years, most of them describe work done in the twenty-one years that have passed since Delbrück gave the first Phage Course at Cold Spring Harbor in 1945. Thus the book may be thought to commemorate also the coming of age of that annual course, which will celebrate its twenty-first birthday this summer.

John Cairns
Gunther S. Stent
James D. Watson

May, 1966.

Table of Contents

I. *Origins of Molecular Biology*

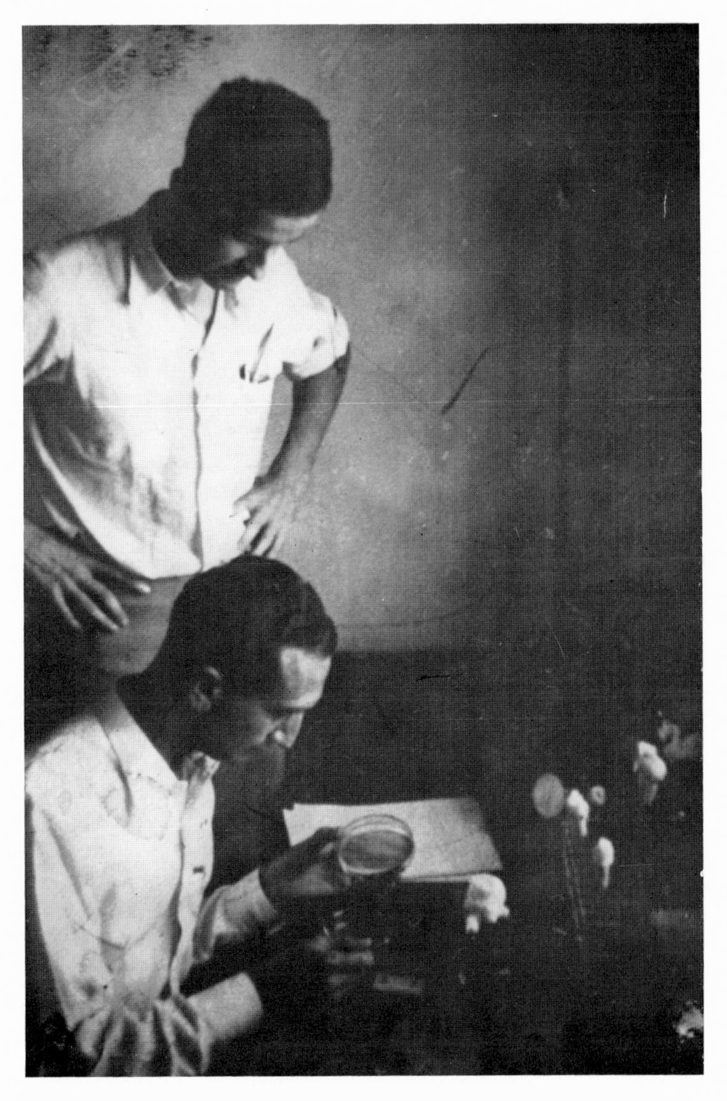

M. Delbrück and S. E. Luria, Cold Spring Harbor, 1941

GUNTHER S. STENT
University of California, Berkeley, California

Introduction: Waiting for the Paradox

Shortly before the end of World War II the great Austrian physicist, Erwin Schrödinger, then living as an anti-Nazi emigré in Ireland, wrote the little book "What is Life?" that was to draw wide attention to the dawn of a new epoch in biological research. Just why this book should have made such an impact was never quite clear. After all, in it Schrödinger presented ideas that were even then neither particularly novel nor original. The book, furthermore, does not make for easy reading; though it seems to be quite clearly written, the clarity turns out to be deceptive, since most readers must have had the uneasy feeling from time to time that perhaps they had not really understood what Schrödinger was trying to get across. On professional biologists the book had probably little or no influence. In so far as they bothered at all to read "What is Life?," they probably considered the title a piece of colossal nerve. At their most charitable, they must have viewed the book with amused tolerance. But its propagandist impact on physical scientists was very great. Their knowledge of biology was generally confined to stale botanical and zoological lore, and having one of the Founding Fathers of the new physics put the question "What is Life?" provided for them an authoritative confrontation with a fundamental problem worthy of their mettle. Since many of these physical scientists were suffering from a general professional malaise in the immediate post-war period, they were eager to direct their efforts toward a new frontier which, according to Schrödinger, was now ready for some exciting developments. In thus stirring up the passions of this audience, Schrödinger's book became a kind of "Uncle Tom's Cabin" of the revolution in biology that, when the dust had cleared, left molecular biology as its legacy.

Schrödinger opens with the comforting statement that "the obvious inability of present-day physics and chemistry to account [for the events which take place in a living organism] is no reason at all for doubting that they can be accounted for by those sciences." Since, as Schrödinger points out next, organisms are large compared to atoms, there is no reason why they should not obey exact physical laws. And even the peculiar quality of living matter, namely that it "evades decay to equilibrium," does not put it beyond the pale of thermodynamics, since organisms evidently feed on "negative entropy," whose ultimate source is the sun. No, the *real* problem in want of explanation is the hereditary mechanism. For, while the genes are evidently responsible for the order that an organism manifests, *their* dimensions are not so very large compared to atoms. How then do the genes resist the fluctuations to which they should be subject? How, wonders Schrödinger, has the tiny gene of the Hapsburg lip managed to preserve its specific structure for centuries while being maintained at a temperature 310° above absolute zero? Schrödinger proposes that genes preserve their structure because the chromosome that carries them is an *aperiodic crystal* whose atoms stay put in energy wells. These large aperiodic crystals are composed of a succession of a small number of isomeric elements, the exact nature of the succession representing the *hereditary code*. Schrödinger illustrates the vast combinatorial possibilities of such a code by an example that uses the two symbols of the Morse code as its isomeric elements.

In considering gene mutation, Schrödinger discusses what he calls "Delbrück's Model," under which alternate isomeric states of the genes correspond to different quantum mechanical energy levels, the probability of transition between states being low because transition requires a high energy of activation. Schrödinger thinks that "we may safely assert that there is no alternative to [this] molecular explanation of the hereditary substance. The physical aspect leaves no other possibility to account for its permanence. If the Delbrück picture should fail, we would have to give up further attempts." But now Schrödinger states an important credo which, as can be inferred from the article "A Physicist Looks at Biology," had been embraced also by Max Delbrück. In fact, this credo probably was the most important psychological incentive for physicists to turn to biology in the first place: "From Delbrück's general picture of the hereditary substance it emerges that living matter, while not eluding the 'laws of physics' as established up to date, is likely to involve 'other laws of physics' hitherto unknown which, however, once they have been revealed will form just as integral part of this science as the former." Thus it was the romantic idea that "other laws of physics" might be discovered by studying the gene that really fascinated the physicists. This search for the physical paradox, this quixotic hope that genetics would prove incomprehensible within the framework of conventional physical knowledge, remained an important element of the psychological infrastructure of the creators of molecular biology.

Schrödinger's book offers no concrete suggestions as to how one might actually go about studying the nature of the gene, but it predicts presciently that "advances in understanding how the hereditary substance works will come from biochemistry, under the guidance of physiology and genetics." "What is Life?" closes with an epilogue on Determinism and Free Will, the traditional philosophical paradox posed by the existence of conscious life. Schrödinger resolves this paradox according to more or less conventional empiricist thought, except that, following the philosophical *Zeitgeist* of the 1940's, he adds an existentialist twist: the conscious "I" is a canvas on which experiences have been collected, and these experiences lead "me" to the belief that "I" control the motion of atoms.

At the time of writing his book, Schrödinger's knowledge of the state of genetics was some years in arrears. He did not seem to have known of the rise of biochemical genetics in the late 1930's and thus did not mention the main doctrinal advance of that epoch, the "one gene-one enzyme" theory, that was to pave the way for molecular genetics. Another symptom of his somewhat outdated information was his emphasis of the fruit fly *Drosophila* as the main genetic object whose hegemony was, by that time, already giving way to micro-organisms, especially to the mold *Neurospora*. A rather ironic gap in Schrödinger's information was that he did not then know that for the past five years Delbrück, whose model of the gene seemed to have inspired Schrödinger's book in the first place, had been working with bacterial viruses, the very experimental material that was eventually to provide the answers to what Schrödinger wanted to know.

Bacterial viruses were discovered in 1915 by the English microbiologist F. W. Twort, and two years later—perhaps independently, perhaps not—by the French-Canadian, F. d'Herelle. In any case, it was d'Herelle who named these viruses *bacteriophages* and it was through his efforts they came to play a glamorous role in the bacteriology of the 1920's. By the middle of the 1930's, however, this glamor had begun to tarnish, since the widely propagandized control of bacterial diseases by means of bacteriophages had failed to materialize. But as interest in the practical application of bacteriophages waned, interest in them as tools for fundamental biological studies waxed. In the early 1930's it was being shown by M. Schlesinger that bacteriophages can be studied by chemical and physico-chemical techniques, and by F. M. Burnet that they are suitable for genetic investigations. At the same time, W. M. Stanley had crystallized the tobacco mosaic virus, an event of great heuristic impact. For that a self-reproducing object like a virus can be crystallized as if it were so much sodium chloride gave momentum to the notion that viruses, as "living molecules," ought to be a most favorable experimental material for unravelling the physical basis of biological self-reproduction.

In the late 1930's three men took up the study of bacteriophages: Alfred D. Hershey, Salvador E. Luria, and Max Delbrück. Their meeting in 1940 marks the origin of the Phage Group whose work and personalities form, in large part, the subject of the retrospective essays of this volume. The members of this group were united by a common goal, namely the desire to understand how during the brief, half-hour latent period the simple bacteriophage particle achieves its hundredfold self-reproduction within the bacterial host cell. The initial growth of this group was so slow that during its first five years only a few recruits joined its ranks. Among these few was T. F. Anderson, one of the first American electron microscopists. But in 1945, Delbrück took a step that was to set off a rapid and autocatalytic growth of the Phage Group. He organized the annual summer phage course at Cold Spring Harbor. The purpose of this course was frankly missionary: to spread the new gospel among physicists and chemists, a purpose that was not exactly hindered by the appearance in that same year of "What is Life?"

The greater number of workers assimilated into the Phage Group through the Cold Spring Harbor course, as well as the easier access to new tools such as radioactive tracers and ultracentrifuges, engendered more rapid progress during the next seven years. In 1952 the fifty or so stalwarts, gathered at the Abbaye de Royaumont near Paris for the first International Phage Symposium, knew by then that the phage DNA is the sole carrier of the hereditary continuity of the virus and that the details uncovered hitherto concerning the physiology and genetics of phage reproduction were to be understood in terms of the structure and function of DNA. In the very next year, the discovery of the Watson-Crick structure of DNA and the proposed mechanism of its replication provided the fundament for that understanding. A period of explosive development now set in, made possible not only by this intellectual breakthrough but also by the sudden increase in government support for biological research. Within another nine years, by 1961, the goal was reached: the mechanism by which the phage DNA replicates and directs the synthesis of viral proteins was more or less understood, and what remained was merely to iron out the details. No paradox had been encountered, no "other laws of physics" had turned up. Making and breaking of hydrogen bonds seems to be all there is to understanding the workings of the hereditary substance.

Granted that the mechanism of reproduction of an organism as simple as a bacteriophage does not seem to embody any hitherto unknown physical principles, what about that of higher forms of life? How can one account for the processes responsible for the orderly morphogenesis of organisms from single germ cells into amazingly complex and highly differentiated multicellular forms? It was with this more formidable question and with the next generation of problem solving already in mind that some phage workers turned to the study of animal cells soon after the discovery of the structure of DNA had made it clear that before long the nature of the gene would be fathomed in molecular terms. These studies on animal cells were to have important practical consequences, particularly in that they were to permit the flowering of quantitative animal virology; and though it still cannot be said that there has been any real breakthrough in understanding differentiation, the rise of molecular genetics brought a radical change to traditional embryology. One special case of differentiation, that of the antibody response of vertebrates, does seem near to its solution now, thanks largely to N. K. Jerne's recognition in 1955, of the selective nature of that response. Now that some reasonable molecular mechanisms for cellular differentiation can be at least *imagined*, the likelihood that the explanation of the development of the embryo will lead to the "other laws" seems to have greatly diminished, and with this denouement diminished also the appeal of embryology as an area for romantic strife.

Round about 1950, Delbrück began to think that bacteriophages were now, as he expressed it, "in good hands." This meant that he was beginning to lose interest in the hereditary mechanism before most of the real breakthroughs had even been made. He could evidently sense already at that time that the quest on which he, Luria, and Hershey had set out would presently lead to the understanding of biological self-reproduction, without the encounter of any paradoxes on the way. Accordingly, he turned toward that remaining major frontier of biological inquiry for which reasonable molecular mechanisms still cannot be even imagined: the higher nervous system. Its fantastic attributes continue to pose as hopelessly difficult and intractably complex a problem as did the hereditary mechanism a generation ago.

Delbrück found sensory perception a suitable point of purchase upon neurobiology. Having in mind the successful precedent of virus as the simplest material exhibiting self-reproduction, he looked about for an analogous simple model system of perception. His first choice was phototaxis of *Rhodospirillum*, a bacterium that translates light and dark stimuli into a motor response. After some initial studies had shown, however, that the nature of this response does not seem to be such a good model for perception after all, Delbrück addressed himself to the light-stimulated growth reaction of the fungus *Phycomyces*. Though they are phylogenetically most remote from the sensory organs of animals, the fungal light receptors do manifest two fundamental aspects of metazoan sense perceptors: accommodation and refractory period. About ten years ago, this work brought into being a Phycomyces Group dedicated to the solution of the fungal growth reaction. Nearly all of its recruits were physicists and, perhaps surprisingly, hardly any besides Delbrück himself were renegade phage workers. As was true for the Phage Group in its formative years, the Phycomyces Group has remained a small band, though Delbrück's recent institution of an annual summer course on *Phycomyces* at Cold Spring Harbor may presently swell its ranks with proselytes. One of the first to join the

Phycomyces Group was W. Reichardt, a physicist who had been studying the optomotor response of a beetle and whom Delbrück persuaded that the behavior of a single fungal cell would be easier to analyze than that of a whole insect. Though Reichardt has remained a part-time phycomycologist, he resumed work on the insects after a few years and showed that, in regard to perception, the insect eye happens to be more tractable than the fungal light receptor.

Now that the success of molecular genetics has made it an academic discipline, one can expect that in the coming years students of the nervous system, rather than geneticists, will form the *avant garde* of biological research. Though it is still far from clear what kind of experimental material is to become the phage of neurobiology, it seems more than likely that it will be a metazoan with nerve cells rather than a unicellular organism. And what Delbrück sensed more than fifteen years ago is now becoming a commonplace: The inability of even imagining any reasonable molecular explanation for such manifestations of life as consciousness and memory still offers some hope that biology may yet turn up some "other laws of physics." But it is also possible that study of the higher nervous system is bringing us to the limits of human understanding, in that the brain, being a finite engine, may not be capable, in the last analysis, of providing an explanation for itself. In that case, the paradox will have been found at last: there exist processes which, though they clearly obey the laws of physics, can *never* be understood.

MAX DELBRÜCK
California Institute of Technology, Pasadena, California

A Physicist Looks at Biology

*Address Delivered at the Thousandth Meeting of the Academy**

A mature physicist, acquainting himself for the first time with the problems of biology, is puzzled by the circumstance that there are no "absolute phenomena" in biology. Everything is time bound and space bound. The animal or plant or micro-organism he is working with is but a link in an evolutionary chain of changing forms, none of which has any permanent validity. Even the molecular species and the chemical reactions which he encounters are the fashions of today to be replaced by others as evolution goes on. The organism he is working with is not a particular expression of an ideal organism, but one thread in the infinite web of all living forms, all interrelated and all interdependent. The physicist has been reared in a different atmosphere. The materials and the phenomena he works with are the same here and now as they were at all times and as they are on the most distant stars. He deals with accurately measured quantities and their causal interrelations and in terms of sophisticated conceptual schemes. The outstanding feature of the history of his science is unification: two seemingly separate areas of experience are revealed, from a deeper point of view, to be two different aspects of one and the same thing. Thus, terrestrial and celestial mechanics, for thousands of years totally separate sciences, were reduced to one science by Newton, at a price, it is true—that of introducing the abstract notion of force acting at a distance and the notions of calculus. Thus, also, thermodynamics and mechanics were shown to be one and the same thing through the discovery of statistical mechanics, as were chemistry and electricity through the discovery of the proportionality between charge transport and mass transport in electrolysis, at a price—that of introducing atomicity into the concept of electric charge. Thus, optics and electromagnetism turned out to be two aspects of one theory, Maxwell's theory of the electro-magnetic field, at the price of introducing further abstract concepts, those of the field vectors. Thus, above all, chemistry and atomic physics were unified through the conceptual scheme of quantum mechanics, at the highest price of all—that of renouncing the ideal of a causal description in space and time.

The history of biology records discoveries of great generality, like those

*Reprinted from the Transactions of The Connecticut Academy of Arts and Sciences, vol. 38, Dec. 1949, pp. 173-190

of the occurrence of sexual reproduction in all living forms, of the cellular structure of organisms, of the ubiquitous presence of closely similar oxidative mechanisms, and many others of lesser generality. It records one great unifying theory, bringing together separate fields: the theory of evolution. To a physicist this is a strange kind of theory. It states, in the first place, that all living forms are interrelated by common descent. This statement is not one that is proved by decisive experiments but one that has become more and more inescapable through centuries of accumulated evidence. At the same time, it is the principle which serves to bring to order vast masses of descriptive taxonomy and geographical distribution. The theory states, further, that evolution has progressed through natural selection of the fittest and the most prolific. The assumption here involved is not whether or not selection occurs. On the contrary, in this respect the theory is tautological, since the organisms best fitted are defined as those which are selected. Rather, the assumption here involved is that the things selected for carry genetic permanence. At the time the theory was proposed this assumption was not proved, in fact in Darwin's time it could not even be stated in clear terms. It gave rise to the basic questions of the new science of genetics: how does heritable variability originate and how is it transmitted? Eventually it gave rise to the new abstractions: genotype and gene. It is most remarkable how late and how slowly these abstractions were established. Actually, today the tendency is to say "genes are just molecules, or hereditary particles," and thus to do away with the abstractions. This, as we shall endeavor to show later on, may be an overstatement of the legitimate claim of molecular physics. On the whole, the successful theories of biology always have been and are still today simple and concrete. Presumably this is not accidental, but is bound up with the fact that every biological phenomenon is essentially an historical one, one unique situation in the infinite total complex of life.

Such a situation from the outset diminishes the hope of understanding any one living thing by itself and the hope of discovering universal laws, the pride and ambition of physicists. The curiosity remains, though, to grasp more clearly how the same matter, which in physics and in chemistry displays orderly and reproducible and relatively simple properties, arranges itself in the most astounding fashions as soon as it is drawn into the orbit of the living organism. The closer one looks at these performances of matter in living organisms the more impressive the show becomes. The meanest living cell becomes a magic puzzle box full of elaborate and changing molecules, and far outstrips all chemical laboratories of man in the skill of organic synthesis performed with ease, expedition, and good judgment of balance. The complex accomplishment of any one living cell is part and parcel of the first-mentioned feature, that any one cell represents more an historical than a physical event. These complex things do not rise every day by spontaneous generation from the nonliving matter—if they did, they would really be reproducible and timeless phenomena, comparable to the crystalization of a solution, and would belong to the subject matter of physics proper. No, any living cell carries with it the experiences of a billion years of experimentation by its ancestors. You cannot expect to explain so wise an old bird in a few simple words.

Perhaps one can hope to understand some of its features—how oxygen is transported, how food is digested, how muscles contract, how nerves conduct, how the senses perceive, etc. One hopes to fix one's attention on features of the greatest generality, features that are an expression of the organization of matter as it is peculiar to all living matter, and to living matter only, to living protoplasm. Do such features really exist? Take, for example, the so-called excitability of living cells. Text books of physiology assure us that excitability is a peculiarity of all living cells. From a physicist's point of view this feature may perhaps best be expressed in the following terms: a living cell is a system in flux equilibrium, matter and energy are taken in from the environment, are metabolized and partly assimilated, partly degenerated, and waste products are given back to the environment. To a first approximation, that in which growth is neglected, this represents a steady state. As long as the environment does not change, the cell does not change even though matter and energy flow through it. If the environment does change slowly, the state of the cell will also change slowly and continuously, but if the environment changes sufficiently rapidly the state of the cell will change abruptly. It will become excited. Thus, upon excitation a nerve cell will discharge an action current, a muscle fiber will contract, a sensory cell will discharge, and similarly all other kinds of cells under proper conditions of stimulation will produce a disproportionately large response to a relatively slight but rapid change in the environment. Do we here deal with a truly general feature of the organization of matter in living cells? More than a hundred years ago Weber went one step further and formulated quantitatively a general law relating the threshold of stimulation with the parameters characterizing the environment. The law can be most easily formulated for situations in which one external parameter, say, the light intensity, changes abruptly from one level to another. Weber's law then states that the threshold change of this parameter which will produce excitation is proportional to the initial value of the parameter. For instance, if at an illumination of 100 foot candles, a change of 5 foot candles to 105 foot candles is necessary to produce excitation, then at a 10 times greater intensity of illumination of 1000 foot candles, a ten times greater change of 50 foot candles to 1050 foot candles will be necessary to cause excitation. The rating which this law of Weber has received in the biological literature has fluctuated from generation to generation, and has reached a very low level in recent times, so low in fact that it is barely mentioned in many of the current text books of physiology. The reasons for this decline furnish an instructive example of our perplexities regarding biological theory. In the first place it is quite clear that the law as stated experimentally can only be expected to hold over a limited range. It must fail at exceedingly low stimulus levels where the proportionate changes in stimulus fall below the size of individual quanta of action, and it must fail at exceedingly high stimulus levels where secondary effects of the stimulus become damaging to the cell. Experimentally, therefore, the validity of the law can only be apparent in a middle region of varying extent. In an ordinary physical law such limitations would not detract from its intrinsic value as a structural unit in the general theoretical edifice. Thus Boyle's law relating pressure and density of a gas fails at high density, where the molecules get crowded for space. Yet Boyle's law has been one of the chief clues to the kinetic theory of gases, and thus to atomic theory. A physicist would not be tempted to belittle a law merely because it is valid only over a limited range. Not so in biology. The biologist is tempted to discard the law altogether and say that it is an artefact. The threshold for stimulation, as a function of stimulus, may not really be proportional to the stimulus, with deviations from this law due to secondary causes at very low and at very high stimulus levels, but may be represented by an ill defined curve, running straight for a shorter or longer stretch. In fact, one can explain even a rather extended straight portion of this curve by appealing to the adaptive value that such a shape of the curve might have. It is perhaps reasonable to say that a proportionality of threshold with stimulus represents an advantageous situation for a cell that has to adjust to a very wide range of stimulus level, and thus to make natural selection responsible for simulating something that looks a physical law, natural selection acting like the overly faithful assistant of a credulous professor, the assistant being so anxious to please that he discards all those data which conflict with his master's theory. Arguments of this kind, combined with our total lack of a proper theory of excitation and the fact that perception in higher organisms anyhow is known to be a very complex business, have led to the present decline in the rating of Weber's law. It remains to be seen whether closer quantitative studies of the law, particularly as displayed by the simplest organisms, will justify this low rating, or whether it will turn out that we have overlooked a powerful clue to the nature of the organization of the living cell. The situation just described is one frequently met with in biology.

If it be true that the essence of life is the accumulation of experience through the generations, then one may perhaps suspect that the key problem of biology, from the physicist's point of view, is how living matter manages to record and perpetuate its experiences. Look at a single bacterium in a large volume of fluid of suitable chemical composition. It assimilates substance, grows in length, divides in two. The two daughters do the same, like the broomstick of the Sorcerer's apprentice. Occasionally the replica will be slightly faulty and an individual arises with somewhat different properties, and it perpetuates itself in this modified form. It is quite easy to believe that the game of evolution is on once the trick of reproduction, covariant on mutation, has been discovered, and that the variety of types will be multiplied indefinitely.

Higher organisms manage the matter of reproduction in a slightly more sophisticated manner, insisting on biparental reproduction, giving each new individual something from each parent, namely, one half of what the parent itself received from its parents, the half selected by an elaborate lottery. The student of evolution appreciates this game of segregation and recombination as a clever trick for trying out heredity in new combinations; it is not the basic thing but a refinement and elaboration of the *really* marvelous accomplishment: ordinary, uni-parental reproduction.

A physicist would like to know how this ordinary reproduction is done. This seems to be the elementary phenomenon of living matter. What sort of a thing is it, from the molecular point of view? What is the most elementary level upon which it can be observed? The answer to this question is that the cellular level is the most elementary level. There is a variety of approaches to this question, but none of these leads essentially below the cellular level.

Take for instance, ordinary classical genetics of some higher organism. We find that heredity is controlled by genes linearly arranged in chromosomes, and it might seem that the problem of reproduction has been reduced to that of the reproduction of the genes, but this is not true because no gene has been observed to reproduce except within the intact functional cell. We have every reason to believe that this dependence on the intactness of the cell is an essential one. The fact that there are many different genes in each cell, and that we have learned to combine different sets of genes in hybridization experiments, teaches us that the thing that is reproduced is a complex thing, but does not teach us how to break down the problem into simpler problems. Or, for instance, take the reproduction of a virus particle, that of a bacterial virus. A single virus particle will enter a bacterial cell and twenty minutes later several hundred virus particles identical with the one which infected the cell may be liberated. At first sight this may seem simpler than cellular reproduction because the individual virus particle is a very much smaller unit than the individual cell and may be analogous to an individual gene or to a small group of genes. In some respects, however, this case is really more complex than that of the reproduction of a cell. In the first place, it, too, requires the presence of a living and functional cell, and in the second place, what is observed is not a reproduction from one to two elements, but from one to several hundred. This feature makes it more complex than an ordinary cell division. Moreover, the complexities of sexual reproduction and of recombination are not eliminated by going to this seemingly elementary level. We have learned recently that when a bacterium is simultaneously infected with two similar but different virus particles, the progeny will contain recombinants in high proportions.[1] From this we draw two conclusions: that the virus particles themselves are complex, and that their reproduction must involve manoeuvers analogous to those occurring in meiosis and conjugation of higher organisms. This is news that is exciting principally by the blow it deals to our fond hope of analyzing a simple situation. Perhaps one might think that it would help to break up the bacterium prematurely, before the end of its natural term. An ingenious technique to do this has actually been developed by Dœrmann.[2] He finds that up to about half time there are no active virus particles, thereafter they increase rapidly in number up to the end of the natural term. Since no active virus particles are found before half time, although at least one was necessary to initiate virus reproduction in the bacterium, the infecting particles must be greatly modified if not actually broken down into subunits. It seems that the real reproduction and recombination has already occurred during the first dark period when no active particle can be detected, because when the first complete particles appear they contain as high a proportion of recombinants as will be found in the ultimate total crop of a full term liberation.[3] Thus, the technique of early inspection did not bring us an essential simplification of the problem.

The hopes associated with the study of bacterial viruses are based on one other feature, the ability to control and vary the experimental conditions under which reproduction occurs, and indeed in this respect we have learned something. We have learned that the reproduction of the virus particles in the bacterial cell requires that the cell be very actively assimilating. In fact, it seems that this is what happens: the virus particle which enters the bacterium commandeers the assimilatory apparatus of the bacterium. The primary products of assimilation, instead of being used to make bacterial substance, are used largely or exclusively to make virus particles of the type that enter.[4] We have no clue whatsoever as to how the virus particle manages to shunt the utilization of assimilatory products into its own channels nor how the reproduction proper is done. Presumably this situation can be analyzed from the biochemical point of view in considerably more detail. Since we know that the virus substance comes largely from assimilates after infection, we could ask more specifically in what order the assimilates acquired are incorporated. In principle this can be done with radio-active tracer experiments by adding or removing labelled compounds at given times after infection of the bacterium. Moreover, by using bacterial strains unable to synthesize one or another major component of the virus particle, say, an amino acid or a purine, one might hope to analyze the pathway by which these constituents are incorporated into the virus particle. It is however not clear to me how experiments of this kind could give a clue to the key problem, the chemical mechanism of replication.

Another way to approach the problem of reproduction is to center one's attention on some of the cellular constituents outside of the chromosomes which seem to have genetic continuity. To this class belong the plastids of green plants and a number of specific structures in protozoa. Here too, however, the genetic continuity is strictly dependent upon the functioning of the whole cell, and since these bodies can only be identified with assurance if they are large enough to be visible under the microscope they do not offer any simplification of the basic problem. There is reason to suspect that genetic continuity is also involved in the production of the enzymes elaborated by the living cells. If this is so, enzymes may be the best material for the study of this key problem but at present the mechanisms involved in enzyme synthesis are not clearly understood. There seem to be four factors involved in the production of any particular enzyme.[5] The first is the external metabolism supplying energy and assimilates. The second is a genetic control of the following kind: one finds that a very specific and limited genetic change, a mutation of a single gene, can be responsible for whether or not a cell can manufacture a given enzyme. A third factor is the substrate upon which this enzyme is supposed to act. In many cases a cell which is genetically able to produce a certain enzyme and is well supplied with nutrients, will not produce this enzyme except in the presence of the particular substance upon which this enzyme can act. One says that the cell responds adaptively to the presence of the substance. It is even possible that every enzyme requires for its formation the presence of its specific substance, only we do not become aware of this necessity because the substance in question may be present within the cell as a normal intermediary metabolite at all times. This adaptive mechanism is obviously of enormous value to the economy of the cell. The cell can thus shift its limited manufacturing capacity into a great variety of channels according to what is offered by a changing chemical environment. While we can readily understand the usefulness of this mechanism, we do not in the least understand its chemical nature. Is it that any substance merely stabilizes corresponding enzyme molecules that are continuously formed even in its absence but would be quickly destroyed if it were not for the stabilization? This notion has been urged by several authors, but it is an unlikely one. An enzyme molecule that is stable only as long as it is held in the claws of its substrate would be pretty useless. It would vanish too quickly if the concentration of the substance is reduced below the level at which the enzyme is saturated with it. Arguments of this kind cannot yet be very well evaluated quantitatively because of the intervention of the fourth factor involved in enzyme synthesis. This factor becomes apparent in experiments in which a cell is offered simultaneously two different substances, for each of which it can form adaptive enzymes which differ from each other. It is then found that these two adaptations compete and interfere with each other. This could not happen if the adaptation merely consisted in the stabilization of enzymes. The phenomenon of interference between different adaptations indicates a power on the part of any enzyme forming process to commandeer supplies into its own channel. In other words, the enzyme forming process, too, involves some kind of self-accelerating mechanism. As yet, however, we are not able to pin down the element responsible for this power. Is it the enzyme itself, an unspecific precursor, a specific region of the enzyme, or any one of these combined with substrate? I believe a great deal of light could be shed on this situation by a much closer analysis than has hitherto been attempted of the adaptation interference phenomenon.

I have been trying to give you an impression of the intellectual uncertainties confronting a physicist entering biology. These are of two kinds. On the one hand, when he thinks he has discovered a law of nature as pertaining to living matter, like Weber's law, he must beware lest he be fooled by natural selection simulating such a law. On the other hand, when he has found a really fine phenomenon universal in the living world and specific to it, like reproduction, he finds that it is indissolubly tied into the enormously complex organization of a living cell. Baffled by these barriers to his usual approach, a physicist may well stop and consider what he may reasonably expect by way of theoretical progress in the application of physical principles to living matter. It is true that physics and chemistry have a firm hold over biology by virtue of two great generalizations: living matter is made up of the same elements as those of the inanimate world, and conservation of energy is valid for processes occurring in living matter, just as it is for all processes in the inanimate world. To the men who first established the validity of these generalizations for living matter, like Helmholtz, for instance, it seemed clear that the processes of living matter must be essentially the same as those of the inorganic world and that there could not possibly exist a biological science ruled by its own laws. For these men the phenomena of life presented essentially mechanical problems. They should be deducible from the laws of Newton's mechanics as movements of particles due to forces originating in other particles. Such a view, more recently slightly modified to incorporate a few quantum concepts, has been the driving force of several generations of bio-chemists. We may quote J. Loeb as fairly representative of this attitude: "The ultimate aim of the physical sciences is the visualization of all phenomena in terms of groupings and displacements of ultimate particles, and since there is no discontinuity between the matter constituting the living and the non-living world the goal of biology can be expressed in the same way." One should add at this point that some of the contemporaries of Newton took quite a different view of the Newtonian principles. To describe motions in terms of forces acting at a distance seemed to them like introducing magic. Moreover, to correlate forces with accelerations, that is, with the second derivative of the function describing the motion, seemed to them the height of abstraction, going beyond what should be permitted to occur in any science, and threatening to remove it from the realm of rational pursuit. Between the times of Newton and Helmholtz, then, a strange inversion took place. What had seemed magical and extravagant at the earlier period, after a century of success had become the *only* way in which one could hope to account rationally and visualizably for the phenomena of nature. Actually, most branches of biology manage to flourish without any recourse to this ideal. All of natural history operates with a system of concepts which has very little contact with the physical and chemical sciences. The habits of animals and plants, their reproduction and development, their relations to their symbionts and to their enemies, can all be described and analyzed with very little reference to the concepts of physics and chemistry. Perhaps the most notable of these independent branches of biology is genetics, which in its pure form operates with "hereditary factors" and "phenotypic characters" in a perfectly logical system, as an exact science without ever having to refer to the processes by which the characters originate from the factors. The root of this science lies in the existence of natural units of observation, the individual living organisms, which in genetics play somewhat the same role as the atoms and molecules in chemistry.

This analogy of the individual living organism with the molecules of chemistry may be valid in more than one sense. In the first place the notion of the individual organism brings into biology discreteness, and number, and identity of type, as the concept of molecule does in chemistry. In the second place, the stability and reproducibility of type in biology we think is a result of the stability and reproducibility of the genes and this stability we think is based upon the stability of certain complex molecular structures. The third and perhaps the most interesting aspect of the analogy is the fact that the individual organism presents an indissoluble unit, barring us, at least at present, from a reduction to the terms of molecular physics. It may turn out that this bar is not really an essential one, but a physicist is well prepared to find that it is essential. This would be similar to the lesson the physicist had to learn most recently in the attempts to develop a proper approach to an understanding of the properties of molecules in terms of their constituent elements. To make this point clearer I want to illustrate it with an historical example. At the beginning of this century enough was known about the existence of atoms and molecules to make the question of their structure a real one. The analysis of the interaction of atoms with light had begun before the constituent elements of the atoms, electron and nucleus had been discovered. It was known that atoms emit radiations of characteristic frequencies which could be determined with high accuracy. The attitude of the physicists of those days was that these characteristic frequencies represent frequencies of vibration of the atoms themselves and as such are expressions of certain properties of the atoms as mechanical systems. Examples of solid objects having characteristic frequencies of vibration had been well worked out and understood in terms of classical physics. The vibrations of the atoms were presumed to be vibrations of electrical matter within the atoms and it was considered the task of future analysis to infer the structure of the atom from a close inspection of the frequencies of vibration and of the conditions under which atoms could be excited in their various modes of vibration. Nothing could have seemed more sane and reasonable a program of research. In fact, many interesting regularities in the system of frequencies characteristic for a given atom were discovered. As long as nothing definite was known of the parts of which atoms are made up, these findings may have seemed a little strange but by no means suggestive that a suitable description in terms of a mechanical model could not be obtained. The situation was changed suddenly when it became clear that every atom is made up of one nucleus and a number of electrons. Very simple and general arguments could be brought forward showing that no system consisting of these elements could possibly have the properties that atoms were known to have. Here was a clear paradox. In Bohr's paper of 1913 this paradox was met by introducing the notions of stable orbits and jumps between these orbits. The frequencies of revolution of electrons in these orbits were unrelated to the frequencies of the light waves emitted. These were very irrational assumptions which shocked and in fact disgusted many physicists of that time.

The crucial point in this abbreviated account of an historical episode is the appearance of a conflict between separate areas of experience, which gradually sharpens into a paradox and must then be resolved by a radically new approach.

As is well known, the resolution of the paradoxes of atomic structure necessitated a revision of our ideals (or prejudices) regarding the description of nature. It was necessary to replace the classical conceptual scheme of particles moving in well defined orbits by the new scheme of quantum states and transition probabilities. Let us consider, for instance, a hydrogen atom in an excited state. Quantum mechanics permits us to calculate the probability that the atom will make a transition to the ground state in a definite time interval, with emission of a light quantum which carries off the difference in energy between the two states. Suppose we wanted to improve the statistical prediction and find out exactly at what moment the transition will take place. To make such a prediction we would need more precise information regarding the state of the atom, beyond the fact that it is in this excited state. We might try to obtain this more precise information by finding out at what point within the atom the electron is located at a certain moment. Such information can be obtained, in principle at least, by using an ideal microscope employing a radiation of very short wave length. Observation with light of such very short wave lengths, however, necessitates the use of an experimental arrangement in which we cannot control the exchange of energy between the atom and the measuring instrument, the light in this case. This uncontrolled exchange of energy will be of such a magnitude that the atom thereafter will be in any one of the excited states. If the first aim, that of obtaining the precise location of the electron, is not to be sacrificed, the microscope has to be used in a fashion which is mutually exclusive with any arrangement designed to give us information about the excited state. While, therefore, obtaining exact information regarding the position of the electron at this particular moment, the uncontrolled interaction between the tool of observation, the short wave light, and the atom introduces an uncertainty in the energy of the atom and we are not better off than before. We have just swapped the knowledge of a stationary state for the knowledge of the location of the electron at one particular moment, and our primary aim of additional information, beyond that of knowing that the atom is in the first excited state, has been frustrated. This state of affairs is at the root of all quantum phenomena. Each process of observation has an individuality which cannot be broken down beyond a certain limit and different types of observation stand to each other in a mutually exclusive, complementary relationship. Different observations will therefore lead to a variety of optimal informations regarding an atomic system, and each such optimal information we describe by the abstract notion of the state of that system. From such knowledge we can make precise statistical predictions regarding the future of the system. The different optimal informations are obtained from mutually exclusive experimental arrangements, and the quantum mechanical formalism is designed to embody this particular feature of complementariness. We believe that quantum mechanics is the final word as regards the behavior of atoms and we base this belief upon the analysis by Bohr and Heisenberg of the possibilities of observation, which shows that the renunciation of the ideal of classical physics for the description of Nature eliminates nothing that could be defined operationally. Therefore this renunciation should not be considered as a loss but as a liberation from unnecessary restrictions and thus as the essential element of our advance which opened the widest possibilities for future developments.

Let us now go back to the situation as it appeared to Helmholtz in the 1870's. For him it seemed that the behavior of living cells should be accountable in terms of motions of molecules acting under certain fixed force laws. How far we have wandered from this ideal! We now know that if we tried to adhere to this ideal we could not even account for the behavior of a single hydrogen atom. In fact, we account for the stationary state of the hydrogen atom precisely at the price of not describing the motion of the electron. But how about the living cell? Should we now perhaps consider it as a super molecule and ideally calculate its stationary states and transition probabilities? Would it be sufficient to modify Helmholtz's views to this slight extent?

Such a view might seem reasonable to the structural chemist, who applies the general concepts of quantum mechanics successfully to more and more complex molecules. Even though his efforts have not yet been successful with any structures of molecular weight higher than a few thousand, a limitation in principle can hardly be expected. However, the experimental analysis of the structural chemist always presupposes the availability of a practically infinite number of molecules of the same kind, and of a practically infinite stability of the molecules if not disturbed.

In the living cell we know that a great deal depends on very fine features of structure. By structure we mean relevant inhomogeneities in the make-up of the cell. These relevant inhomogeneities go right down to the atomic scale. It would certainly not be possible, even in an ideal experiment, from observation of an *individual* living cell to gain knowledge of these details sufficient to make quantum mechanical calculations of the development of the cell. To make structural observations of such extreme finesse again a practically infinite number of cells in *identical quantum states* would be required. Such a requirement can certainly not be met in practice and it seems likely to me that also it cannot be met in principle. This leads to the point of the third analogy between individual living organisms and molecules. It may turn out that certain features of the living cell, including perhaps even replication, stand in a mutually exclusive relationship to the strict application of quantum mechanics, and that a new conceptual language has to be developed to embrace this situation. The limitation in the applicability of present day physics may then prove to be, not the dead end of our search, but the open door to the admission of fresh views of the matter. Just as we find features of the atom, its stability, for instance, which are not reducible to mechanics, we may find features of the living cell which are not reducible to atomic physics but whose appearance stands in a complementary relationship to those of atomic physics.

This idea, which is due to Bohr, puts the relation between physics and biology on a new footing. Instead of aiming from the molecular physics end at the whole of the phenomena exhibited by the living cell, we now expect to find natural limits to this approach, and thereby implicitly new virgin territories on which laws may hold which involve new concepts and which are only loosely related to those of physics, by virtue of the fact that they apply to phenomena whose appearance is conditioned on *not* making observations of the type needed for a consistent interpretation in terms of atomic physics.

I would like to explain this point of view a little further by a brief reference to the problem of spontaneous generation.[6] We know that life originated on earth within less than a billion years after conditions had become reasonable, perhaps in much less time and perhaps not once but repeatedly. We do not know the chemical environment of those days, but it can hardly have been anything that could not be reproduced today nor is it likely to have been anything very specific. Therefore, spontaneous generation should be an experimentally reproducible phenomenon. Imagine that we knew these conditions. We could then prescribe a certain synthetic medium and temperature and predict that spontaneous generation should occur with a certain probability per time unit, and it would seem that this would bring life completely into the domain of chemistry. Or would it? Conceivably spontaneous generation would be a very gradual thing, going through lengthy stages of inaccurate and poorly defined replications, perhaps evolving from something like crystallization from supersaturated solutions. In that event the algebra of natural selection and of statistical fluctuations would enter decisively into the kinetics of spontaneous generation and would limit the predictability of the outcome, just as it does in the later stages of evolution. The creatures emerging in such a test tube experiment presumably would be extremely different, biochemically, from those of our world, and not being adapted to our world would have no chance of survival in it. This implies that we could not tell the precise chemical make-up of a living thing from its mode of chemical origin. Nor can we tell it, as pointed out before, from a structural analysis of its actual state. The detailed chemical structure of a living cell is thus operationally undefinable and the concept therefore meaningless. It remains to be seen whether a retrenchment in the demand for a detailed chemical description for such phenomena as replication, chromosome movements, excitation, active transport, etc., will facilitate giving a coherent account of these phenomena.

Perhaps you will think that such speculations and arguments as here presented are very dangerous: they seem to encourage defeatism before it is necessary, and to open the door to wild and unreasonable speculations of a vitalistic kind. I sympathize with this criticism and want to justify the presentation I have given by saying that Bohr's suggestion of a complementarity situation in biology, analogous to that in physics, has been the prime motive for the interest in biology of at least one physicist and may possibly play a similar role for other physicists who come into the field of biology. Biology is a very interesting field to enter for anyone, by the vastness of its structure and the extraordinary variety of strange facts it has collected, but to the physicist it is also a depressing subject, because, insofar as physical explanations of seemingly physical phenomena go, like excitation, or chromosome movements, or replication, the analysis seems to have stalled around in a semidescriptive manner without noticeably progressing towards a radical physical explanation. He may be told that the only real access of atomic physics to biology is through biochemistry. Listening to the story of modern biochemistry he might become persuaded that the cell is a sack full of enzymes acting on substrates converting them through various intermediate stages either into cell substance or into waste products. The enzymes must be situated in their proper strategic positions to perform their duties in a well regulated fashion. They in turn must be synthesized and must be brought into position by manœuvers which are not yet understood, but which, at first sight at least, do not necessarily seem to differ in nature from the rest of biochemistry. Indeed, the vista of the biochemist is one with an infinite horizon. And yet, this program of explaining the simple through the complex smacks suspiciously of the program of explaining atoms in terms of complex mechanical models. It looks sane until the paradoxes crop up and come into sharper focus. In biology we are not yet at the point where we are presented with clear paradoxes and this will not happen until the analysis of the behavior of living cells has been carried into far greater detail. This analysis should be done on the living cell's own terms and the theories should be formulated without fear of contradicting molecular physics. I believe that it is in this direction that physicists will show the greatest zeal and will create a new intellectual approach to biology which would lend meaning to the ill-used term biophysics.

REFERENCES

1. DELBRÜCK, M. 1949. Internat. Symposium of the Centre Natl. de la Recherche Scientifique, *8*: 91–104.
2. DOERMANN, A. H. 1948. Carnegie Institution of Washington Yearbook, *47*, 176–185.
3. ——— *Personal communication.*
4. COHEN, S. S. 1949. Bacteriological Reviews, *13*: 1–24.
5. MONOD, J. 1949. Internat. Symposium of the Centre Natl. de la Recherche Scientifique *8*: 181–199.
6. PIRIE, N. W. 1948. Modern Quarterly *3* (new series): 82–93.

GEORGE W. BEADLE
The University of Chicago, Chicago, Illinois

Biochemical Genetics: Some Recollections

I have often thought how much more interesting science would be if those who created it told how it really happened, rather than reported it logically and impersonally, as they so often do in scientific papers. This is not easy, because of normal modesty and reticence, reluctance to tell the whole truth, and protective tendencies toward others.

My first exposure to genetics came during the summer after my second year at the University of Nebraska. Professor F. D. Keim had just returned from a leave of absence at Cornell University, where he worked on a Ph.D. thesis on the genetics of spelt-club wheat hybrids. I was employed at the rate of thirty cents an hour to take data on the progeny—kernel color, size, and weight; rachis internode length; glume texture and shape; plant height; and an array of other traits.

To find out what genetics was all about, I spent my spare time reading H. E. Walter's textbook *Genetics*. It was a simple book—just about right for me—and I was fascinated. I decided then and there to investigate wheat genetics on my own. But I did little more than make some crosses in the greenhouse.

Keim, no longer living, was a truly remarkable man. I do not suppose anyone would claim he was a scholar of special distinction. Nor was he a brilliant teacher in the usual sense. He could not always solve the problems in the book we used as a text in his course, and he sometimes asked me if I could. The flattery was effective, though I suspect not premeditated. He won respect through his honesty and lack of pretense. He had an abundance of infectious enthusiasm and a fabulous understanding and judgment of students. During his career as a Professor of Agronomy he advised hundreds of students and, so far as I knew, never made a mistake. He sent George F. Sprague, now one of the world's top corn breeders, into genetics and plant breeding. He urged me to go to Cornell to take graduate work, rather than returning to the farm as I had intended—and arranged for an assistantship to make it possible. He sent Earl Patterson and Francis Haskins to the California Institute of Technology to become geneticists, and convinced Adrian Srb that he would make a better geneticist than English scholar.

As I look back, I wonder if an understanding of people may not be more important in a teacher than knowledge of the subject or the brilliance with which he teaches it.

After going through phases during which I wanted to be an English major, an entomologist or a plant ecologist, I ended up as a graduate student at Cornell University working with Rollins A. Emerson and Lester W. Sharp on the genetics and cytology of maize. Those were exciting times, for Barbara McClintock, Marcus Rhoades, Ian Phipps, H. W. Li, T. H. Shen, H. S. Perry and a number of others were students at the same time.

Unlike A. C. Fraser who taught the beginning genetics course, which rated very high with students, Emerson did little or no formal teaching. Fraser was a superb teacher, in the sense of organizing the facts of genetics known at the time and presenting them clearly and logically. He was little concerned with what remained to be discovered, and I have often suspected that was the reason few, if any, of his many hundreds of students were inspired to become geneticists.

Emerson, on the other hand, was little interested in what was already known but fascinated with what remained to be discovered. He infected his students and colleagues with his own enthusiasm for adding to known knowledge. During corn pollinating season we all worked from dawn to dark, with Emerson setting the pace and presiding over lunch and rest-period intellectual bull sessions. In my book, he was a truly great teacher—not unlike Max Delbrück, in many respects.

In working with genes for pollen sterility, I discovered the three genetic types of abnormal chromosome behavior: asynapsis, polymitosis and sticky chromosomes. I shall never forget the incomparable thrill of discovering the asynaptic character. My enthusiasm was shared—so much so in the case of Barbara McClintock that it was difficult to dissuade her from interpreting all my cytological preparations. Of course she could do this much more effectively than I.

In 1928 or 1929 Bernard O. Dodge, then of the New York Botanical Garden, gave a seminar at Cornell University on Neurospora. He had dissected in order and grown ascospores of crosses between a normal strain and one lacking asexual spores. He had found 4:4 segregations, with first division separation of the two types, in most asci. But there were some asci showing alternate pairs of spores of the two types. This was at a time when many cytologists and geneticists regarded the first meiotic division as reductional and the second as equational or mitotic. Dodge was at a loss for an explanation of the apparent second division segregation. Several of us, who had just been reviewing Bridges' and Anderson's evidence for four-strand crossing over in Drosophila, suggested the explanation—crossing over between the centromere and the segregating gene pair. This was my first introduction to Neurospora.

A year or so prior to this, Dodge was insisting to T. H. Morgan, then at Columbia University, that Neurospora was a more favorable organism for genetic studies than was Drosophila. Finally, after much persuasion, for Dodge was a most enthusiastic and persistent persuader, Morgan was induced to take some Neurospora stocks and try them. He carried them to the California Institute of Technology, on occasion transferring them to new culture media, as instructed by Dodge.

Soon thereafter, there appeared a young man who proposed to do graduate work in Morgan's new Division of Biology at Caltech. It was Carl C. Lindegren. Morgan asked his background, and when Lindegren said "Bacteriology," it was suggested he work on the genetics of Neurospora. In 1931, when I arrived at the California Institute of Technology, this work was in full swing.

On nearing the completion of my graduate work at Cornell, I had applied for, and was awarded, a National Research Council Fellowship. But there was a condition. I had proposed to remain at Cornell, for I wanted to continue with corn cytogenetics and thought Cornell was the best possible place to do so. But the Chairman of the Fellowship Committee, then Professor C. E. Allen of the University of Wisconsin, had other views. He and his Committee believed a change of institution was good in principle and said I could have the fellowship if I would elect my second choice, namely, the California Institute of Technology.

I have had many occasions to thank that Committee, not because Cornell was not a good place, but because a change of laboratories was the best thing that could have happened in my case. Ever since, I have argued that moves to different institutions after undergraduate work, after graduate work, and after postdoctoral work are much to be desired, other things being anywhere near equal.

Before leaving Cornell, H. S. Perry and I had worked out what we—I, at least—thought was a marvelous theory of crossing over. In brief, it held that chiasmata arise by pairwise separation of the four chromatids in alternate planes—always reductional at the centromere and equational or reductional elsewhere—with subsequent breakage of the chiasmata at anaphase and recombination of the broken chromatids.

We corresponded with E. G. Anderson, then at Caltech, about it. He replied that he did not know about the hypothesis, but he was pleased to know that we at least understood the genetic facts of crossing over. Ours was a fine hypothesis, later elaborated and published by Karl Sax of Harvard University. Unfortunately it proved to be wrong.

On arriving at Caltech in early 1931, I was anxious to talk with C. B.

Bridges about our hypothesis. I was almost overwhelmed at the thought of approaching such a distinguished and brilliant geneticist, for I thought he would be so quick of mind that I would not be able to keep up my end of the discussion. It was with much amazement and relief that I discovered Bridges to be not only most friendly but quite slow and methodical in our discussion. He insisted on going over each point carefully and fully before going on to the next one. With Sturtevant, Dobzhansky, Emerson, and Schultz, on the other hand, I was often quite lost, for those were the days of the so-called scute subgenes, Oenothera analysis, and translocations in Drosophila. My head often buzzed on listening in on conversations about scute-8, scute-4, gaudens-velans, hookeri and so on, but finally I learned the new jargon.

The early 1930s were depression days and laboratory work was done at minimal expense. I recall that Sterling Emerson and I once needed a Harvard trip balance—cost about ten dollars. We were certain that Morgan would not approve such an expenditure, for he knew that geneticists did not need elaborate equipment of that kind. So we persuaded James Bonner to request it and then give it to us. He was a plant physiologist and really did need things of that kind.

I soon discovered that Morgan was likely to be in his most agreeable mood on Sunday mornings, when he could work in his small downstairs laboratory on Ciona. It was on such an occasion that I got him to approve the purchase of a new 90x achromatic oil immersion objective for the research microscope I was using. That was indeed a triumph.

Speaking of Morgan, I well recall the occasion on which my morale was given a substantial boost. I had shown a manuscript on sticky chromosomes to Sturtevant. After reading it, he asked, "May I show this to the Boss?" To me this was a high compliment, for Morgan, who was affectionately known as "the Boss" by close associates both at Columbia University and at Caltech, did not by any means have every manuscript referred to him. As an indication of the depth of the affection with which the term "the Boss" was held, several years after Morgan's death a postdoctoral fellow at Caltech who had been a graduate student in our laboratories at Stanford continued to refer to me by that term; he was promptly told, "That term is *never* to be used in this laboratory for anyone other than Doctor Morgan!"

In 1934, Boris Ephrussi came to Caltech to work on embryology and genetics of mice, using transplantation technics. These were times of much talk about gene action, with such questions being asked as whether all genes act all the time. Morgan had just published a book, *Embryology and Genetics*. I remember Ephrussi once commenting that the difficulty was just that—embryology *and* genetics—for the organisms investigated by the two branches of biology were not the same. The classical organisms of embryology—sea urchins, for example—were not favorable for genetic study and those favored for genetics, like Drosophila, were difficult embryologically.

We thought something should be done about it and finally proposed that we each gamble up to a year of our lives trying to do it. I would try this in Ephrussi's laboratory in Paris. Morgan inquired informally of the Rockefeller Foundation officers whether they might consider supporting me during such a venture. The reply was that this could not be done on the required short notice, for none of the fellowship officers had met me, and that was a requirement. Morgan thereupon said Caltech would pay my salary—the welcome sum of $1,500 per year—if I could manage. By leaving my family in a house renovated by me and provided rent-free by Caltech at the "corn farm," I could manage, by living on approximately two dollars per day in Paris. I do not know, but I have often suspected that Morgan personally provided the $125 per month.

I arrived in Paris in May, 1935, and we immediately began to attempt to culture Drosophila tissues. This proved technically difficult; so, on Ephrussi's suggestion, we shifted to transplantation of larval embryonic buds destined to become adult organs. We sought advice from Professor Ch. Perez of the Sorbonne, who was a widely recognized authority on metamorphosis in flies. He said we had selected one of the worst possible organisms and that his advice was to go back and forget it. But we were stubborn and before many weeks had devised a successful method of transplantation. The first transplant to develop fully was an eye. It was the occasion for much rejoicing and celebration at a nearby café. In this way, we confirmed the nonautonomous development of eye-pigmentation that Sturtevant had earlier discovered for the character vermilion eye in gynandromorphs.

Before long, we had established the existence of two diffusible eye-pigment precursors that we believed to be sequential in the formation of the brown component of eye pigment, according to the scheme

$$\longrightarrow v^{+}\text{ substance} \longrightarrow cn^{+}\text{ substance} \longrightarrow \text{pigment brown}$$

$$\uparrow \qquad\qquad \uparrow$$

$$v^{+}\text{ gene} \qquad cn^{+}\text{ gene}$$

This was the beginning in our minds of the one-gene-one-enzyme concept, but we did not then use that expression.

At the end of 1935, Ephrussi and I returned to Caltech and continued our transplantation work in collaboration with C. W. Clancy. In the fall Ephrussi returned to Paris and I moved to Harvard University as an assistant professor. We both attempted to determine the origin and nature of the diffusible pigment precursors, which we then called hormones because they were produced in one part of the body and had their effects elsewhere.

It was a busy year, for I not only gave the lectures in the botany semester of the general course in biology but also managed to get a respectable amount of research done. In March of 1936 I was invited to accept a professorship in the School of Biological Sciences at Stanford with superb provisions for research support. The decision whether to accept was drawn out over a period of weeks. During this period, I sought the advice of Edward M. East, Professor of Genetics. I shall never forget the discussion.

"The fellows at Stanford seem to think they want me to join them," I said.

"Well, what do you want me to say about it?" East asked.

"I just want to know what you think about it."

"Oh, you want to know what I think of it? Well, I'll tell you what I think of it. Stanford never was any good, it isn't any good, and it never will be any good. That's what I think about it."

A month later, after the decision to move, East came to my laboratory and said, "Beadle, I *knew* you were going to Stanford." East was a fine friend and colleague. As those who had the privilege of knowing him appreciate, his gruffness and extravagant manner of speech were purely synthetic, deliberately cultivated to spur students and colleagues to greater efforts.

My nine years at Stanford were both pleasant and productive. Charles V. Taylor, Dean of the School, possessed boundless energy, enthusiasm and confidence. He had in fact promised support for research beyond his financial resources. But his confidence stood him in good stead, for after I accepted he immediately applied to the Rockefeller Foundation for the research support he had promised. I have often wondered what he would have done had the Foundation declined to make the grant. I am sure he would have found a way out. With the funds provided we built laboratories, equipped them and induced Edward L. Tatum to accept a position as Research Associate.

We spent three years trying to identify eye-pigment precursors and came pretty close to doing so. Tatum had in fact obtained an active substance as a product of bacterial metabolism of tryptophan. It was kynurenine esterified with sucrose, but because of the molecular weight of the complex we failed to identify it. Butenandt in Germany did so by testing known relatives of tryptophan, an approach that is obvious in retrospect.

During the course of this work Tatum's late father, Arthur Lawrie Tatum, then Professor of Pharmacology at the University of Wisconsin, visited Stanford. On this occasion he called me aside and expressed concern about his son's future. "Here he is," he said, "not clearly either biochemist or geneticist. What is his future?" I attempted to reassure him—and perhaps myself as well—by emphasizing that biochemical genetics was a coming field with a glowing future and that there was no slightest need to worry.

In 1940 we decided to switch from Drosophila to Neurospora. It came about in the following way: Tatum was giving a course in biochemical genetics, and I attended the lectures. In listening to one of these—or perhaps not listening as I should have been—it suddenly occurred to me that it ought to be possible to reverse the procedure we had been following and instead of attempting to work out the chemistry of known genetic differences we should be able to select mutants in which known chemical reactions were blocked. Neurospora was an obvious organism on which to try this approach, for its life cycle and genetics had been worked out by Dodge and by Lindegren, and it probably could be grown in a culture medium of known composition. The idea was to select mutants unable to synthesize known metabolites, such as vitamins and amino acids which could be supplied in the medium. In this way a mutant unable to make a given vitamin could be grown in the presence of that vitamin and classified on the basis of its differential growth response in media lacking or containing it.

There was never any slightest doubt that this approach would be successful—in my mind at least—for we had complete confidence in the one-gene-one-enzyme hypothesis. The only question in doubt was the frequency of the kinds of mutations we were in a position to produce and identify with the metabolites and methods available to us. The mutants came, first one requiring thiamine, then pyridoxine, and soon paraaminobenzoic acid.

These results were reported at the 1940 American Association for the Advancement of Science meetings in Dallas, Texas. I recall vividly the discussion period, where a fellow scientist we knew well commented that either we were incredibly industrious or he would have to be skeptical. I strongly suspect he did not believe the results. The hypothesis seemed much too simple to him, a view I shall have occasion to refer to again.

It was clear to us that we had an approach capable of adding rapidly to our understanding of the relation of genes and specific known chemical reactions. It was equally clear that we would require additional financial support. The need for such support posed an interesting problem. Three years earlier, the experimental biology group at Stanford headed by C. V. Taylor had applied for a large grant from the Rockefeller Foundation. It was made: $200,000 for ten years, with a stipulation we were not to apply for more for a period of ten years. The Foundation knew Taylor well!

That meant we could not properly ask for further support from the Rockefeller Foundation. The Research Corporation seemed a logical possibility. So in 1941 I went to New York to see Frank Blair Hanson of the Rockefeller Foundation to tell him that I knew of the ten-year pact and to ask if there were any objection to my going to the Research Corporation for additional funds. Of course he said there was no objection.

On the same day, I went to see Howard Andrews Poillon at the Research

Corporation and told my story. He called in Robert E. Waterman. Not knowing he was the Waterman of Williams and Waterman vitamin B-1 fame, I retold the story. Waterman was ahead of me and tremendously enthusiastic. He and Poillon said they would give support—$10,000 right off. At that point there was a telephone call for me. It was Hanson, who said he had been thinking that, since the Rockefeller Foundation had got us started, they might give added support, despite the pact. Poillon and Waterman said that was only reasonable and proper, and that I should by all means go back to Hanson. They said I should send the Research Corporation a carbon copy of our application to the Rockefeller Foundation and that, if the grant were not made, they would promise then and there to make the grant. That was a marvelous example of foundation flexibility and speed of decision. The Rockefeller Foundation made the grant without delay. This plus further grants enabled us to invite David M. Bonner, Norman H. Horowitz and Herschel K. Mitchell to join our group and to move forward with dispatch.

In 1945 I held a Sigma Xi National Lectureship under which I gave lectures on biochemical genetics and the one-gene-one-enzyme interpretation. I was much impressed with the resistance to this notion, especially in agricultural colleges where workers were familiar with the genetics of such characteristics as egg production, and milk production in dairy cattle. They were sure gene action could not be generally described in the simple way we had postulated. It seems to me the status of the concept dropped to an all-time low at the Cold Spring Harbor Symposium of 1951. In rereading the volume on those meetings, I have the impression that the number whose faith in one-gene-one-enzyme remained steadfast could be counted on the fingers of one hand—with a couple of fingers left over.

I have several times been asked when the one-gene-one-enzyme hypothesis was first proposed and by whom. I have thought about this question many times and have reread a number of papers to see if I could discover the answer. I have not been successful. The first reference in so many words that I know of is in a review of biochemical genetics that I wrote for the 1945 volume of *Chemical Reviews*. I am sure Ephrussi and I had the concept in mind. Clearly Tatum and I also had when we planned our Neurospora work. I have read it into the papers of A. R. Moore, L. T. Troland, R. B. Goldschmidt, S. Wright, J. B. S. Haldane and others. Prior to any of these, A. E. Garrod was quite clear in his view that an enzyme concerned with the cleavage of the ring of 2,5-dihydroxyphenylacetic acid (homogentisic acid) is controlled by a gene. One must remember that the protein nature of enzymes was not clearly demonstrated until 1926 when J. B. Sumner crystallized urease. Even then the evidence was not widely accepted for several years.

Why was Garrod's work unappreciated and forgotten? It was well known to Bateson and he referred to it in his 1909 *Mendel's Principles of Heredity*. W. E. Castle's textbook, *Genetics and Eugenics*, 1920 edition, lists Garrod's 1902 paper on alcaptonuria in the bibliography but does not refer to it in the text. Thereafter it seems to have been dropped from all standard textbooks of genetics, until 1942 when J. B. S. Haldane in *New Paths in Genetics* discussed alcaptonuria in the context of phenylalanine-tyrosine metabolism and its genetic control.

I referred to this remarkable situation in a seminar at the University of California some years ago and pointed out that these derelict textbooks included Sturtevant and Beadle's 1939 *An Introduction to Genetics* as well as Goldschmidt's 1938 *Physiological Genetics*. Following my presentation, Professor Goldschmidt explained to me that he could not understand how this could have happened in his case, for he had known of Garrod's work and had referred to it in his own earlier works. It seems clear that Goldschmidt, like the many others who had read Garrod's papers and books, simply failed to see the significance of his view. He was in good company, for as Bentley Glass has recently pointed out, Muller referred to the analysis of alcaptonuria in 1922 without recognizing its basic significance.

It seems to me that, like Mendel, Garrod was so far ahead of his time that biochemists and geneticists were not ready to entertain seriously his gene-enzyme-reaction concept. Like Mendel's, Garrod's work remained to be rediscovered independently at a more favorable time in the development of the biological sciences. I strongly suspect that an important common component of the unfavorable climate for receptiveness in these two instances is the persistent feeling that any simple concept in biology must be wrong. Here are some classical examples of such concepts:

The synthesis of urea by Wöhler.
Darwin's theory of evolution.
Mendel's principles of inheritance.
Garrod's gene-enzyme-reaction concept.
The crystallization of urease by Sumner.
The crystallization of tobacco mosaic virus by Stanley.
The transformation of pneumococcus by Avery, MacLeod, and McCarty.
The Watson-Crick structure of DNA.

In connection with the last of the examples listed, I have a vivid recollection of a 1956 discussion with two distinguished biochemists, X and Y, about the significance of the Watson-Crick structure of DNA. I was not making much of an impression. Finally I asked X, who I assume must be the "Old Chemist" in Erwin Chargaff's incredible essay *Amphisbaena* (in

Essays on Nucleic Acid, Elsevier, 1963), "Do you believe that the Watson-Crick structure is essentially correct?" The amazing answer: "Yes, I think it is correct, but I don't think it has anything to do with replication."

Moral: Do not discard an hypothesis just because it is simple—it might be right.

K. G. ZIMMER
Institut für Strahlenbiologie, Kernforschungszentrum Karlsruhe, Karlsruhe, Germany

The Target Theory

1. INTRODUCTION

To write about work done more than thirty years ago, about the views and aims that gave rise to certain experiments, is not an easy task. In a recent review by another contributor to this volume (Stahl, 1959) we find the statement ". . . the primary aim in employing radiation in the study of phage is to elucidate the *normal* state of affairs." Well, this is exactly what I did *not* have in mind when starting to work with Timoféeff-Ressovsky about a year or two before we came into contact with Delbrück. The problem that fascinated me (and, by the way, it still does) was to find out as much as possible about the primary physico-chemical processes produced in elementary biological entities by ionizing radiation. At that time, genetic changes in Drosophila were about the most elementary and the most clearly defined biological reactions available and formed, therefore, the system of choice for such work—not to speak of the brilliant personality of Timoféeff-Ressovsky which made team-work an exciting adventure. Of course, the two views, i.e. the one held by Stahl and the other one that I prefer, are not, eventually, distinguishable. They rather form different approaches to one problem: one cannot use radiations for elucidating the normal state of affairs without considering the mechanisms of their actions, nor can one find out much about radiation induced changes without being interested in the normal state of the material under investigation.

2. THE STATE OF THE ART IN 1932

Nevertheless, I joined the team with the intention of using Drosophila to investigate the actions of ionizing radiations, and I should state clearly what attracted me to do so. In 1932 quantitative radiobiology had just become of age; that is, relevant experiments had been done for about 20 years. The early observations had revealed the need for quantitative analysis and formulation of hypotheses concerning the mechanism of action. This need arose because two observations were made, the explanation of which was by no means obvious.

The first of the two puzzling observations was that such remarkably small amounts of energy can be followed by biological effects when the energy is delivered to the biological material by ionizing radiation. To give this point force, many comparisons have been suggested, e.g. that the amount of energy absorbed by drinking a cup of tea would be fatal to man if delivered as X irradiation instead of heat.

The second of these puzzling observations stands in close connection to the research which gave rise to it. This was mostly carried out by irradiating populations of biological objects (such as bacteria) and determining the fraction of individuals which showed a given effect after a given dose. It had been previously recognized that such experiments could be of value only if the population was as homogeneous as possible in respect to all biological parameters, such as size and age of the individuals. Cor-

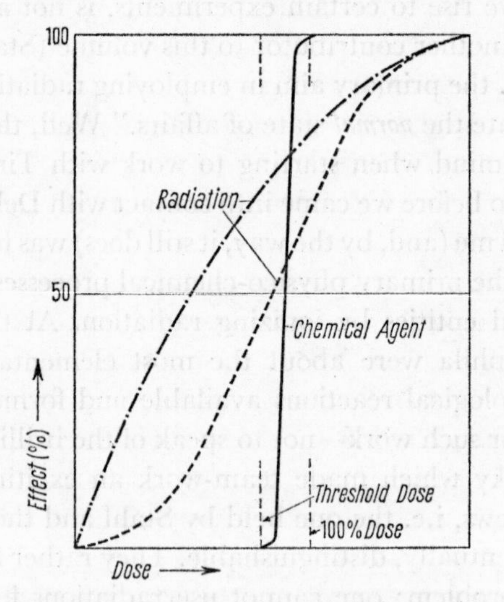

FIGURE 1. Diagram illustrating dose-effect curves for action of poisons and of radiation.

responding experiments with poisons (chemical agents) had mostly given dose-effect curves of a kind which showed practically no effect up to a "threshold dose" and then climbed steeply up to 100% effect (Fig. 1). But experiments with radiations led in many cases to dose effect curves rising slowly to 100% and in which no threshold could be recognized. Concerning the experiments with chemical agents, the difference between threshold value and 100% dose was generally regarded as an effect of unavoidable biological variability (scatter of sensitivity). The results of irradiation, however, were scarcely amenable to an explanation of this sort, since application of an analogous line of thought seemed to demand a wholly unusual variability in the biological parameters. In more recent times, more detailed investigations have raised doubts as to whether these arguments of old-time radiobiology were really sound. These doubts will have to be considered later (cf. section 4). It was the discovery of the form of a dose-effect curve for which a plausible explanation did not seem available that led to an entirely new line of thought: the application of the concepts of quantum physics to biological problems. These concepts in a generalized form have been well justified as a working hypothesis, since there is no doubt but that in this way modern physical concepts came into contact with biology, and that the synthesis of the two specialties so initiated has been remarkably fruitful. At this point our interest lies primarily in the beginnings of this development, that is, in the first hypothesis by way of which modern physical concepts were introduced some forty years ago into radiobiology, and thereby into biology: the "hit" theory (Dessauer, 1922; Blau and Altenburger, 1922). According to this view, the form of the observed dose-effect curve is due to the fact that absorption of radiation is not a continuous but a quantized process which follows the statistical principle that bears the name of Poisson. According to the mathematical formulation of this idea, the observed effect should appear in a member of a population having received macroscopically homogeneous irradiation when a minimal number of absorption events (called "hits") have happened in this individual. A strong resemblance could be shown to exist between the observed dose-effect curve and the curve calculated for the probability of occurrence of given numbers of absorption events for a given dose. In these calculations, the inevitable biological variability of the test material was, at first, admittedly neglected.

A very important development came some years later which not only independently extended the concepts of the hit theory but led further to the "target" theory (Crowther, 1924, 1926, 1927). The significant part of this work was that it offered the possibility of calculating from the dose-effect curve a volume, the target, within which the required number of absorption events must occur during irradiation, with a given probability. Comparison of the hit and target theories shows that the hit theory is to a large degree formal and very similar to the theory of chemical kinetics. The target theory, in contrast, demands that a well defined physical event must be chosen as the "hit." The three-dimensional target volume postulated in the original version of the theory can be computed only if the dose, stated as the number of absorption events per unit volume or mass, is given: here, choice can be made between a whole series of physical processes as absorption events, e.g. ionizations, excitations, primary ionizations, etc. Similarly, a two-dimensional target (reaction cross-section) can be worked out if the dosage is given in terms of the number of particles crossing a unit area. In any case, in applying the target theory, it is always necessary that the hit process should be clearly defined; a necessity which, as shown by later developments (Timoféeff-Ressovsky and Zimmer, 1947), has been often very helpful for further analysis.

A very important application of the target theory was made a few years later (Holweck and Lacassagne, 1930). Under the name of statistical ultramicrometry, a process was suggested of applying techniques of irradiation to determine the size of biological objects or structures, on the assumption that the targets calculated from experiments with irradiation in general corresponded to biological structures or functional units possibly too small to be measured in other ways. It must be admitted that these hypotheses of statistical ultramicrometry assumed implicitly that each hit on a target is followed by the observed effect with unit probability: but nothing is known *a priori* about this probability, and it is difficult to determine.

The hit theory thus appears to offer an explanation of the radiobiological dose-effect curve, which had seemed at first sight so difficult to understand. Beyond that, the target theory opens the way to understanding why irradiation is so efficient in inducing biological effects. For, if these effects result from transfer of energy to small targets, and not from release of energy into the whole bulk of the material, it is easy to understand the difference between the action of radiation and of heat. Because of the statistical nature of the absorption of radiation, introduction of small increments of energy into the material as a whole could mean that small targets receive relatively large amounts. This conception is well expressed by the word "point-heat" coined by Dessauer. Obviously, the question should be raised as to what kind of event can result from this localized release of energy. The idea at first associated with point-heat was local denaturation of protein, whereas other authors regarded chemical reactions induced by local energy release as more likely.

3. THE "GREEN PAMPHLET" OR "DREIMÄNNERWERK"

In order to apply the ideas outlined in the preceding section to the process of radiation-induced mutation, the already existing experimental results, mainly in Drosophila, had to be analyzed and extended. In the course of about two years, Timoféeff-Ressovsky, Zimmer and their collaborators succeeded in filling in several gaps in the understanding of the effects of various radiations other than ordinary X rays. The problems to be solved were concerned mainly with suitable sources of radiations and with accurate dosimetry. There is no need to describe this work in any detail here, though it may be pointed out that our use of 1 gm of radium as the source was something quite unusual in radiobiology, and that an exact evaluation of the energy absorbed by Drosophila from a field of gamma-rays had not been attempted before.

At about the time these studies reached completion, Delbrück became interested in our line of work: however hard I try, I cannot remember exactly how the contact was established, but I do remember vividly the discussions that followed. Two or three times a week we met, mostly in Timoféeff-Rossovsky's home in Berlin, where we talked, usually for ten hours or more without any break, taking some food during the session. There is no way of judging who learned most by this exchange of ideas, knowledge and experience, but it is a fact that after some months Delbrück was so deeply interested in quantitative biology, and particularly in genetics, that he stayed in this field permanently.

As an outcome of these discussions, a joint paper had been completed "Über die Natur der Genmutation und der Genstruktur" (N. W. Timoféeff-Ressovsky, K. G. Zimmer, and M. Delbrück, 1935) which was published in the *Nachrichten der Gesellschaft der Wissenschaften zu Göttingen* in the form of a little pamphlet with a bright green cover. Consequently, its friends and its critics used to refer to it as the "Green Pamphlet" or, somewhat deprecatingly, as the "Dreimännerwerk" ("Three-men-paper"): team work was not very usual in Germany thirty years ago, and inter-disciplinary team work appeared rather strange to some scientists. Nevertheless, the paper met with considerable interest and became widely known in many countries. If the main idea of the work can be described most succinctly by the words "to develop a 'quantum mechanical' model of the gene" (Stent, 1963), its later and more important sequelae impressed a geneticist of our days as follows "... in the years immediately preceding World War II, something quite new happened: the introduction of ideas (not techniques) from the realm of physics into the realm of genetics, particularly as applied to the problems of size, mutability, and self-replication of genes. ... Though this first application of physical ideas to a particular set of problems did not work out too well, the whole outlook in theoretical genetics has since been perfused with a physical flavour. The debt of genetics to physics, and to physical chemistry, for ideas began to be substantial then ..." (Pontecorvo, 1958).

At this point we might as well remember that I myself entered the field rather by the complementary approach, i.e., aiming to find out how radiations bring about biological effects. For this facet of the general problem (Zimmer, 1961) a recent appraisal of the "Green Pamphlet" and of its late effects is available too: "The 'hit' and 'target' theories were first brought into prominence in the late 20's ... The important development of this concept really has come through the publications of three investigators: Timoféeff-Ressovsky; Zimmer ...; and Delbrück. ... It is unfortunate that the 'hit' and 'target' theories have been so much neglected in the last few years. Both are very useful and helpful for interpreting radiation effects. ... They have not, however, always proved to be the most useful, especially with the entrance of biochemical approaches to modern radiation biology." (Hollaender, 1961).

Obviously, the "Green Pamphlet" is considered by some to have served a dual and useful purpose by initiating a new line of research in genetics and by stimulating others to apply similar ways of reasoning to radiobiology. It may be of interest, therefore, to mention some more recent results closely related to the subject matter of the "Green Pamphlet" and exemplifying what pitfalls these borderline fields hold in store for us.

4. SOME LATER DEVELOPMENTS

Using the terminology of the "hit" and "target" theories, as given in Section 2, the results forming the basis of the "Green Pamphlet" can be stated as follows: (i) The fraction of sex-linked lethal mutations in an irradiated population of Drosophila rises with the dose D of X rays according to the equation $N^*/N_0 = 1 - \exp(-vD)$, thus indicating a one-hit-process. (ii) The formal volume v of the target, as calculated from the same equation, is (within certain limits) independent of the spatial density of ionization (linear energy transfer), if the doses D are counted in numbers of ionizations per unit volume of Drosophila. Consequently, one ionization within the formal target may be considered a hit. (iii) Taking into consideration additional data on temperature dependence of radiation-induced, as well as of spontaneous mutation, a "quantum-jump" may be regarded as the physical process produced by a hit in a target and leading to mutation.

There is no reason to discuss (iii) in any detail here, but (i) and (ii) seemed certainly well established in 1935. Later on, further experiments lent additional support. In fact, the deviations of the target volumes v shown in Table I from the weighted mean value of $\bar{v} = 1.77 \cdot 10^{-17}$ cm^3 are so small that the data form one of the most carefully tested cases of a one-hit curve in radiobiology. Nevertheless, as time passed I became worried about the approximation inherent in this reasoning: the complete neglect of a possible biological variability. At first there was no sign of its existence in the material under investigation, but theoretical analyses showed that hit-curves can be badly distorted by a quite moderate variability in target volume, hit number, or multiplicity of targets (Zimmer, 1941). Some years later (in fact, incited by unpublished experiments on the action of X rays on the eggs of some water-snail whose name I have forgotten) I investigated graphically the possibility that approximate single-hit curves could arise through superposition of multi-hit curves. Thence it appeared that, based on different, quite plausible assumptions, curves can be obtained which wrap sinously round exact single-hit curves, within the limits of the accuracy obtainable in radiobiological experiments (Zimmer, 1950; and unpublished). About a decade later a really comprehensive investigation of these possibilities was carried out at our suggestion (Dittrich, 1960). An instructive case of deception by an apparent single-hit curve is shown in Fig. 2.

Table 1

FORMAL TARGET VOLUMES FOR INDUCTION OF SEX-LINKED RECESSIVE LETHALS IN *DROSOPHILA MELANOGASTER*, AS CALCULATED FROM EXPERIMENTS BY TIMOFÉEFF-RESSOVSKY, ZIMMER ET AL. (ZIMMER, 1943).

Dose[1] D in ion pairs per cm³	$N/N_0 = 1 - N^*/N_0$	$v = -\frac{\ln N/N_0}{D}$	N_0	vN_0	$\bar{v} = \frac{\Sigma v_i N_{0i}}{\Sigma N_{0i}}$
		Radium-β rays			
$2.33 \cdot 10^{15}$	0.9621	$1.67 \cdot 10^{-17}$	1872	$3.13 \cdot 10^{-14}$	
$4.67 \cdot 10^{15}$	0.9082	$2.06 \cdot 10^{-17}$	1531	$3.15 \cdot 10^{-14}$	
$7.24 \cdot 10^{15}$	0.8847	$1.68 \cdot 10^{-17}$	1214	$2.04 \cdot 10^{-14}$	
$9.33 \cdot 10^{15}$	0.8512	$1.73 \cdot 10^{-17}$	1057	$1.83 \cdot 10^{-14}$	$1.78 \cdot 10^{-17}$ cm³
		Radium-γ rays			
$2.17 \cdot 10^{15}$	0.9645	$1.66 \cdot 10^{-17}$	1642	$2.72 \cdot 10^{-14}$	
$4.34 \cdot 10^{15}$	0.9188	$1.96 \cdot 10^{-17}$	1293	$2.53 \cdot 10^{-14}$	
$7.24 \cdot 10^{15}$	0.8800	$1.77 \cdot 10^{-17}$	1184	$2.09 \cdot 10^{-14}$	
$8.69 \cdot 10^{15}$	0.8653	$1.67 \cdot 10^{-17}$	822	$1.37 \cdot 10^{-14}$	$1.76 \cdot 10^{-17}$ cm³
		X rays, 160 kV			
$0.97 \cdot 10^{15}$	0.9828	$1.76 \cdot 10^{-17}$	3082	$5.43 \cdot 10^{-14}$	
$1.93 \cdot 10^{15}$	0.9669	$1.76 \cdot 10^{-17}$	5020	$8.83 \cdot 10^{-14}$	
$3.86 \cdot 10^{15}$	0.9375	$1.68 \cdot 10^{-17}$	3948	$6.65 \cdot 10^{-14}$	
$5.79 \cdot 10^{15}$	0.8966	$1.88 \cdot 10^{-17}$	3504	$6.66 \cdot 10^{-14}$	
$7.82 \cdot 10^{15}$	0.8739	$1.75 \cdot 10^{-17}$	3107	$5.43 \cdot 10^{-14}$	$1.77 \cdot 10^{-17}$ cm³
		X rays, 70 kV			
$1.21 \cdot 10^{15}$	0.9793	$1.74 \cdot 10^{-17}$	9346	$16.3 \cdot 10^{-14}$	
$2.42 \cdot 10^{15}$	0.9575	$1.78 \cdot 10^{-17}$	16467	$29.3 \cdot 10^{-14}$	
$4.42 \cdot 10^{15}$	0.9185	$1.92 \cdot 10^{-17}$	3466	$6.65 \cdot 10^{-14}$	
$4.73 \cdot 10^{15}$	0.9141	$1.86 \cdot 10^{-17}$	11738	$21.9 \cdot 10^{-14}$	
$6.03 \cdot 10^{15}$	0.8815	$2.09 \cdot 10^{-17}$	2064	$4.32 \cdot 10^{-14}$	
$7.23 \cdot 10^{15}$	0.8768	$1.81 \cdot 10^{-17}$	6442	$11.7 \cdot 10^{-14}$	
$9.65 \cdot 10^{15}$	0.8412	$1.79 \cdot 10^{-17}$	9116	$16.3 \cdot 10^{-14}$	$1.81 \cdot 10^{-17}$ cm³
		X rays, 10 kV			
$1.75 \cdot 10^{15}$	0.9698	$1.71 \cdot 10^{-17}$	3338	$5.71 \cdot 10^{-14}$	
$2.42 \cdot 10^{15}$	0.9588	$1.74 \cdot 10^{-17}$	2731	$4.75 \cdot 10^{-14}$	
$3.51 \cdot 10^{15}$	0.9395	$1.76 \cdot 10^{-17}$	2124	$3.75 \cdot 10^{-14}$	
$4.83 \cdot 10^{15}$	0.9201	$1.72 \cdot 10^{-17}$	1816	$3.13 \cdot 10^{-14}$	
$7.00 \cdot 10^{15}$	0.8869	$1.71 \cdot 10^{-17}$	1641	$2.82 \cdot 10^{-14}$	$1.73 \cdot 10^{-17}$ cm³

1. Assuming the production of $1.61 \cdot 10^{12}$ ion pairs per cm³ of Drosophila-tissue and per R of X rays, β rays and γ rays.

In the meantime, the influence of the stage of Drosophila germ cells on their X-ray induced mutability became more clearly recognized. The usual way to investigate this phenomenon had been to work out a so-called "brood pattern" giving the rate of mutation for germ cells of various stages after irradiation with a given dosage of radiation. Comparing such brood patterns obtained at various doses made the single-hit curve described above appear quite unbelievable (Fig. 3). In order to elucidate the meaning of this apparent contradiction, the tiresome task was undertaken in my present laboratory to obtain statistically significant dose-effect curves for various stages of germ cells (Traut, 1962, 1963). An example of the results is given in Fig. 4. Here, it must be emphasized that the strange form of the curves

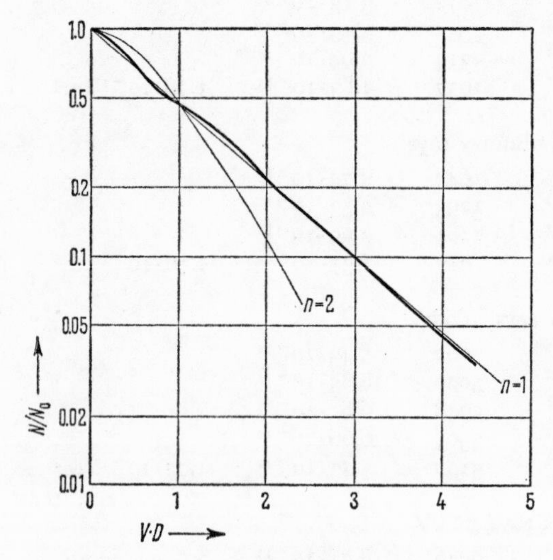

FIGURE 2. Approximate agreement of a single-hit curve with a two-hit mixed curve formed by super-position of 4 two-hit curves with differently sized targets (Dittrich, 1960).

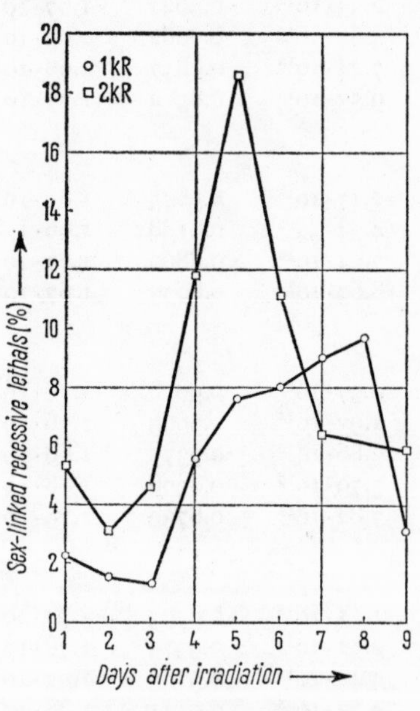

FIGURE 3. The dependence of radiation-induced lethal rate on stage sensitivity after irradiation of 3-4-day-old B-males; 9 one-day broods. Spontaneous rate subtracted (Traut, 1963).

is significant, as shown by careful statistical analysis using an electronic computer. Nevertheless, summing up arithmetically all the mutations obtained in the various broods for given doses resulted in a dose-effect curve closely approximating the single-hit curve (Fig. 5) which, thirty years ago, formed one of the starting points of the "Green Pamphlet." Complete agreement cannot be expected, as arithmetic summation is, of course, not identical with neglecting stage-dependence of mutability, which may be considered as a process of "biologic summation." Thus, the old single-hit curve turned out to be quite reproducible and to hold for all practical purposes (such as problems of protection from radiation damage). But it became clear also that though the dose-effect relation of mutation induction has the form of a single-hit curve, it certainly does not have the meaning of a single hit.

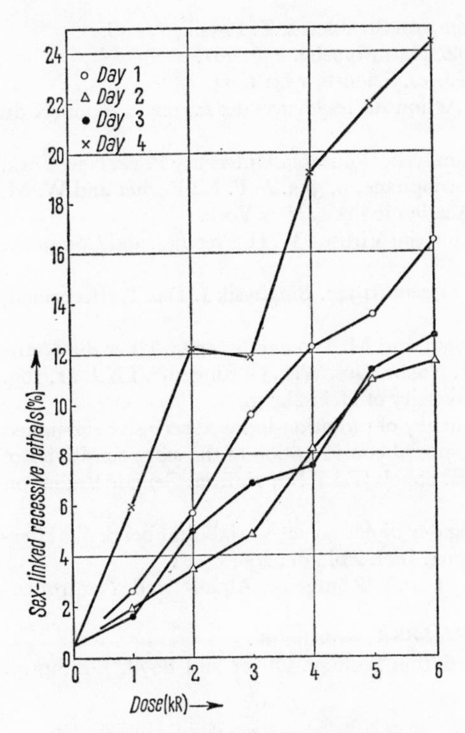

FIGURE 4. Dose effect curves for lethals induced in stages with different sensitivity. 3-4-day-old B-males were irradiated (Traut, 1963).

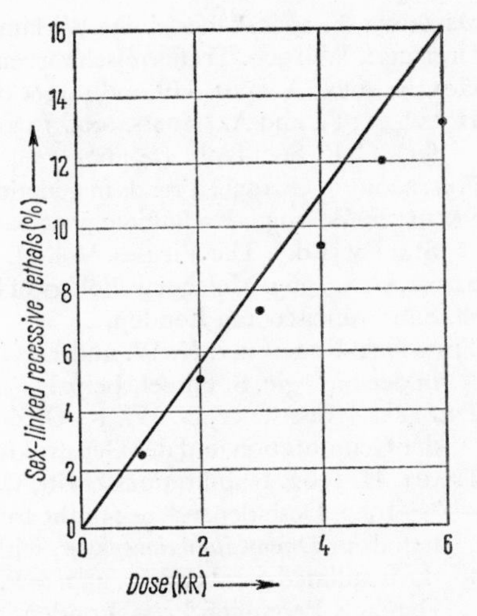

FIGURE 5. The solution to the apparent contradiction. • : mean frequencies of sex-linked recessive lethals integrated arithmetically over the first four days after irradiation (Traut, 1963). ——: dose-effect curve as used for the "Green Pamphlet." This curve was obtained neglecting stage dependency of mutability i.e. by "integrating biologically" over the first days after irradiation (Timoféeff-Ressovsky, Zimmer, and Delbrück, 1935). Abscissa linear, ordinate logarithmic.

This result removes one of the foundation-stones of the "Green Pamphlet." Strangely enough, that does not seem to matter any more, for two reasons: (i) the concept of the gene and modern trends in genetic research as well as in radiation biology have changed considerably during thirty years, as will undoubtedly become evident from the subsequent papers in this book, and (ii) the "Green Pamphlet" has served a useful purpose by helping to initiate exactly these modern trends.

REFERENCES

BLAU, M. and K. ALTENBURGER. 1922. Über einige Wirkungen von Strahlen II. Z. Physik, *12*: 315.

CROWTHER, J. A. 1924. Some considerations relative to the action of X-rays on tissue cells. Proc. Roy. Soc. (London), *B 96*: 207.

———1926. The action of X-rays on Colpidium colpoda. Proc. Roy. Soc. (London), *B 100*: 390.

———1927. A theory of the action of X-rays on living cells. Proc. Camb. Phil. Soc., *23*: 284.

DESSAUER, F. 1922. Über einige Wirkungen von Strahlen I. Z. Physik, *12*: 38.

DITTRICH, W. 1960. Treffermischkurven. Z. Naturforschg. *15b*: 261.

HOLLAENDER, A. 1961. Hit and target theories. Science, *134*: 1233.

HOLWECK, F., and A. LACASSAGNE. 1930. Action sur les levures des rayons X mous (K du fer). C. R. Soc. Biol., *103*: 60.

PONTECORVO, G. 1958. Trends in genetic analysis. Columbia University Press, New York.

STAHL, F. W. 1959. Radiobiology of bacteriophage, p. 353. *In* F. M. Burnet and W. M. Stanley [Ed.], The Viruses. Vol. II. Academic Press, New York.

STENT, G. S. 1963. Molecular Biology of Bacterial Viruses. W. H. Freeman and Company, San Francisco and London.

TIMOFÉEFF-RESSOVSKY, N. W. and K. G. ZIMMER. 1947. Biophysik I. Das Trefferprinzip in der Biologie. S. Hirzel, Leipzig.

TIMOFÉEFF-RESSOVSKY, N. W., K. G. ZIMMER and M. DELBRÜCK. 1935. Über die Natur der Genmutation und der Genstruktur. Nachr. Ges. Wiss. Göttingen, VI N.F., *1*, 189.

TRAUT, H. 1962. Habilitationsschrift, University of Heidelberg.

———1963. Dose-dependence of the frequency of radiation-induced recessive sex-linked lethals in *Drosophila melanogaster*, with special consideration of the stage sensitivity of the irradiated germ cells, p. 359. *In* F. H. Sobels [Ed.], Repair from Genetic Radiation Damage. Pergamon Press, London.

ZIMMER, K. G. 1941. Zur Berücksichtigung der biologischen Variabilität bei der Treffertheorie der biologischen Strahlenwirkung. Biol. Zbl., *61*: 208.

———1943. Statistische Ultramikrometrie mit Röntgen-, Alpha- und Neutronenstrahlung. Physikal. Z., *44*: 233.

———1950. Otchet. Fondy Ural. Fil. A. N. SSSR, Sverdlovsk.

———1961. Studies on Quantitative Radiation Biology. Oliver and Boyd, Edinburgh and London. 124 p.

HERMAN M. KALCKAR
Department of Biological Chemistry, Harvard Medical School, and Biochemical Research Laboratory, Massachusetts General Hospital, Boston, Massachusetts

High Energy Phosphate Bonds: Optional or Obligatory?

Delbrück introduced me to quantitative biology and genetics through his first phage course, held at Cold Spring Harbor in 1945. At that time, few biochemists thought along the lines of Poisson distributions, although there were some notable exceptions (*vide* the treatise by Linderstrøm-Lang and Fizz-Loony on "The Thermodynamic Activity of the Male Housefly" [Linderstrøm-Lang, 1962]). It is impossible, however, to grasp the fundamental aspects of natural selection and population dynamics—so beautifully illustrated by the Luria-Delbrück fluctuation test—without paying attention to Poisson distributions, exponential equations, etc. And so Delbrück, who is uncompromising when something crucial is at stake, set an entrance examination on such matters for all who applied to take his course.

When it was first suggested to me to discuss here the development of the concept of phosphorylation, my response was ambivalent because it has been a quarter of a century now since I was actively concerned with this problem. But my reluctance waned once I put my mind to this task within the context of this series of essays, and I set out to try to reconstruct the *esprit* of the biochemical *avant garde* of those days. Foremost among the evocative expressions that were then coined is "high energy bond," which is a type of forceful slang introduced by Lipmann in 1941—the year when the symbol "~P" was also born. The expression was first used to distinguish certain types of phosphoric ester bonds which, upon hydrolysis, release free energy on the order of 10 kcal, from ordinary phosphate ester bonds, the hydrolysis of which releases only 2 to 3 kcal of free energy (Lipmann, 1941). This notion of making any such distinction was entirely novel to most American biochemists at that time.

While Lipmann was preaching the gospel of phosphorylation and the "high energy phosphate bond" on the eastern seaboard a quarter of a century ago, I was beginning my missionary work at the same time in the "Pacific triangle"—Caltech, Berkeley, and Stanford. As I was in the vicinity of the chief architects of modern physical chemistry, G. N. Lewis and his disciples, I meticulously tried to avoid any type of slang in my colloquia on energetics coupling in fermentative or oxidative phosphorylation. Although I am not sure that I succeeded in eliciting much interest for my own work, oxidative phosphorylation, I was not without luck with my discourses on phosphorylation during my 1939 stay at Caltech. The new notions of acid anhydrides and "dry" fission of acylphosphates had just been introduced by Lipmann (1940) and by Warburg and Negelein (Negelein and Brömel, 1939), and the Coris had discovered dry fission, i.e. phosphorolysis of glycogen (Cori and Cori, 1936). It was possible, therefore, to make a rather forceful case out of substrate phosphorylation, and these monumental problems elicited lively discussions, particularly with Delbrück, Borsook, Coryell, and Pauling.

Perhaps one can say that the discovery of the coupling of the biosynthesis of acylphosphates to the "dry" dehydrogenation of carbonyl compounds marked the beginning of molecular biology, at least as far as explaining the mechanism of generation of energy in the living cell. I particularly recall some remarkable discussions with Charles Coryell, who at that time was part of the Linus Pauling "resonance team." We had fun trying to use notions like "opposing resonance" for the powerful phosphoryl donors like phosphoguanidine pyrophosphates and acylphosphates (Kalckar, 1941).

The realization of the existence of phosphoric esters characterized by a high ΔH of hydrolysis goes back to Meyerhof and Lohmann (1927). The successful application of this knowledge to one cellular function, muscle contraction, was formulated by Lundsgaard in 1930 on the basis of his discovery of alactic muscle contraction (Lundsgaard, 1930).

Neither oxidative phosphorylation of yesterday (Kalckar, 1938, 1939) nor of today has yet reached the conceptual level of molecular biology, in spite of many brilliant efforts. The most decisive event was perhaps the discovery of the mitochondrial matrix as an integral part of oxidative phosphorylation (Kennedy and Lehninger, 1949). Apart from this, the most spectacular progress has been in elucidating the role of phosphorylation in the energetics of muscle action. In any event, it seems that phosphorylation is the fundamental biological "coinage" for transfer and transition of free energy.

In chemical evolution, phosphate may well be considered irreplaceable. Wald (1962) lists a number of subtle physical features of phosphate and phosphoric esters. According to Cruickshank (cited by Wald, 1962), phosphoric esters show asymmetries of their phosphate tetrahedral ions, in that the P-O bond of the carbon or nitrogen phosphorus is "stretched" whereas the other P-O bonds tend to "contract." This feature, in connection with the presence of unoccupied 3d orbitals in P, provides a foothold for nucleophilic agents, such as ROH, RNH_2, or water.

Inorganic phosphate and inorganic pyrophosphate can of course also act as nucleophilic agents themselves (Cori and Cori, 1936; Kalckar, 1949; Kornberg, 1950). In fact, in the coupling process, sulfhydryl of the protein or of a specific coenzyme usually plays a primary role, and the high ΔF of the acyl-mercapto linkage is preserved by "dry" fission, using inorganic orthophosphate as a nucleophilic agent (Lynen, 1953; Racker, 1954). The acyl-phosphate formed might subsequently "trade in" with ADP by transferring the "dry phosphoryl" group to ADP forming ATP. In these types of "phosphoryl" transfer, the terminal phosphate of the P-acceptor, ADP, is the nucleophilic agent. The bond lengths of acyl-phosphate or ATP probably follow the general Cruickshank rule of phosphate ester bonds outlined above. The "Cruickshank features" might be even more extreme in the case of phosphoric anhydrides. The hydrolytic splitting of ATP illustrates this feature in an interesting way. Thus hydrolysis of adenosine triphosphate in $H_2{}^{18}O$ leaves all the ^{18}O in the liberated phosphate (Koshland and Levy, 1964). Likewise in phosphorylation, the P-O linkage of the anhydride bond is the bond which encourages an attack by a nucleophilic group like R-OH or R-NH_2 etc.

In the origin of life, phosphate may have been "tailor made" to serve together with hydrogen, ammonia, and carbon dioxide in launching the first successful ensemble of autocatalytically replicating organic molecules. In any event, that phosphate played a role in early chemical evolution seems most likely.

Oxidative phosphorylation is a later development in evolution. The coupled phosphorylation encountered here or in photophosphorylation may, however, be generated in quite a different way than the previously discussed mechanism. It has been customary to ascribe the splitting of ATP by adenosine triphosphatase as an energy-yielding reaction which can be used to drive ions across a membrane against a gradient. However, Mitchell's interesting suggestion (Mitchell, 1961) that the oxygen cytochrome system releases OH^- or $H_3O_2{}^-$ ions on one side of the membrane and the dehydrogenase system (starting with succinate) releases H^+ ions on the opposite side ("Chemiosmosis") would pave the way for an alternative and perhaps more appealing formulation of mitochondrial phosphorylation. The polarity and structure of a membrane containing an operating respiratory chain might well suffice to create a powerful ion potential which in turn can be used for formation of $\sim$P.

The Mitchell theory might also explain another curious phenomenon in oxidative phosphorylation. The oxygen of inorganic phosphate is able to undergo a rapid exchange with the OH^- of water, provided intact mitochondria are performing oxidative phosphorylation (Cohn, 1951). This important but rather puzzling observation can be explained if one assumes that the trapped H^+ drives OH^- out of orthophosphate creating an electrophilic "phosphoryl" group.

If the phosphoryl group is located close to a nucleophilic acceptor like adenosinediphosphate, perhaps embedded in a membrane ATP_{ase}, formation of ATP would ensue. The reverse reaction, ATP splitting, releasing phosphoryl which can pick up $^{18}OH^-$ from $H_2{}^{18}O$, might in fact directly contribute to raise the potential again. Mitchell and Moyle have been able to support this thesis recently (Mitchell and Moyle, 1965). As can be seen, the chemiosmotic phosphorylation is quite a different type of coupling than that discovered in fermentation; perhaps it is also more differentiated.

It should be pointed out that separation of ions might well proceed independent of $\sim$P. Likewise, utilization of this ion potential for various cellular functions may find channels other than that of "phosphate bond energy." One could well raise the question whether oxidative phosphorylation may not be an entirely optional pathway for many types of cells. A dramatic illustration of the possibility of creating "high energy bonds" without phosphorylation has, in fact, been provided through Mayne and Clayton's (1966) recent paper (communicated by none other than Delbrück to the National Academy of Science). It was demonstrated in that work that chloroplasts can elicit a flash of light, having the spectrum of chlorophyll fluorescence, upon induction of brief acid-base transitions or oxidation-reductions. Phosphorylation does not seem to come into play in this type of chemiluminescence since the presence or absence of ATP or phosphate acceptors has no detectable effect on the phenomenon.

My interest in cell physiology during my apprentice years was driven by my impatience with classical physiology because it seemed to avoid any attempt to account for the mechanisms of active cell functions. "Bioenergetics," as it is called nowadays, was therefore a field which attracted me immensely. Delbrück, having then recently come from the field of atomic physics and quantum mechanics, was much more impatient with the problems with which *he* wanted to concern himself. He wanted to bypass all the phenotypic diversities of the living cell and go straight to the problem of gene replication and gene action. Was it possible to find fundamental natural laws for gene action and gene replication?

Right from the outset, twenty-seven years ago, when Delbrück initiated his search for such laws by the concentrated study of the T-phages, diversities cropped up plentifully in this system (Delbrück, 1946). One of the major complexities to be encountered was the discovery of 5-hydroxymethylcytosine in the DNA of T-even phages (Wyatt and Cohen, 1953), a pyrimidine whose existence would indeed be difficult to explain by invoking chemical evolution based on structural chemical characteristics. Subsequent pursuit of this problem (cf. Hattmann and Fukasawa, 1963) made it evident, however, that it was what we could call "social" incompatibilities between phage and host which may have governed the evolutionary history of hydroxymethylcytosine and its glucosylated DNA derivatives. Such social encounters may also underlie the origin of using compounds which are not confined to a few organisms but are almost ubiquitous. For instance, the capacity of organisms for synthesis of compounds like uridine diphosphoglucose (UDP-glucose) and UDP-galactose seems to be as widespread as their ability to synthesize ATP or UTP. Mutant organisms which have lost their ability to synthesize ATP or UTP would be nonviable under almost any circumstances, especially if that loss derived from an impairment of phosphorylation. In contrast to the certain death of ATP-less or UTP-less mutants, we know of several viable mutants unable to synthesize UDP-glucose or UDP-galactose (Fukasawa et al., 1962; Sundararajan et al., 1962) and yet able to grow perfectly well without any supplements except their usual nutrients.

The widespread occurrence of galactose compounds furnishes an important example of an evolutionary development which cannot be based on chemical evolution. Pathways for the synthesis of UDP-glucose and UDP-galactose have been developed in bacteria, plants, and animal cells (Frydman and Neufeld, 1963; Kalckar et al., 1959; Kalckar, 1960). Yet, galactose compounds play no role in the function or viability of living cells as far as chemical evolution is concerned. Although Salmonella mutants lacking 4-epimerase lose all the galactose of their polysaccharides, their generation time and general viability in galactose-free nutrient media are unaffected (Fukasawa and Nikaido, 1959). The galactose-free creatures are even protected from attack by the phage P-22 (Fukasawa and Nikaido, 1960), because they have lost the superficial phage receptor sites whose specificity depends on galactose and the deoxyhexoses which nest on galactose.

One could ask why any cell should waste the "high energy phosphate" resident in UTP for the sake of synthesizing a compound like UDP-galactose which seems superfluous, and even hazardous. One answer to this question may be that perhaps the hazards are outweighed by the prospects of novel biological combinations. Phage P-22 is a temperate phage which offers a chance to acquire new genetic information by grace of lysogeny to a small fraction of those potential host cells that contains accessory sugars like galactose or deoxyhexoses. Perhaps this innovation, or "transfiguration" of the minority offers advantages that greatly outweigh the heavy toll of death —the fate of the majority of the population.

Finally we must bear in mind the possibility that the development of pathways for the biosynthesis of some of these accessory sugars may be completely optional and carry neither positive nor negative selective values. This touches on the problem of whether we are always entitled to invoke the concept of natural selection when discussing biochemical evolution.

If we grant to the social encounters of living cells a more important role in their biochemical evolution, we may better understand the many confusing metabolic features of the living cell. My own early enthusiasm for the recognition of the "$\sim$P" is undiminished but tempered by the facts of "Life" and living communities. Delbrück expressed the essence of many of these problems in his characteristic way in a memorable speech at the thousandth meeting of the Connecticut Academy of Arts and Sciences in 1949, which is included in this book. (Paul Hindemith was one of the other distinguished scholars who addressed that body on that occasion.)

The curiosity remains, though, to grasp more clearly how the same matter, which in physics and in chemistry displays orderly and reproducible and relatively simple properties, arranges itself in the most astounding fashions as soon as it is drawn into the orbit of the living organism. The closer one looks at these performances of matter in living organisms, the more impressive the show becomes. The meanest living cell becomes a magic puzzle box full of elaborate and changing molecules, and far outstrips all chemical laboratories of man in the skill of organic synthesis performed with ease, expedition, and good judgment of balance. The complex accomplishment of any one living cell is part and parcel of the first-mentioned feature, that any one cell represents more an historical than a physical event. These complex things do not rise every day by spontaneous generation from the nonliving matter—if they did, they would really be reproducible and timeless phenomena, comparable to the crystallization of a solution, and would belong to the subject matter of physics proper. No, any living cell carries with it the experiences of a billion years of experimentation by its ancestors. You cannot expect to explain so wise an old bird in a few simple words. (Delbrück, 1949.)

These reflections of 1949 go beyond the realms of molecular biology—but why not?

REFERENCES

CORI, C. F. and CORI, G. T. 1936. Mechanism of formation of hexosemonophosphate in muscle and isolation of a new phosphate ester. Proc. Soc. Exp. Biol., N.Y., *34*:702.

COHN, M. 1951. A study of oxidative phosphorylation with inorganic phosphate labeled with oxygen-18. *In* W. B. McElroy and B. Glass [Ed.], Phosphorus Metabolism, Baltimore, The Johns Hopkins Press, vol. I, p. 374.

DELBRÜCK, M. 1946. Experiments with bacterial viruses (bacteriophages). Harvey Lectures, Springfield, Ill., Charles C. Thomas, vol. 41, p. 161.

———1949. A physicist looks at biology. Trans. of Conn. Acad. of Arts and Sciences, *38*: 173.

FRYDMAN, R. B., and E. F. NEUFELD. 1963. Synthesis of galactosylinositol by extracts from peas. Biochem. Biophys. Res. Comm., *12*: 121.

FUKASAWA, T., and H. NIKAIDO. 1959. Galactose-sensitive mutants of *Salmonella*. Nature, *184*: 1168.

———, ———1960. Formation of phage receptors induced by galactose in a galactose-sensitive mutant of *Salmonella*. Virology, *11*: 508.

FUKASAWA, T., K. JOKURA, and K. KURAHASHI. 1962. A new enzymic defect of galactose metabolism in *Escherichia coli* K-12 mutants. Biochem. Biophys. Res. Comm., *7*: 121.

HATTMANN, S., and T. FUKASAWA. 1963. Host-induced modification of T-even phages due to defective glucosylation of their DNA. Proc. Natl. Acad. Sci., *50*: 297.

KALCKAR, H. M. 1938. Fosforyleringprocesser i dyrisk vaer. Dissertation, Copenhagen, Nyt Nordisk Forlag, Arnold Busck.

———1939. The nature of phosphoric esters formed in kidney extracts. Biochem. J., *33*: 631.

———1941. The nature of energetic coupling in biological syntheses. Chem. Rev., *28*: 71.

———1949. Enzymatic reaction in purine metabolism. Harvey Lectures, Springfield, Ill., Charles C. Thomas, vol. 45, p. 11 (Publ. 1952).

———1960. Hereditary defects in galactose metabolism in man and microorganisms. Fed. Proc., *19*: 984.

KALCKAR, H. M., K. KURAHASHI, and E. JORDAN. 1959. Hereditary defects in galactose metabolism in *Escherichia coli* mutants, I. Determination of enzyme activities. Proc. Natl. Acad. Sci., *45*: 1776.

KENNEDY, E. P., and A. L. LEHNINGER. 1949. Oxidation of fatty acids and tricarboxylic acid cycle intermediates by isolated rat liver mitochondria. J. Biol. Chem., *179*: 957.

KORNBERG, A. 1950. Reversible enzymatic synthesis of diphosphopyridine nucleotide and inorganic pyrophosphate. J. Biol. Chem., *182*: 779.

KOSHLAND, D. E. Jr., and H. M. LEVY. 1964. Evidence for an intermediate in ATP hydrolysis by myosin. *In* J. Gergeley [Ed.], Biochemistry of Muscular Contraction, Boston, Little, Brown and Co., p. 87. Retina Foundation, Institute of Biological and Medical Sciences, Monographs and Conferences, vol. II.

LINDERSTRØM-LANG, K. 1962. Selected papers. Danish Science Press, Copenhagen; Academic Press, New York.

LIPMANN, F. 1940. A phosphorylated oxidation product of pyruvic acid. J. Biol. Chem., *134*: 463.

———1941. Metabolic generation and utilization of phosphate bond energy. Advances in Enzymology, New York, Interscience Publishers, vol. 1, p. 99.

LUNDSGAARD, E. 1930. Weitere Untersuchungen über Muskelkontraktionen ohne Milchsäurebildung. Biochem. Z., *227*: 51.

LYNEN, F. 1953. Acetyl coenzyme and the "fatty acid cycle." Harvey Lectures, New York, Academic Press, vol. 48, p. 210.

MAYNE, B. C., and R. C. CLAYTON. 1966. Luminescence of chlorophyll in spinach chloroplasts induced by acid-base transition. Proc. Natl. Acad. Sci., *55*: 494.

MEYERHOF. O., and K. LOHMANN. 1927. Über den Ursprung der Kontraktionswärme. Naturwiss., *15*: 670.

MITCHELL, P. 1961. A chemiosmotic hypothesis for the mechanism of oxidative and photosynthetic phosphorylation. Nature, *191*: 144.

MITCHELL, P., and J. MOYLE. 1965. Stoichiometry of proton translocation through the respiratory chain and adenosine triphosphatase systems of rat liver mitochondria. Nature, *208*: 147.

NEGELEIN, E., and E. BRÖMEL. 1939. R-diphosphoglycerinsäure, ihre Isolierung und Eigenschaften. Biochem. Z., *303*: 132.

RACKER, E. 1954. Formation of acyl and carbonyl complexes associated with electron-transport and group-transfer reactions. *In* A Symposium on the Mechanism of Enzyme Action, W. D. McElroy and B. Glass [Ed.], Baltimore, The Johns Hopkins Press. Sponsored by the McCollum-Pratt Institute of the Johns Hopkins University.

SUNDARARAJAN, T. A., A. M. C. RAPIN, and H. M. KALCKAR. 1962. Biochemical observations on *E. coli* mutants defective in uridine diphosphoglucose. Proc. Natl. Acad. Sci., *48*: 2187.

WALD, G. 1962. Life in the second and third periods; or why phosphorus and sulfur for high-energy bonds? *In* M. Kasha and B. Pullman [Ed.], New York, Academic Press, p. 127.

WYATT, G. R., and S. S. COHEN. 1953. The bases of the nucleic acids of some bacterial and animal viruses: the occurrence of 5-hydroxymethylcytosine. Biochem. J., *55*: 774.

II. *The Phage Renaissance*

M. Delbrück and S. E. Luria, Cold Spring Harbor, 1953

EMORY L. ELLIS
Devcom Associates, LaHabra, California

Bacteriophage: One-Step Growth

"Is such a virus living? The question is unanswerable without a definition of 'life'. And any definition of life must be arbitrary." (BEADLE, 1964)

I. PREFACE

In the mid-1930's Thomas Hunt Morgan was the chairman of the Biology Division of the California Institute of Technology, and the leader of its group of geneticists. In another part of the Institute, physicists were studying high-voltage X rays, and physicians associated with them were examining effects of these radiations on human cancers. In the Chemistry Division, whose chairman was Linus Pauling, the tools for elucidation of molecular structure were being applied to proteins. In this small institution, these lines of research did not proceed in isolation from each other.

The importance of inter-disciplinary communication is more generally appreciated today than thirty years ago, but it has always been a feature of research at Caltech. It played a part in the initial interest at Caltech in bacteriophage, which came about through studies in the biochemistry of malignant tumors. Since some tumors were known to be transmissible by cell-free filtrates, it appeared that more knowledge regarding the nature of filtrable viruses would be helpful in understanding these, and perhaps other malignancies. Bacteriophage was the filtrable virus chosen for study as the model system.

This chapter is an effort to summarize the early work and to reconstruct the circumstances which surrounded and influenced it. Perhaps it is presumptive for someone like myself, who has been away from biochemical research for more than twenty years, to contribute a chapter to a book on that subject. Obviously, that contribution will have limits—and certainly it cannot relate closely to the current state of the art. But I shall attempt it anyway, in particular since it has been my pleasure to re-read some of the work of the thirties, and to consider this early work in the context of the later developments, as set forth in the excellent books of Stent (1963) and of Adams (1959). What I find most impressive is how much progress has been made meanwhile in the understanding of life processes, and how exquisite are the techniques and how powerful are the instruments now used in this research, compared to the simple and crude experiments of the 1930's.

A most pleasant aspect of the preparation of this chapter is the opportunity to record the important early contributions made by Max Delbrück. It is most appropriate that this volume should be prepared and dedicated to recognizing the role he played in illuminating the mysteries surrounding the replication of genetic material.

Before proceeding further, the reader should be warned that some of the discussion will be speculative. Speculation leads to hypotheses, which are, in turn, important steps toward planning productive experimentation. Perhaps recounting some of the speculations of the thirties would be of interest to the reader, and may even provide him with some stimulation as well. If so, time spent in preparation of this paper is well rewarded.

2. DISEASES INDUCED BY FILTRABLE VIRUSES

By the late 1930's it had been shown that specific filtrable viruses are the causative agents of some diseases in plants, of some cancerous growth in animals, and of the lysis of some bacterial species. One of these viruses, that causing the tobacco plant mosaic disease, had been isolated in crystalline form and shown to contain nucleic acid and protein.

We speculated about the process by which cancerous growth is produced in animals by filtrable viruses. Logically, this seemed to consist of two main steps: (a) infection of one or more animal cells with the virus, and (b) proliferation of these infected cells on the surrounding substrate. The infected cells behave as if they have been released from some general control of their division existing before infection, or as if infection resulted in stimulation of cell division.

How do these virus-induced malignant growths differ from those not produced by filtrates? Some of the latter seemed to result from protracted mechanical or chemical irritation (e.g., prolonged contact with methyl cholanthrene). Others seemed to be the consequence of hereditary factors. Still others appeared to happen more or less at random. Are all of these cancers also virus-induced, but the viruses involved so hedged in by conditions limiting infectivity that it takes long irritation, or favorable hereditary factors, to produce a finite probability that one or more cells will be susceptible to the postulated virus? The incidence of such cancers would then involve a compound probability: a low probability that the right virus is present, and another low probability that the required cell susceptibility occurs during its presence. To produce the tumor, both conditions would have to be met in the same cell and at the same time.

3. BACTERIOPHAGE, FERTILIZATION, AND VIRUS DISEASES

There appeared to exist some formal similarities in the processes of bacteriophagy, fertilization of egg-cells by sperm, and infection in virus diseases. If these do indeed have common aspects, even though taking place in substrates as different as man and bacteria, then study of the process in the system lending itself to quantitative study seemed likely to be the most rewarding. To make even a very approximate assay of a virus suspension which must be tested against animals (e.g., rabbit papilloma virus) required a large animal colony, with all its attendant problems and expense. In addition to the cost, a long period of time was required for a single assay, compared to the few hours required for an assay of a phage suspension. Work with plant viruses, such as the tobacco mosaic virus, was intermediate between animal viruses and phage in respect to time and magnitude of laboratory facilities and support required. Clearly, then, bacteriophage was by far the best material from these points of view. It made good sense, therefore, to try to learn all that one could from this easily managed experimental subject before progressing to the more difficult viruses requiring plant or animal substrates for their assay.

The wide variation in the structure of different molecules is a result of variation in the manner of assembly of only a few kinds of atoms. Thus, with virus infection, egg fertilization, and bacteriophagy, it was thought that the differences might result from a differing combination of fundamental reactions, relatively few in number. Penetration of the bacterial cell wall by bacteriophage and penetration of the egg membrane by the fertilizing sperm might not be fundamentally very different. Unmasking of certain enzymes might be a very similar process for normally developing cells and for the cell providing the substrate for virus multiplication. Replication of the genetic material is, of course, fundamental for cell division, as it is also for filling a cell with virus particles—the major difference being just which genetic material is replicated.

Fertilization of an egg cell, infection of a bacterium by a phage particle, and virus-induced conversion of a normal cell into a malignant cell seemed to have much in common. Certainly, in all three cases, penetration of the cell by a piece of replicating genetic material is involved. In the egg, this replication keeps pace with the cell division process. In the bacterium, phage replication goes on without concomitant cell division, to a catastrophic end for the bacterium. However, in lysogenic bacteria, the analogy to the egg seemed somewhat closer, since here the phage seems to replicate fast enough to maintain its presence in successive sub-cultures of the bacterium, but not fast enough to kill (lyse) the accommodating host. The case of conversion of a normal animal cell into a malignant cell by a tumor virus is different from phage infection. Obviously, this cell does not lyse as a result of unbridled multiplication of some virus infecting it. The analogy with the fertilized egg seemed to be closer—infection stimulates cell division, and the virus replicates along with this division.

These similarities and differences provided the reasons for commencing study in detail of the process of bacteriophagy. We hoped that once we understood it, we would be in a better position to understand virus-induced malignancies. It was this argument which led us to start work on bacteriophage. But our studies were soon broadened by the interest and collaboration of Max Delbrück, who brought to us his basic interest in genetics and his skills in physics and mathematics.

4. PHAGE REPLICATION

The history of the discovery of bacteriophage is told by Stent (1963). Here I want to mention only that for many years the process of lysis of bacteria by phage was called the Twort-d'Herelle phenomenon, but that it was Felix d'Herelle who had immediately recognized the particulate nature of phage:

> During my residence at the University of Leiden, in discussing this question with my colleague, Professor Einstein, he told me that, as a physicist, he would consider this experiment as demonstrating the discontinuity of the bacteriophage. I was very glad to see how this deservedly-famous mathematician evaluated my experimental demonstration, for I do not believe that there are a great many biological experiments whose nature satisfies a mathematician.
>
> —d'Herelle, 1926, p. 83

D'Herelle had also realized that its life history must include attachment to the bacterium serving as host, multiplication in that bacterium, and subsequent lysis of it, setting the daughter virus particles free into the solution.

> . . . the increase in the number of corpuscles does not take place in a continuous progressive fashion, but by successive liberations. It may be pertinent to observe that in order to clearly observe this phenomenon it is essential that the experiments be performed in such a way that the course of the reaction is not obscured . . .
>
> —d'Herelle, 1926, p. 117

The temptation was great for the early workers to turn immediately to the practical applications of these discoveries, and much of the early work on bacteriophage was aimed at the cure of diseases in man. D'Herelle was fully aware of possible clinical applications of phage, and spent some of his effort in these areas. However, his interests were broad enough that he was aware also of its value as a material for more basic studies. The important opportunity that phage presented was in understanding the replication process, an opportunity of which we decided to avail ourselves. As our work progressed, it became ever clearer that one can study separately and quantitatively the main parts of the phage growth process (adsorption and growth steps). We saw also that in addition to experiments with large populations of phage particles and bacteria, which give a picture of the average behavior, one can also make quantitative measurements with a single phage particle.

5. GROWTH EXPERIMENTS

The image of the process of multiplication of bacteriophage given by d'Herelle (1926) was reasonable, and in accord with the experimental observations, but seemed to attract more opponents than supporters. Considering that the powerful present-day experimental tools (electron microscope, analytical centrifugation, and radioactive tracer techniques, among others) were not available to d'Herelle, his description of the growth process struck remarkably close to the picture we have today.

A. *Step Growth*

The masterful book published in 1926 in both English and French (d'Herelle, 1926) described a three-step process for the life history of the bacteriophage virus: (1) attachment to the susceptible bacterium, (2) multiplication in the cell, and (3) disintegration of the cell to set free the progeny virus particles and attachment of the progeny to other susceptible bacteria, if such are present. These notions derived from quantitative work based on the plaque-count, and dilution methods of assay that he had invented.

Our work at Caltech on bacteriophage commenced with the isolation of a phage active against an available *Escherichia coli* strain. This was done by plating this strain with sewage ultra-filtrates, and selecting a phage from a plaque of a size which would be readily seen, but small enough to allow counts of 50 or more on a petri plate culture. This phage was then purified by serial culture. Replicate plaque counts were made on serial dilutions to test the linearity of the method of assay. Counts proved to be linear with dilution, and reproducible.

From the beginning, our interest was in growth processes, and not in classifying different "races" of phage, or in finding medical uses for them. It was fortunate that the phage isolated was one which d'Herelle would have called "virulent." Its adsorption to bacteria in young dense cultures was rapid, making it possible to start growth of a parent phage population with reasonable simultaneity, so that their progeny might "graduate" from the infected bacterial cells also with a reasonable degree of simultaneity. Thus, stepwise growth curves were readily demonstrated, if one took the precaution to obtain the required simultaneous initial adsorption, or attachment.

By a series of experiments performed with different bacterial species I am convinced that with bacteriophage races of low potency it is impossible to demonstrate the process of fixation to the bacteria. At least, it is impossible to demonstrate that the corpuscles disappear from the liquid. The reason for this is obvious, for with such races fixation takes place slowly, and instead of the process occurring almost simultaneously with practically the entire number of inoculated corpuscles, as is the case when the bacteriophage is very active, the time of fixation varies enormously among the different corpuscles. It thus happens that a great number have not been fixed when those which are first fixed have already commenced to reproduce. Consequently, the fixation remains undetected.

—d'Herelle, 1926, p. 107

To obtain rapid adsorption, one needs not only a virulent strain of phage and a young rapidly growing bacterial culture but also a high concentration of bacteria. To avoid the ambiguity that could arise from attachment of more than one phage particle to a single bacterium, the phage particles were added in amounts which left them outnumbered by the bacteria by about 1,000 to 1. In order to reduce further the ambiguity which might result from attachment of the phage over an extended period of time, the culture containing 2×10^8 bacteria per ml was allowed three minutes to adsorb the phage and then diluted fifty-fold. Less than seven percent of the phage are not adsorbed at the end of the three minute period, and the dilution decreases the adsorption rate of the remaining free particles to a value at which forty minutes more would be required for half of them to be adsorbed. With our phage strain, this procedure provided an excellent demonstration of the step-wise growth described by d'Herelle.

Soon after Max Delbrück arrived in the Caltech Biology Division, intent on discovering how his background in physical sciences could be productively applied to biological problems, I showed him some step-growth curves. His first comment was "I don't believe it." He soon dispelled this initial reaction of disbelief by his own analysis of the phenomenon, and promptly joined in the work with enthusiasm, bringing to it his training in mathematics and physics, his intense interest in genetics, and the many important ideas that he contributed freely. In the work which followed, the step-wise growth curves were repeated and refined, and experiments were devised to study separately the features of the life cycle of the phage (Ellis and Delbrück, 1939). These growth curves provided a measure of the average burst size, and the spread in time during which the bursts occur for the first step. After this first step, variation in time of reinfection partly obscures the values of these factors for the second and third steps, which were nevertheless clearly visible.

B. *Adsorption*

The rate of adsorption of free phage (P_f) on living bacteria was determined, so that this factor could be taken into account in the interpretation of the growth curves. This was done by sampling a suspension of phage and bacteria at intervals, centrifuging the bacteria out, and making a phage assay of the supernatant liquid. Adsorption was found to proceed according to the equation:

$$\frac{-d(P_f)}{dt} = k_a(P_f)(B)$$

where (B) is the bacterial count, and (P_f) is the concentration of free phage. The value of the rate constant k_a, is 1.9×10^{-9} at 25°C, when concentrations are expressed in virus particles and bacterial cells per milliliter.

This result provided the necessary assurance that our experimental conditions avoided ambiguity as to average burst size, and time of rise, measured from first adsorption of the phage on the bacterial cell.

C. *Absolute Assay*

To have full confidence in the plaque count method of assay, it is important to determine unambiguously the real number of viable phage particles in a solution. We believed that losses occur in the plating technique of assay, even under the best conditions. Some phage particles in a suspension added to a bacterial culture might be inactivated by substances present in the culture (Ellis and Spizizen, 1941). The fraction so inactivated would depend on the relative concentrations of susceptible bacteria and the inactivating agent. The phage particles which do not become adsorbed and, remaining free in suspension, do not multiply, could diffuse away when the suspension is plated on agar. The plating technique itself might result in loss. Thus, if some fraction of the bacteria were not able to survive the change from broth to agar plate, and if this were true also for infected bacteria, then it is clear that the infected bacterium which "died" would not produce its plaque, unless its death also released the contained phage. Whatever the losses, they would be known if there were an absolute assay method for viable phage.

Such an absolute method does exist, based on the statistics of random sampling. Max Delbrück was jubilant when his search for the original appropriate reference to this statistical formula turned out to be a paper by Poisson published in 1837, more than a hundred years prior to our paper. One selects a concentration of phage such that among many small samples taken from the suspension about half contain no phage particle. After addition of bacteria to and incubation of the samples, lysis occurs in those samples that initially contained a viable phage particle, and turbidity, due to bacterial growth, occurs in those samples that contained no phage. The original phage concentration can then be reckoned from the fraction of samples containing no phage, from the relationship:

$$n = ln\ P_o$$

where P_o is the fraction of samples containing no particle, and n is the average number of particles per sample. The determination is the more accurate the greater the number of samples taken.

This absolute assay, when compared with a plaque-count assay of the same suspension, shows that for the phage and the experimental procedures and conditions that we used the plaque count method detects about 40 percent of the viable phage particles present. This capacity of the plating technique to register phage (i.e. the fraction of phage giving plaques) was called the "efficiency of plating." It seemed to be relatively constant for the various conditions tested.

D. *Single-Step Growth*

Knowing the adsorption rate and the efficiency of the plaque count assay, it was possible to proceed with measurement of the principal parameters of the phage growth process. These parameters are the time elapsed between adsorption and lysis of the bacterium, the average burst size, and the time-spread of the burst among the infected bacteria.

Since the phage adsorption rate depends on the concentration of the bacteria, it is feasible first to obtain rapid adsorption of most of the phage to a dense bacterial suspension, and then to slow the process by dilution to a point where subsequent adsorption of any phage remaining unadsorbed makes an insignificant contribution to the ensuing phage growth. In the experiments reported (Ellis and Delbrück, 1939), phage was allowed ten minutes' contact with bacteria at a concentration of 2×10^8 organisms per cc. The infected culture was then diluted 1:10,000 and incubated further. Another 1:10 dilution was made at the time of the start of the rise in phage titer, further restricting adsorption of the progeny whose release by lysis of infected bacteria was commencing.

In these experiments, a single rise in phage titer is observed, making it possible to examine with more assurance the time of start and completion of the rise, the average burst size, and the effects of temperature on these characteristics. It turns out that the temperature influences the length of the latent period and rise period, but not the burst size. Growth curves for three temperatures are shown in Fig. 1.

A prominent feature of the growth curve is the initial rise, which is an increased probability of plaque formation (increased efficiency of plating), and not an actual increase in phage number. This was established by an experiment in which bacteria grown in broth were compared with bacteria grown on agar and suspended in broth as the plating cultures. The initial rise was absent in the latter case, the values of the assay here being initially at the higher level reached in the former case after the initial rise. Thus, it is apparent that the efficiency of plating on the broth-grown bacterial culture increases during the initial rise (from forty per cent to sixty-five per cent), whereas the efficiency on agar-grown bacteria is already at the higher value.

E. *The Individual Burst*

We found that the same technique of multiple small samples containing on the average less than one phage particle, used for the absolute assay, can also be used to examine the size of *individual* bursts. The experiment is performed very simply by plating the entire sample after incubation to the first burst. Some samples will have had more than one phage particle initially, and so will represent two or more bursts. The fraction of samples with one, two, or more phage particles can be estimated from the Poisson relation:

$$Pr = \frac{n^r e^{-}}{r!}$$

where Pr is the fraction of the samples containing r particles, and n is the average number of phage particles per sample. Of course, one cannot determine which individual samples had two phage particles.

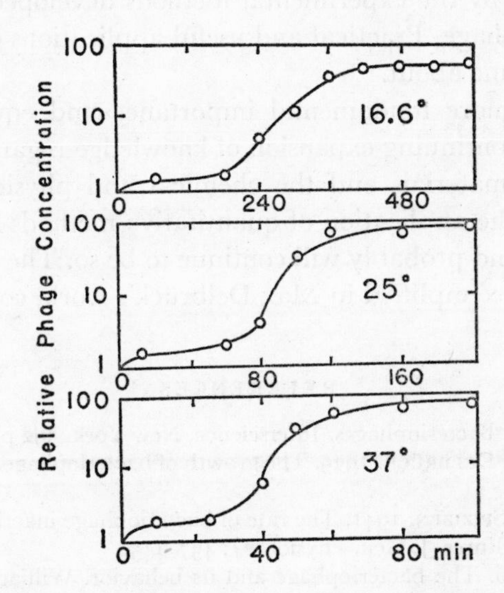

FIGURE 1. One step growth curves of a coliphage at three temperatures.

With this single burst experimental procedure we found that the average burst of sixty phage particles observed in the one-step experiments is the result of a summation of bursts yielding widely varying numbers of phage particles. The variation ranges from a few particles to values of about twice the average. Other experiments showed that the burst size is not related to the time of burst.

CONCLUDING REMARKS

Phage is a material of great convenience for studying many aspects of virus behavior. Quantitative assay of both phage and bacteria are rapid and more accurate than most biological assays (although not as accurate as most chemical analyses). One can study with ease either the individual behavior of a single phage particle or bacterium or the average behavior of a larger population, things that are much more laborious for analogous studies with plants or animals. One can readily achieve near-absolute control over the environment, including control of some variables which are not controllable in animal or plant experiments. Phage has been a prime material for elucidation of the processes taking place in the replication of genetic material. The electron microscope, the techniques of isotope marking of molecules in order to trace their history and fate, and modern methods of chemical analysis have contributed markedly to this—today's knowledge would not be possible without them. The study of other viruses and the development of a knowledge of their natural histories will be (and already has been) aided by the experimental methods developed, and the results obtained, with phage. Practical and useful applications of this knowledge are certain to come about.

Possibly of more fundamental importance and equal practical importance is the continuing expansion of knowledge regarding the duplication of genetic materials and the chemical and physical factors which influence this. The application of quantitative methods has been basic to all of this work and probably will continue to be so. The talent for quantitative thinking, exemplified in Max Delbrück's work, contributed greatly to this progress.

REFERENCES

ADAMS, M. H. 1959. Bacteriophages. Interscience, New York. 592 p.

ELLIS, E. L., and M. DELBRÜCK. 1939. The growth of bacteriophage. J. Gen. Physiol. *22*: 365–384.

ELLIS, E. L., and J. SPIZIZEN. 1941. The rate of bacteriophage inactivation by filtrates of *Escherichia coli* cultures. J. Gen. Physiol. *24*: 437–445.

D'HERELLE, F. 1926. The bacteriophage and its behavior. Williams and Wilkins Co., Baltimore. 629 p.

STENT, G. S. 1963. Molecular biology of bacterial viruses. W. S. Freeman and Co., San Francisco. 474 p.

THOMAS F. ANDERSON
Department of Molecular Biology, The Institute for Cancer Research, Philadelphia, Pennsylvania

Electron Microscopy of Phages

During my three years (1937–1940) at the University of Wisconsin, I must have read many scientific papers in the Medical Library, but today I can remember only three—one on growth of bacteriophage by Emory Ellis and Max Delbrück (1939) and two by Delbrück (1940a, b) on adsorption and one-step growth. The experiments were beautifully designed and reported in an elegant style that was new to me. The three papers carrying the Delbrück label formed a little green island of logic in the mud-flat of conflicting reports, groundless speculations, and heated but pointless polemics that surrounded the Twort-d'Herelle phenomenon.

Little did I know then that in a few years I would casually visit the tiny island that Delbrück was building and spend most of the next twenty-five years helping an ever increasing number of workers to enlarge and beautify it with pretty pebbles and bits of coral. Eventually almost all of our offerings fell into place to form a beautiful arrangement, but in the early days one of us would occasionally come up with a smelly skeleton or a piece of rotting driftwood that Delbrück would fling away as being unworthy of his paradise. Other, more fortunate offerings could be assigned only tentative locations, for in the early darkness, assignments had to be made by the smell, taste or feel of things that were too small to be observed meaningfully by ordinary light.

In 1940, when I first heard of the electron microscope which was said to have been developed in Germany it almost seemed to be a hoax perpetrated on the rest of the world by the Nazis. It should be recalled that in 1940 our relations with Germany were so strained that it was difficult to obtain current literature from that country. The few articles on the electron microscope that I was able to find in Madison were not impressive. However, the National Research Council planned to administer a fellowship for the Radio Corporation of America which would pay someone $3000 per year to explore the biological applications of the instrument. Reasoning that a company with RCA's resources would surely have other more interesting things for me to explore if the electron microscope project were unproductive, I applied for and received the fellowship.

On arriving at the RCA Manufacturing Company in Camden, N.J., in September, 1940, I was pleasantly surprised to find that the microscope was not a hoax after all—in Dr. V. K. Zworykin's department, Dr. L. Marton had already built an instrument that was capable of 50 A resolution. With Stuart Mudd, David Lackman, and Katherine Polevitsky, Marton had taken micrographs of bacteria that revealed structural details that were quite invisible in the light microscope. Even the techniques were not too difficult to learn. In a few weeks I myself was able to take reasonably good pictures and for a year I took pictures of specimens supplied by outstanding workers in many fields. At first we worked with Marton's microscope and later with the commercial Model B instrument that James Hillier and A. W. Vance completed with Zworykin in November or December of 1940.

IDENTIFICATION OF PHAGE PARTICLES

Ruska (1940) and Pfankuch and Kausche (1940) were the first to study bacteriophage preparations with the electron microscope, and later Ruska (1941) saw sperm-shaped particles in them. He suggested that these particles should be interpreted either as the virus itself or as bacterial constituents.

Enter Luria

It was early in November 1941, in the second year of my fellowship, that the young Italian scientist Salvador E. Luria came to RCA. He wanted to discuss the possibility of checking the sizes of some bacteriophages which he and Exner had just estimated from the cross sections for X-ray killing (Luria and Exner, 1941). It sounded like a reasonable proposal to me, and so he promised to prepare phage stocks of high titer and return with them in a month, provided that security clearance could be obtained for him to enter the RCA laboratory, which was busy with defense contracts at the time. He did get clearance and he did show up with stocks of three different phages on Monday, December 8, 1941, the day after Pearl Harbor. We didn't accomplish much that day, or even that week, because the titers of his stocks were too low. However, during the week of March 2, 1942, Luria returned with stocks that assayed as high as 10^{10} phages/ml! Now we could really see the phage as tadpole-shaped particles, whose heads ranged from 600 to 800 A in diameter, depending on the species. We wasted no time getting those results into print. In a simple, straightforward paper communicated March 12, 1942, Luria and I noted that: "(*a*) [characteristic] particles are always present in highly active phage suspensions and missing in any control suspensions (media, bacterial cultures, bacterial filtrates, etc.); (*b*) they are readily adsorbed by the bacterial cells of the susceptible strain and fail to be adsorbed by other bacteria; (*c*) the size from a given strain is uniform and corresponds essentially to measurements by indirect methods; (*d*) the structure of both the 'head' and 'tail' is characteristic of the strain of phage; (*e*) preliminary experiments on the lysis process seem to demonstrate the liberation of these particles from the lysing bacteria" (Luria and Anderson, 1942).

There was no doubt in our minds that we had found the phage particles, but certain other workers remained skeptical for years. I remember particularly the reaction of Alfred Hershey's teacher, kindly old Professor J. J. Bronfenbrenner, who had worked on bacteriophages for many years at Washington University in St. Louis. He had inferred from diffusion experiments that the PC (now T2) phage we had micrographed was only 160–180 A in diameter, with a still smaller "active" phage fraction only 30 to 40 A in diameter. When he first saw our pictures of PC phage he clapped the palm of his hand to his forehead and exclaimed, "*Mein Gott!* They've got tails!" On April 23, 1942, he wrote me, ". . . if these tails should represent organs of locomotion, it is quite possible that the diffusion of the particles might be speeded up to such an extent as to nullify the results of our diffusion experiments . . ." At various times Bronfenbrenner had his technicians send me preparations of phage, but we were never able to get good pictures of his phage particles. In each case control cultures kept in St. Louis mysteriously lost titer by the time I was able to micrograph the samples. He became more skeptical, and as late as July 22, 1943, he wrote: "A couple of months ago both Doctor Delbrück and Doctor Luria visited St. Louis and gave very interesting reports. We have naturally discussed the significance of the electron microscope findings. I do not think it will surprise you when I say that I am still unconvinced as to the nature of the sperm-like structures recorded by the electron microscope . . ."

WHAT ARE THEY—REALLY?

Of course, for all we knew then, phages might be motile, and certainly a large proportion of the pictures we obtained (and intuitively suppressed) showed phage particles oriented with their heads pointing toward the bacteria, as though, when the specimen was dried, they had been caught in the act of swimming toward their host cells. It took me eleven years to expose this apparent action at a distance as an artifact of drying (Anderson, 1953).

Enter Delbrück

Max Delbrück came into the picture-taking business during the summer of 1942, when RCA installed an electron microscope for me in the biologists' summer camp, the Marine Biological Laboratory at Woods Hole. Toward the end of that summer Luria brought Delbrück and his wife Manny to Woods Hole for two weeks' further study of bacteriophages.

Significant results were obtained particularly from an examination of the adsorption and multiplication of two phages α and γ (the present-day T1 and T2) both active on the now familiar strain B of *E. coli*.

Our work confirmed some of Delbrück's (1940a, b) earlier results, in that it demonstrated the adsorption of the phages on the host cell and, at the predicted time, the lysis of the host cell and the expected liberation of a hundred or so daughter particles from each cell. Evidently, the particles multiply inside the cells, rather than at their surfaces, for until lysis occurs the number of particles visible at the surface remains constant. This constancy also meant that very few, if any, of the particles enter the cell. This observation seemed to be of the "greatest consequence" to Delbrück and led him to write the remarkable paragraph that concluded the discussion of our paper (Luria, Delbrück, and Anderson, 1943). Before I quote that paragraph it should be recalled that we were then groping to find a place for viruses in the order of living (or non-living) things. With the aid of Einstein's logic, as purveyed by d'Herelle, most workers were convinced that viruses are at least particulate, rather than non-particulate (whatever *that* may be), and our electron micrographs amply confirmed this conviction. W. M. Stanley's crystallization of tobacco mosaic virus caused many to consider crystallizable viruses as molecules. Our pictures were not used to refute this view, but we wrote, "while no harm is done by calling viruses 'molecules,' such a terminology should not prejudice our views regarding the biological status of the viruses, which has yet to be elucidated."

Enzymes? No!!

The pictures did, however, disprove the ideas of A. P. Krueger, according to which (as stated in a preliminary version of our paper) the "cell contains a precursor of the virus particle which, upon infection of the cell with a virus particle is converted into virus. . . . According to this theory an uninfected bacterium of the strain here considered should contain 140 precursor particles of virus α and 135 precursor particles of virus γ. The pictures show clearly that this is not the case; since bacteria lysed under the influence of virus γ show no evidence of particles of virus α and vice versa." While we were preparing the manuscript, I objected in a letter that this statement might be too strong: that "the possibility still exists that the two viruses have the same precursor which is converted into the virus which infects the bacterium, the precursor having possibly a different shape than either virus. It seems a bit premature to go very far into definite theories of virus growth at this time when it should be possible to obtain definite evidence on this matter in the near future."

In answer to my objection, I received this Shavian reply from Delbrück: "I am willing to omit this paragraph but I don't see why we should. . . .

The possibility that the two viruses have the same precursor I am not willing to concede. If that were so, the precursor would have to be very different from either one or the other virus. If one admits such a possibility the precursor idea looses its meaning. Why not call the bacteria a precursor of the virus, or those chemicals that will go into the making of the virus? I am sure that Krueger will eventually choose this road of withdrawal but it seems to me a very silly one. . . ."

Sperm? Maybe?

The last paragraph in our discussion bothered me more than the attack on Krueger. Here is a preliminary version:

Finally, a point may be mentioned which seems to us perhaps of the greatest consequence. We have seen that the new virus is liberated from within the cell. On the other hand, the pictures of bacteria infected with virus γ and taken at fifteen minutes showed that the adsorbed virus particles, or at least most of them, do not penetrate into the interior of the cell but remain on the outer surface of the membrane. This observation creates a difficulty in interpreting virus growth. How do the infecting particles reproduce if they remain outside while the new virus is generated in the interior of the cell? One might assume either that the infecting particles act through the membrane, or that only one particle can enter the cell. The latter idea seems attractive in the light of results of growth experiments on multiple and mixed infection. These had shown that a bacterium always reacts as though only one particle of virus had been effective. The pictures here reproduced, if interpreted on the assumption that one virus particle enters the cell, would indicate that the entry of one virus particle bars the entry of other virus particles by making the bacterial membrane impermeable to them. The highly peculiar phenomenon of mutual exclusion between virus particles attacking a cell could thus be explained by a mechanism alternative to that proposed in a previous discussion (Delbrück and Luria, 1942). An interpretation of this kind, suggesting an analogy with the fecundation of monospermic eggs, would lend support to those theories of the systematic position of virus which consider it as phylogenetically related to the host rather than as a parasite.

To this I objected in a letter to Delbrück:

p. 16, last paragraph: I have just read parts of Hadley's review and am indeed impressed by the number of theories. I should have gone over this paragraph more closely the first time. The first sentence seems to give more weight to the new idea than you, who have just proposed [cf. Delbrück and Luria, 1942] the "competition for a key enzyme theory," might wish. The analogy with fertilization seems to me to be amusing and can be carried for some way as I have done in the appended note, but I would not want to take it seriously until I knew a lot more of what goes on in the phage reaction, or for that matter, in the egg. If we could show that phage can give rise to the bacteria, we might have a stronger case.

Is there any chance that the phage particles still outside the cells in the micrographs could be adsorbed but inactive phage particles? [This question was ignored in the correspondence that followed.]

It seems to me that there is a contradiction between the last two paragraphs.

If I try to define carefully what I mean by "systematically related," I come close to identifying Kreuger's virus "precursor" in the bacterial cell. What is meant by "systematically related"? Does it mean "derived from" in an evolutionary sense only? If so, the phage must be a parasite. Or does it mean that the virus is like a sperm in relation to the bacterium—that the bacterium and the phage have a gene or more in common? If so, these genes would be virus precursors. Systematic relationships have been demonstrated serologically [in other systems], but the bacterium and phage do not seem to have common antigens. Bacteria are usually classified physiologically by the reactions produced in certain nutrient media, but phage has not been shown to produce any reactions in media comparable to those produced by growing bacteria.

Could you lead me by easy stages over the logic involved in the statement of the last sentence, i.e., assuming that the exclusion of sperm from the egg is accomplished by the lifting off of the fertilization membrane and that the bacteria accomplish the exclusion of all but one phage particle in the same way, how this (assumed) fact would lend support to theories of the systematic position of bacteria and virus? I have a vague feeling for it, but it *is* vague.

Hoping to show that the instantaneous reaction of most biologists to our pictures was ridiculous I also sent Luria and Delbrück the following critical note appended to the above letter.

ANALOGY BETWEEN PHAGE-BACTERIUM AND SPERM-EGG

The superficial aspects of the observations obtained so far are amusing. The phage particles with heads and tails look so much like sperm that some of the pictures of phage and bacteria could easily be mistaken for lower magnification pictures of sperm and egg. Just how far this analogy can be pushed is not clear, but we can list the points of similarity and difference. As in the case of sperm and egg a rather high degree of specificity is shown—sperm will adsorb on eggs of closely related species only, phage will adsorb on bacteria containing only the proper surface antigen. Part of the sperm enters the egg and in most cases only one sperm reacts in this way—the course of development of the egg is independent of the number of sperm which are simply adsorbed; we do not know yet whether any part of any phage particle enters the bacterium (the pictures show that many phage particles do not), but the course of events following adsorption (lysis) is apparently unaffected by the multiplicity of infection as though only one phage particle had reacted. In the egg, the exclusion of all but one sperm is associated with the lifting of a membrane from the surface of the egg; we do not know the exclusion mechanism in the case of bacteria. Eggs fertilized with damaged sperm are spoiled and will not develop normally even though good sperm be added; bacteria infected with damaged phage are also spoiled, their normal multiplication is hindered as is the production of phage from them when good phage is added. In the case of eggs it is difficult to experiment with the effects of adding simultaneously more than one kind of sperm; but both α and γ phage react with coli B and when added to the bacteria simultaneously, lysis gives rise only to γ particles although the time of lysis is longer when γ alone is added than when α alone is added. Of course the analogy breaks down completely when we consider the products of the two reactions for phage-infected bacteria are lysed with the production of more phage while the fertilized egg has never been observed to break down with the production of more sperm.

A statement like this was usually sufficient to end serious speculation along the sperm-egg line by most audiences I had faced, but it had no effect on a pair like Luria and Delbrück. Luria wrote to me:

I didn't answer your letter before because I waited [for] the answer to a series of questions I had asked to Max concerning it. We seem to be in good agreement, and hope you will agree too.

I think we should leave the paragraph about Krueger's precursors as it is: there are hundred of reasons to *be sure* that he is wrong, and this of the el. mics. is only a simple and direct one. Do not doubt that everyone but Krueger will agree.

For the last paragraph, which should remain in its main part, Delbrück has a suggestion of a mild transformation, which he has probably sent to you, and which would make it clearer by omitting the "phylogenetically" and adding a reference to the big Hadley's paper, which contains the whole story. . . .

Within a few days I received this letter from Delbrück:

p. 16 "precursor." The situation is this. Krueger is a former assistant of Northrop and while in Northrop's lab he developed a very misleading method of assay for phage, based on preconceived analogies of phage with enzymes. Later, when he was in Berkeley, and after Northrop and his people had discovered pepsinogen, trypsinogen, chymotrypsinogen, and the conversion of these "precursors" into the enzymes in vitro, Krueger decided that the phage growth reaction was the same thing. First he published the idea. Then he published a preliminary note saying that he had accomplished conversion of phage from precursor in vitro, i.e. he claimed to have extracted phage precursor from the bacteria. When he found that he could not reproduce this he published experiments in which he claimed to prove conversion of precursor into phage within the bacteria, in the cold, and in the absence of nutrient. These experiments . . . made a great impression on people who know nothing about phage technique. For criticism of these (ten!!) papers see my review article. From all this it is clear what Krueger's precursor was meant to be, namely a protein molecule, similar to the phage-protein, which, without the intervention of other enzymes in a spontaneous autocatalytic reaction, is converted into phage. My position is this: firstly Krueger has never given evidence in favor of this view, since the experiments are based on a simple error of interpretation, secondly our present pictures show among the bacterial debris nothing similar to the phage-to-be. I think the second fact is worth pointing out, both as an observational fact by itself and as point against Krueger whose notions have made a widespread if not deep impression.

A conducted tour in easy stages to the last paragraph follows shortly. It is written and in the process of being revised.

In due time I received the following essay:

A CONDUCTED TOUR BY EASY STAGES TO THE LAST PARAGRAPH

FOR T. F. A.

I will take for granted that phages and viruses are similar things, in the sense that the relation between phage and bacterium is essentially the same as that between virus and plant or animal cell. If proof of this assumption was needed,

I believe that the occurrence of interference in both cases has sealed it. This assumption is helpful in limiting the possibilities for theories of the nature of phage.

If one wants to talk about the nature of phage (or of virus) the first problem seems to be this: is the relation between virus and host cell analogous to that between parasite and its host, between, say, hookworm and man? In such a parasitic relation the partners are systematically unrelated. In a very general sense, all living things are, of course, phylogenetically related, even hookworm and man, but it is not this relationship that brings them together and the common ancestors can be disregarded when the interaction is discussed.

In contrast to the parasitic relation, the virus-host relation might be analogous to any of the following relations:

(a) Tumor cells are transplantable within a very limited range of species. The original tumor cell of such a transplantable strain is always from an organism whose phylogenetic relationship with the host is essential for the "take" reaction. White blood cells are a particular kind of tissue that may grow wild in tumor-like fashion and give rise to the transmissible disease of leukemia.

(b) Sperm and egg represent a very specialized relationship developed in the higher, bisexual, living world and modified in a great variety of ways in the two higher kingdoms. The analogy, if any, with the virus-host relation can be at best only a very remote one and may lie in this; that the virus may in some cases be a necessary part of the life cycle of the host and that research has not yet been able to track down this aspect, just as the abnormal cases of heredity were the first and for a long time the only ones that were studied.

(c) The genes are particles of virus size which grow and divide in step with the cell and which may or may not be essential to the cell. The synchronization of their growth is not altogether perfect since there are specialized polyploid cells and since polyploidy may also be induced artificially. In general, the whole chromosome set of a cell duplicates synchronously. This harmony, which is essential to the functioning of the higher organisms, is insured by the complicated maneuvers of mitosis. In aberrant cases any part of any chromosome is capable of duplication out of turn. Transmissible genes and really wild growing genes are not known, but some people have suspected that the viruses are just this group of missing cases.

All of these cases furnish examples of materials which grow with a varying degree of independence of the "host," but for which the close relationship with the host is an essential feature.

For the viruses now, we do not know whether to consider them as degenerated or aspiring bacteria or as runaway genes. If the latter is the right point of view, then it follows that there must be a "grand cut" somewhere along the line that seems to run smoothly from the bacteria through the rickettsia to the viruses. In saying this I have taken for granted that the bacteria are not related to the hosts which they may parasitize.

How can one now go about deciding between the two alternative views regarding the nature of virus? It has been claimed that phages can be derived from normal bacteria. Others have countered this argument by saying that the cultures in question had been carriers or had been contaminated. The balance of evidence is nil. Hadley and others have claimed the existence of filterable stages of bacteria. I believe their filters were bad. It has been said that some of the bacterial antigens are responsible for the adsorption of phage and that the reaction between the two is similar to that between antigen and antibody or to that between sperm and fertilizin. This is probably true and, in my opinion, a strong argument in favor of the relatedness between phage and bacterium. On the other hand, nothing of this kind is known regarding the relation between plant and animal viruses and their hosts.

Finally, let us consider interference. This was discovered first for plant viruses and was there interpreted as an immunity reaction. People thought that the plant perhaps produces antibodies which prevents the second virus from establishing itself. This was soon shown to be wrong. People then thought that it was due to a competition for some substrate which is needed as building material for the viruses and which is present in limited amount. This Luria and I proved to be wrong since it does not at all fit the details of our interference experiments. We then proposed the "key-enzyme" idea, assuming that the cell contains a sort of mysterious innermost sanctum at which the virus reproduction takes place. All the theories mentioned approach the problem of interference from the parasite viewpoint! The cell is considered as a prey to two invaders, but happens to be so constructed that only one of them can succeed. The cell does not *do* anything to bring about interference. The view mentioned in the last paragraph of our paper is totally different from the previous ones in that it assumes a quick and specific reaction of the cell to the attack. Moreover, this reaction is similar to the reaction of monospermic eggs to sperm. The quick and specific reaction of the cell suggests that the cell is not unprepared for the attack and the particular type of reaction suggests further that the attack may have a functional meaning which remains yet to explore.

SAY A LITTLE TOO MUCH

Luria responded as follows:

p. 16, II par. (Krueger). As you saw, also Max thought that it is necessary, or at least useful to maintain it in order to destroy Krueger's fantasies. I think however that we should add at the third line the word "promptly" between "is" and "converted." This fits better Krueger's idea, and the aspect of it which our pictures do not support.

p. 17. Here too, we apparently of the same idea with Max. I personally think that it is always better to say a little too much, if one is not too affirmative, that [than] too little. After all, one should think of the possible interpretations and propose them for general discussion. I think Max's "conducted tour," which I received a few days ago, explains very well the situation as it seems to be now. Let's go ahead with the paragraph modified according to Max's suggestions. . . .

By the way, I think I would like this time to put the "micron" line only on the pictures of the phage alone, not on the others. People should begin to think in terms of 30,000 x magnification.

As a good electron microscopist I saw to it that the micron mark was on every picture when our article was sent to the *Journal of Bacteriology* on February 17, 1943. But the last two paragraphs remained essentially as they were before our exchange. It was two against one in a democracy; however the last sentence of our discussion, with the addition of the sheltering clause I have italicized, had become more discreet: "An interpretation of this kind, *for the correctness of which the experiments offer as yet hardly more than a hint*, would suggest an analogy with the fecundation of monospermic eggs, and would lend support to those theories of the systematic position of virus which consider it as ["phylogenetically" omitted] related to the host rather than as a parasite (cf. Hadley, 1928)."

Delbrück and Luria were certainly right to use our electron microscopic evidence to help smash the ready-made precursor theory; but to abandon the key-enzyme theory they had just proposed (and stoutly defended when I attacked it verbally at Woods Hole in 1942) in favor of the sperm-egg analogy still seems to require an explanation. Many of us were later involved in a plot to interest the "real plant and animal virologists" in bacteriophages by calling them "bacterial viruses." Was the key-enzyme theory designed to interest enzymologists and the sperm-egg idea intended to get all biologists stirred up over still another possibility for which there was as yet admittedly "hardly more than a hint"? Actually, the hint has turned out to be not far wrong, for exclusion of superinfecting phage seems to be mediated by changes induced in the bacterial wall by the primary infection (see Stent, 1963 for a discussion); but this still has little to do with "relatedness" between virus and host. And today, when one considers some of the temperate instead of the virulent phages, one wouldn't be too far off in proposing the sperm-egg analogy; for what is transduction but a sort of sexual communication between bacteria mediated by sperm-shaped phage particles like λ or Pl?

THE BASIC QUANDARY

What Delbrück and Luria did seize upon as being very strange was the observation that few if any particles seem to penetrate the host. As it turned out, it required the use of radioactive tracers by Hershey and Chase (1952) to clarify the reason for this apparent lack of penetration. Similarly, tracer techniques were required to sink once and for all the precursor theory of phage replication.

A PERSONAL CHOICE

My RCA Fellowship ended in September 1942 and it was not difficult for me to decide that of the many fields that had been opened by the electron microscope the study of bacteriophages offered the most interest and excitement. I joined the staff of the Johnson Foundation at the University of Pennsylvania, where L. A. Chambers had obtained an RCA Model B electron microscope from the American Philosophical Society to study influenza virus and Rickettsia.

Evidently, Ellis had become interested in things other than phage by this time, and no one had kept him informed of the developments in electron microscopy. In response to my request for reprints, I received the following letter which seems somewhat wistful.

Dear Mr. Anderson:

I am glad to hear of your interest in bacteriophage. The reprints you requested are coming under separate cover. I took the liberty of including two by Max Delbrück of which I happened to have extras—since they represent a continuation of a line of thought with which I would like to continue to be associated but for other circumstances.

With best wishes for the success of your work—I am

Sincerely yours,

Emory L. Ellis

P.S. Delbrück is at Vanderbilt U., Nashville, Tenn.—He has some other papers in which you would be interested.

THE T-SET

In the summer of 1944 the phage workers under the influence of Delbrück made an important decision. Previously, almost every investigator who worked with bacteriophages had his own private collection of phages and host bacteria. It was therefore almost futile to compare results of different workers, or even to gather a satisfying amount of information about one system. Delbrück insisted that we concentrate our attention on the activity of a set of seven phages on the same host, the now famous *E. coli* strain B and its mutants, in nutrient broth at 37°C. He wanted everyone to work under these standard conditions, but it will be seen that some of us were somewhat rebellious. The set of approved phages had been collected by Demerec and Fano (1945) for their studies of the patterns of mutation of strain B to resistance to the phages. They were chosen from a large number of phages as being "well-behaved," i.e. they give easily countable plaques, and the strains of B that are phage-resistant can be freed of the phage to which they are resistant. Thus there was to be no adventurous nonsense about lysogeny. This and the other phenomena shown by temperate phages, such as transduction, were to be exploited later by mavericks who had not joined in the pact but could nevertheless build on the solid foundation of the relatively simple virulent behavior of the seven phages of the T system.

CLASSIFICATION BY MORPHOLOGY, SEROLOGY, OR HOST RANGE?

I set out to determine the morphologies of those phages which had been arbitrarily labelled T1 to T7. This would have been a simple matter if Delbrück had not objected violently when the samples of T5 and T6 that had been sent to me appeared to have the same morphologies. The matter was corrected when new samples showed T5 to be distinctive, while T6 looked like the other even-numbered phages, T2 and T4.

It developed that the seven phages can be classified into four morphological groups: (T1), (T2, T4, T6), (T3, T7), and (T5). Sensibly enough, this classification agreed with Delbrück's serological classification (Anderson, Delbrück, and Demerec, 1945), which was doubtless the reason he objected to the morphology found originally for T6. It should be concluded that the morphological and serological relationships between the phages are the biologically (i.e. phylogenetically) valid ones, but as I recall, this did not seem completely obvious, to me at least, until genetic recombination between related phages were discovered by Hershey (1946) and by Delbrück and Bailey (1946). Even then Delbrück was strangely guarded in interpreting the genetic results and brought up red herrings that initially cast doubt on an uninhibited interpretation. But that should be someone else's story.

It now seems clear that the patterns of mutation of strain B to resistance to the various phages bears no relation to the morphological and serological classification. Thus the mutation of strain B to B/1,5 does not relate T1 and T5 to each other, but rather relates the host's receptor sites for T1 and T5 in some way that is still not clear. Similarly, mutation to B/3,4,7,pr relates the receptor spots for T3, T4 and T7 to the host's mechanism for the synthesis of proline. Incidentally, it still seems odd to me that none of the one-step mutations of strain B render it unsuitable as a host for a group of phylogenetically related phages in a way that does not involve receptor spots.

IN THE LAND OF SERENDIP

As everybody knows, most discoveries in science are necessarily quite unexpected: during a planned study that is more or less routine one encounters a surprising result that must be explained. I would like to describe two or three such observations that I stumbled upon at about that time and which eventually helped us to understand some of the mysteries that surrounded phage structure and growth.

Strange Behavior of T4 in a Synthetic Medium

In 1943 I had undertaken a program to determine whether all the host cell machinery is required to support phage growth, or whether, as proved later to be the case (Anderson, 1948a), the bacterium supports T2 growth even after it had been heavily damaged by ultraviolet light. A second, related experiment was to determine the kinetics of UV-inactivation of the virus-host cell complex at various times after infection, in order to determine how the number of replicating phage units increases during the latent period (cf. Luria and Latarjet, 1947). To perform these experiments properly I felt that I should study phage replication in a medium that is transparent to the UV light. I chose Friedlein's ammonium lactate (F) medium, which supports growth of strain B very well. I soon found that T2 has a high efficiency of plating on F medium agar, but that T4 and T6 have very low plating efficiencies. So, I started to look for the substance(s) present in nutrient broth that enable T4 and T6 to form placques.

Adsorption Cofactors

To make a long story short, it turned out that certain strains of T4 and T6 do not attach to strain B bacteria unless the particles themselves have been activated by an aromatic *L*-amino acid cofactor like *L*-tryptophan (Anderson, 1945, 1948b).

Naturally, Delbrück was skeptical. But once he had repeated my results he became an avid cofactor fan; for, coming on the heels of the work of Beadle and Tatum on the biochemical genetics of *Neurospora*, the cofactor mutants seemed to offer some promise of similar developments in bacteriophages. In addition, these new genetic markers would add to the list of those available for genetic studies. Delbrück isolated a number of new cofactor mutants, including one requiring Ca^{++} in addition to L-tryptophan; and unlike my mutants, his cofactor-requiring strains are all inhibited by indole (Delbrück, 1948). He and his students spent many enthusiastic years trying to solve the mechanism of cofactor activation. Unfortunately, the cofactor mutants have so far not been very useful in phage genetics. However, they are becoming fashionable again, for, as pointed out by Kellenberger et al. (1965), the cofactor phenomenon represents the first directly observed example of "allosterism," a magic word in enzymology today. The electron microscope shows that in cofactor-requiring strains, the tail fibers are not available for attachment to the host until the cofactor is added. Then the tail fibers seem to be released from a connection somewhere on the tail sheath and are free to attach the phage particle to receptor spots on the host cell.

Osmotic Shock

The discovery of cofactor requirements in the T-even phages also led to the localization of the DNA in these phage particles (Anderson, 1950). For, from experiments designed to determine how the activation of T4 by tryptophan depends on salt concentration, it transpired that T4 is inactivated if it is incubated in concentrations of NaCl higher than 2M and then rapidly diluted in a medium of low osmotic pressure. The electron microscope showed that particles inactivated in this way have empty heads and R. M. Herriott (1951), who was much better equipped than I to do the necessary chemistry, showed that such empty-headed phage particles consist of protein and contain none of the DNA present in normal T4. Evidently, osmotic shock liberates the phage DNA from the phage heads. In this rather roundabout way an accidental start was made on elucidating the functional anatomy of the T-even bacteriophages.

Heads or Tails

There remained the question of whether phage particles attach head first to the host cell, as sperm attach to eggs, or whether they attach tail first, as nothing else I can think of except *Caulobacter*, or stalked bacteria. I have already mentioned that in some of our electron micrographs the particles seemed to attach tail first, while in others the particles ringed the bacteria with all heads pointed toward the host cell. It seemed obvious that some surface tension artifact of specimen drying was responsible for these ambiguities, but it was not until 1949 that the critical point method for eliminating such artifacts in specimen preparation finally occurred to me. I brought a few of the resulting stereoscopic pictures of adsorbed T2 to the *Viruses: 1950* meeting at Caltech. Later, Delbrück looked at many more such pictures in my lab in Philadelphia and agreed the phages are adsorbed by the tips of their tails and that none of the particles seem to enter the bacteria or their ghosts (Anderson, 1952).

I still preferred to discount the significance of this observation, because at this time we had no idea what proportion of the particles in a phage preparation are inactive and which might therefore linger indefinitely on the surface. Shortly thereafter however, Luria, Williams, and Backus (1951) showed that in the best T2 phage preparations, as many as 50 per cent of the particles visible in the electron microscope are able to form plaques. If the particles do penetrate the bacterial ghosts they must become unrecognizable. However, it also seemed possible they do not penetrate; and indeed, one could see empty headed phage ghosts on the bacterial surface. I remember in the summer of 1950 or 1951 hanging over the slide projector table with Hershey, and possibly Herriott, in Blackford Hall at the Cold Spring Harbor Laboratory, discussing the wildly comical possibility that only the viral DNA finds its way into the host cell, acting there like a transforming principle in altering the synthetic processes of the cell.

THE DENOUEMENT

In the hands of Hershey and Chase (1952) this joke proved to be not only ridiculous but true: the T-even bacteriophage particle *is* like a tiny disposable protein syringe loaded with viral DNA and designed to attach itself specifically to a host cell and then to inject its DNA into it. The identification of DNA as *the* essential viral constituent, the determination of its structure, and the recognition of its informational content opened a new branch of molecular biology, which one calls molecular genetics nowadays to distinguish it from studies of the coordinated service functions other molecules perform in the maintenance and proliferation of living things.

When we finally recognized this dichotomy during the Cold Spring Harbor Symposium of 1953, it was like finally hearing the hilariously improbable punch line at the end of a long preposterous tale.

RÉSUMÉ

Looking back on this era, it seems to me that the electron microscope's initial role was to give us concrete objects to think about rather than abstract symbols like T1, or T2. It soon showed us that there are many types or families of bacteriophages, each with its morphological individuality. The electron microscope thus removed the mystery of what phage particles look like, only to substitute the deeper mysteries of how the particles are organized, what the function of each part might be, and why the particles appear to remain on the surface of the host instead of diving into it like a respectable parasite. The resolution of these mysteries has been shown to require the intelligent application of additional methods of research—the microscope can only suggest solutions, not confirm them.

In the early days I had no ultimate goal in research except to see how life processes work. Once started with the bacteriophages, I continued research on them because it seemed that here one could do more interesting experiments than with any other material. Furthermore, the engaging group of people working on these viruses was highly compatible; among ourselves we saw, and still see, very little competitive secrecy and backbiting: we were all interested in cooperating with each other to promote the work as a whole.

At each phase in our groping toward discovery, Max Delbrück seemed to be present not so much as a guide, perhaps, but as a critic. To the lecturer he was an enquiring, and sometimes merciless, logician. If one persevered, he would be fortunate to have Max as conscience, goad, and sage.

REFERENCES

Anderson, T. F. 1945. The role of tryptophane in the adsorption of two bacterial viruses on their host *Escherichia coli*. J. Cellular Comp. Physiol., *25*: 17–26.

———1948a. The growth of T2 virus on ultraviolet-killed host cells. J. Bacteriol., *56*: 403–410.

———1948b. The activation of the bacterial virus T4 by L-tryptophan. J. Bacteriol., *55*: 637–649.

———1950. Destruction of bacterial viruses by osmotic shock. J. Appl. Phys., *21*: 70.

———1952. Stereoscopic studies of cells and viruses in the electron microscope. Amer. Naturalist, *86*: 91–100.

———1953. The morphology and osmotic properties of bacteriophage systems. Cold Spring Harbor Symp. Quant. Biol., *18*: 197–203.

Anderson, T. F., M. Delbrück, and M. Demerec. 1945. Types of morphology found in bacterial viruses. J. Appl. Phys., *16*: 264.

Delbrück, M. 1940a. Adsorption of bacteriophages under various physiological conditions of the host. J. Gen. Physiol., *23*: 631–642.

———1940b. The growth of bacteriophage and lysis of the host. J. Gen. Physiol., *23*: 643–660.

———1948. Biochemical mutants of bacterial viruses. J. Bacteriol., *56*(1): 1–16.

Delbrück, M., and W. T. Bailey, Jr. 1946. Induced mutations in bacterial viruses. Cold Spring Harbor Symp. Quant. Biol., *11*: 33–37.

Delbrück, M., and S. E. Luria. 1942. Interference between bacterial viruses. I. Interference between two bacterial viruses acting upon the same host, and the mechanism of virus growth. Arch. Biochem., *1*: 111–141.

Demerec, M., and U. Fano. 1945. Bacteriophage-resistant mutants in *Escherichia coli.* Genetics, *30*: 119–136.

Ellis, E. L., and M. Delbrück. 1939. The growth of bacteriophage. J. Gen. Physiol., *22*: 365–384.

Hadley, P. 1928. The Twort-d'Herelle phenomenon. A critical review and presentation of a new conception (homogamic theory) of bacteriophage action. J. Infect. Diseases, *42*: 263–434.

Herriott, R. M. 1951. Nucleic-acid-free T2 virus "ghosts" with specific biological action. J. Bacteriol., *61*: 252–254.

Hershey, A. D. 1946. Spontaneous mutations in bacterial viruses. Cold Spring Harbor Symp. Quant. Biol., *11*: 67–77.

Hershey, A. D., and M. Chase. 1952. Independent functions of viral protein and nucleic acid in growth of bacteriophage. J. Gen. Physiol., *36*: 39–56.

Kellenberger, E., A. Bolle, E. Boy de la Tour, R. H. Epstein, N. C. Franklin, N. K. Jerne, A. Reale-Scafati, J. Séchaud, I. Bendet, D. Goldstein, and M. A. Lauffer. 1965. Functions and properties related to the tail fibers of T4. Virology, *26*: 419–440.

Luria, S. E., and T. F. Anderson. 1942. The identification and characterization of bacteriophages with the electron microscope. Proc. Natl. Acad. Sci., *28*: 127–130.

Luria, S. E., M. Delbrück, and T. F. Anderson. 1943. Electron microscope studies of bacterial viruses. J. Bacteriol., *46*: 57–77.

Luria, S. E., and F. M. Exner. 1941. The inactivation of bacteriophages by X-rays—influence of the medium. Proc. Natl. Acad. Sci., *27*: 370–375.

Luria, S. E., and R. Latarjet. 1947. Ultraviolet irradiation of bacteriophage during intracellular growth. J. Bacteriol., *53*: 149–163.

Luria, S. E., R. C. Williams, and R. C. Backus. 1951. Electron micrographic counts of bacteriophage particles. J. Bacteriol., *61*: 179.

Pfankuch, E., and G. A. Kausche. 1940. Isolierung und übermikroskopische Abbildung eines Bakteriophages. Naturwiss., *28*: 46.

Ruska, H. 1940. Die Sichtbarmachung der Bakteriophagen Lyse im Übermikroskop. Naturwiss., *28*: 45–46.

———1941. Über ein neues bei der Bakteriophagen Lyse auftretendes Formelement. Naturwiss., *29*: 367–368.

Stent, G. S. 1963. Molecular Biology of Bacterial Viruses. W. H. Freeman and Co., San Francisco. 474 p.

A. H. DOERMANN
Department of Genetics, University of Washington, Seattle

The Eclipse in the Bacteriophage Life Cycle

INTRODUCTION

The premolecular biologists of twenty years ago considered bacterial viruses to be parasitic microorganisms whose life cycles were only superficially understood. The organismic view found rather forceful support from the early studies of virus mutation by Luria (1945) and by Hershey (1946) and from the first demonstration by Delbrück and Bailey (1946) of genetic recombination among these forms. In the meantime, extraordinary progress has been made so that, in the molecular age of today, a phage particle is recognized as a package of genetic information encoded in the length of a nucleic acid molecule carried in an expendable injection apparatus. The organization of this genetic material has by now been scrutinized in considerable detail; so much so, in fact, that to many people a T2 or T4 particle represents little more than a specifically ordered sequence of nucleotides. Small wonder then that it is today difficult for the molecular biologist to carry himself back to that earlier age.

It is difficult to appreciate that in 1946, on the thirtieth anniversary of the discovery of bacteriophages, one of the main difficulties in studying the life cycle was our complete ignorance about the state and number of intracellular phage. Thus Delbrück, in a Harvey Lecture delivered in that year, admitted "... to analyze the multiplication of a virus within the cell of its bacterial host, it should be our first aim to develop a method of determining the number of virus particles which are present within a bacterial cell at any one moment. Here I, and those who have been associated with me in this work, have to make the first admission of failure." The possibility that intracellularly viruses might exist in a condition different from the state in which they are recognized extracellularly was, of course, appreciated. Nine years earlier, in fact, the Wollmans (1937) had already proposed that intracellular phage may be in a non-infectious phase, because they failed to recover particles capable of plaque formation following lysozyme disruption of infected *Bacillus megaterium* cells. More recently Pirie (1946) argued that at least part of the intracellular virus is present in a form which is complexed with host cell material. And Luria (1947) proposed, as a mechanism for multiplicity reactivation of ultraviolet-inactivated phages, that once having entered bacteria, the phages break down into several independently reproduced subunits, one copy of each subunit being finally incorporated into every mature infectious virus particle. Nevertheless, a direct assay of intracellular phage, without which these speculations were difficult to test, was still lacking.

The general awareness of the problem of quantitating intracellular phage led to a number of experiments to solve that problem, but little success attended these early efforts. For example, Delbrück and Luria (1942) attempted mixed infections using the short-latent-period virus T1 together with T2 (named α and γ respectively at that time) which has a longer latent period. They entertained the hope that the lysis which terminates the short latent period of T1 would liberate any T2 present in the cells prior to the normal end of the longer T2 latent period. The purpose of this experiment was not achieved, but instead they discovered the phenomenon of mutual exclusion: namely that infection of T2 completely prevents growth of T1. Not even the infecting particles of T1 were, in fact, recovered. Another approach was that of Luria and Latarjet (1947) using ultraviolet radiation and X rays (Latarjet, 1948). Through target analysis they attempted to learn something of the fate of the infecting phage particle and of the kinetics of virus multiplication. But the interpretation of those surprising target results is not yet clear even today. In general it may then be said that next to nothing was known at that time about the fate of the infecting phage particle. That virus multiplication might occur while phages are appreciably altered from their extracellular infectious state was little more than conjecture.

THE CYANIDE-LYSIS METHOD OF LIBERATING INTRACELLULAR BACTERIOPHAGE

Observations not originally intended for the purpose finally led to a method for titrating intracellular phage. Microscopic studies of lysis, which recognized "lysis from without," were reported by Delbrück as early as 1940, and these were extended with turbidimetric experiments in his laboratory at Vanderbilt University. In a study of lysis and lysis inhibition in T-even bacteriophages, nephelometric measurements showed that high multiplicities of infection with some phages (notably wild-type T6) promptly caused a large reduction in turbidity (Doermann, 1948). Figure 1, taken from those experiments, shows turbidity curves for *Escherichia coli* infected with T6 at multiplicities ranging from 4.2 to 39 phage particles per cell. The fact that the drop in turbidity was quite precipitous, particularly at the highest multiplicity employed, suggested that lysis from without might be used to disrupt phage-infected cells prematurely and effect liberation of their viral contents.

The selection of the phage strain to be used for induction of lysis was not difficult. The turbidimetric data indicated that wild-type T6 was the most effective lysing agent. Furthermore, mutants of the bacterial strain *Escherichia coli* B existed which are completely resistant to T6 and for which no host-range mutants had been found. These same strains, when used as plating bacteria, assayed each of the other T-phages with an efficiency equal to that of strain B, making possible precise assays of even a small number of phage particles of these types in the presence of very large numbers of T6 particles.

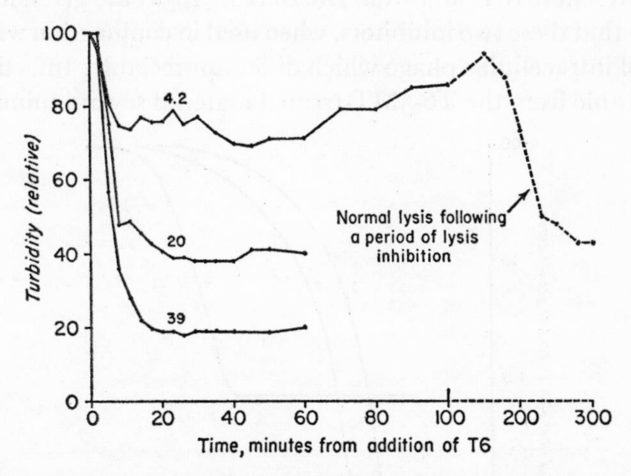

FIGURE 1. Lysis of bacteria by high multiplicities of T6. The relative turbidity of T6 infected bacteria was measured in a nephelometer of the type described by Underwood and Doermann (1947). The number by each curve indicates the ratio of phage to bacteria at time zero. The data come from Doermann (1948) Figure 1 (for ratio 4.2) and Figure 8 (ratios 20 and 39).

The selection of a phage whose intracellular titer could be readily estimated proved to be more difficult. It appeared at first to be advisable to select the phage with the longest latent period, and T5 was tried initially (40 minute latent period in nutrient broth at 37°C). It was indeed found that T5 could be liberated prematurely by T6-induced lysis from without (Doermann, 1946). Phage T5 was, however, an unfortunate choice because precise quantitative experiments were made difficult by its very slow rate of adsorption to strain B, coupled with the unavailability of a good T5 antiserum for eliminating unadsorbed phages. In both of these respects, T4 was superior to T5; and, in spite of its shorter latent period (23 minutes in nutrient broth at 37°C), it proved to be a better choice.

The first experiments of disrupting T4-infected cells with T6 were, nevertheless, only a slight improvement over the T5 experiments. Lysis from without by T6, judging from turbidimetric measurements, seemed slow compared to the speed with which intracellular T4 virus accumulated, leading to the suspicion that a significant increment of virus was occurring between the time of addition of T6 to the culture and the time of rupture of the majority of the T4-infected cells. In spite of Delbrück's warning that there would be too many "messifying" factors involved, attempts were made to find a metabolic inhibitor which would promptly stop intracellular phage multiplication but not interfere with the induction of lysis from without by T6. Out of a number of inhibitors which were tried, the most extensive experiments were made with 5-methyltryptophan (5MT) and with cyanide (CN). Representative results from Doermann (1952) are given in Figure 2. They show that these two inhibitors, when used in conjunction with T6 give estimates of intracellular phage which differ appreciably. Infectious phage are recoverable from the T6-5MT-treated material several minutes earlier

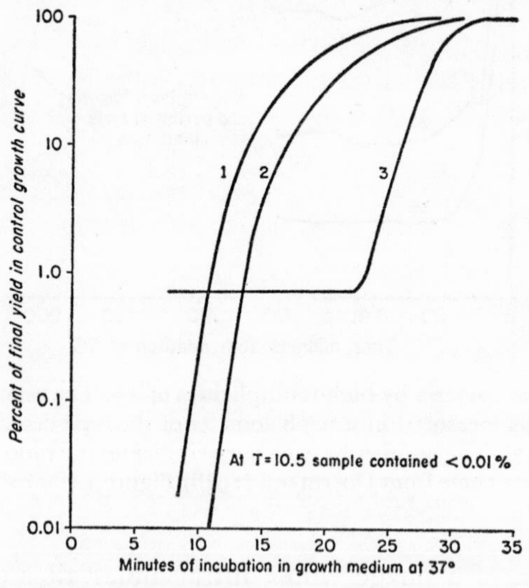

FIGURE 2. The appearance of mature T4 particles within infected bacteria. Samples of cells mixed with T4 at time zero (single infection) were subsequently subjected to high titer T6 together with inhibitory concentrations either of 5-methyltryptophane (curve 1) or cyanide (curve 2). The third curve represents one-step growth measured on aliquots of the same cultures from which curve 2 was derived. Each curve is constructed from several experiments. The data are detailed in Doermann (1952) Figure 1 (giving curves 2 and 3) and Figure 4 (giving curve 1).

than from the T6-CN-treated samples. The lag in phage recovery from the latter samples persists throughout the latent period. Although it was reasonable to suppose that the inhibitor giving the smaller yield at any particular time of treatment is acting more promptly and therefore reflects the instantaneous intracellular titer more faithfully, these experiments gave little assurance that use of either inhibitor was, in fact, providing an accurate measure of the titer of intracellular phage present at the time of treatment.

To give confidence that it does, corroboration with a different method, especially for stopping phage development, seemed to be required.

ALTERNATIVE METHODS OF FREEING INTRACELLULAR PHAGES

Chilling infected cultures to 6°C or lower proved to be an adequate method of stopping phage growth. At this low temperature, however, lysis from without is too slow a process to be an effective agent for disrupting cells. A physical method seemed to be called for, and treatment with sonic vibration was tried (Anderson and Doermann, 1952). It had been shown earlier (Anderson, Boggs, and Winters, 1948) that, although T4 and the other large phages are quite sensitive to sonic energy, the phages T1, T3, and T7 are resistant to sonic treatments which disintegrate most of the cells of the bacterial host. To test the method of chilling and sonication, T4, which had been used in the T6-CN procedure, was therefore replaced by T3. Results obtained by T6-CN treatment were compared with data obtained from chilling and sonication of aliquots from the same T3-infected

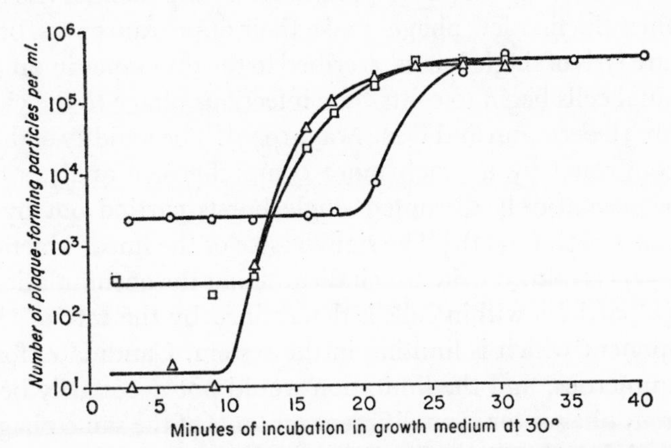

FIGURE 3. Comparison of intracellular T3 estimates obtained by two different methods. *Escherichia coli* B was infected with T3 (single infection) and samples of the infected bacteria disrupted at intervals with a T6-cyanide mixture (triangles) or with sonic treatment (squares). The circles show normal one-step growth measured in aliquots from the same culture. Data are from Anderson and Doermann (1952) Figure 3a.

culture. The comparison showed that there is very little difference in the estimates of intracellular titer obtained by the two techniques. Figure 3 gives a representative experiment. Because the results from two radically different procedures gave nearly identical determinations, it was concluded that these methods do provide means of measuring accurately the number of infectious intracellular virus at any time in the latent period.

In the next few years a number of other systems were devised for rupturing infected bacteria. Fraser (1951) equilibrated cell suspensions with high pressures of nitrous oxide and used rapid decompression to rupture them. This procedure was used in electron microscopic studies of intracellular phage development by Levinthal and Fisher (1952). Another combination which was adopted by DeMars (1955) for immunological studies makes use of glycine (Kay, 1952) to disrupt infected cells while phage growth is inhibited by cyanide. In some ways the most convenient technique is simply to shake suspensions of infected bacteria with a small volume of chloroform (Séchaud and Kellenberger, 1956) which causes immediate lysis of cells which have already elaborated the phage lysozyme. These methods, as well as others not discussed here, all give more or less the same result as the T6-CN procedure so that today a variety of alternatives are available which can be adapted to many specialized objectives in the analysis of events taking place in the infected bacterium at any particular moment.

Two major observations characterize the early data. One is that, contrary to what might have been expected for an intracellular parasite, the kinetics of accumulation of infectious phage particles are not exponential, but are more or less linear. [The approximately exponential rise during the interval when the first few phages make their appearance was, on the basis of premature lysis of single bursts, ascribed to the difference in time at which the individual cells begin to contribute infectious phage to the yield from a mass culture (Doermann and Dissosway, 1949). The validity of this hypothesis was confirmed by a much more comprehensive analysis of a large number of prematurely disrupted single bursts carried out by Bentzon, Maaløe, and Rasch (1952).] The significance of the linear kinetics is, even today, not yet certain. It evidently indicates that the accumulation of infectious phage particles within cells is determined by the rate of synthesis of some component which is limiting in the system. Candidates for this role could be numerous, and the limitation would not necessarily be identical with different phages (or even different mutants of the same phage), in different bacterial strains, or under different culture conditions.

GENETIC ANALYSIS OF THE FIRST-MATURED PHAGES

More significant was the second observation, namely that those particles which infect the cell are, for the first half of the latent period at least, not recoverable. The infecting phages enter a so-called *eclipse*. What the nature of the eclipse is, was not immediately understood. At first, following the ideas of Pirie (1946), it seemed plausible to suppose that the eclipse might result from a temporary association of the infecting virus with some of the cell components. Luria's (1947) interpretation of his multiplicity reactivation data, however, pointed toward the possibility that the infecting particle may be reduced to subunits upon entering the bacterium, and that the reappearance of infectious phages depends on reassembly of complete sets of such elements. What was needed was a technique whereby the first appearing phages emerging from the eclipse could be compared with the infecting phages. Just at this time, Hershey and Rotman (1949) showed that genetic recombinants for a pair of markers are randomly distributed among the individual mixedly infected bacteria, each cell producing about the same proportion of recombinant phages. This result made it possible to use a genetic test to compare the first phages emerging from the eclipse with those originally introduced into the cell. Thus bacteria were infected on the average with several phages of each of two parental genotypes which differed in two genetic loci. Table 1, taken from Doermann (1951), shows the data from three of the first experiments of this kind. It is evident that among the first phage particles recovered after the eclipse, genetic recombinants occurred with a frequency almost as great as in the normal cross lysates.

Table 1

OCCURRENCE OF GENETIC RECOMBINANTS AMONG THE PROGENY PHAGES OBTAINED FROM PREMATURELY LYSED PHAGE-HOST COMPLEXES.

Strain B of *Escherichia coli* was mixedly infected with T2H r_{13} and T2H *h* using multiplicities of several phages of each type. The infected bacteria were induced to lyse by T6-cyanide treatment at several times in the latent period. The progeny obtained were plated with a mixture of B and B/2 which permits identification of all four genotypes. Data are from Doermann (1951).

Experiment No.	*Time of T6-cyanide Treatment*	*No. of Phage per Cell*	*Total No. of Recombinants Observed*	*Per cent Recombinants*
1	20	2.3	10	2.0
2	20	6.0	97	2.2
1	25	36	8	1.6
3	27	67	81	2.4
2	29	102	34	2.3
1	30	113	19	3.0
1	Normal burst	288	47	3.0
2	Normal burst	516	131	3.5
3	Normal burst	365	27	2.3

Clearly, the infectious phages which appear immediately after the eclipse are genetically different from the phages used originally to infect the culture. These results were later confirmed in other experiments in which the marker combinations used give higher recombination values, permitting more precise experiments. Although these later experiments uncovered an upward drift in the proportion of recombinants inside cells as the latent period progresses, genetic recombinants were invariably found with a frequency at least half that found in the normal lysate. That was the case even when samples were disrupted so early that only a single phage was found for every twenty bacteria lysed (Doermann, 1953).

The genetic experiments showed unambiguously that upon entering a cell, the state of the phage was altered. This new state was called the *vegetative state*. Since genetic recombination had been shown to occur in that state, judging from the early appearance of recombinants, the further conclusion could also be drawn that the replication of genetic material must also occur in that state, as the following argument shows. For, if the recombinants found early had been formed prior to replication of the genetic material, they would, during subsequent replication, have grown into large clones. Hershey and Rotman (1949), however, showed that recombinants are randomly distributed among mixed bursts of normal size. Replication must therefore have preceded or occurred simultaneously with recombination.

CONCLUSION

The molecular interpretation of the eclipse was not immediately obvious either from the lysis experiments or from the simple genetic experiments done at that time. Clarification awaited the demonstration by Hershey and Chase (1952) of the true nature of the early steps in infection. By labeling the protein and the DNA of phage separately, they showed that the viral DNA is injected into the host cell, and after that is accomplished the protein moiety, consisting mainly of a microsyringe, is largely, or entirely expendable. The vegetative state was thus shown to begin with the introduction of the phage DNA molecule into the bacterial cell. From this discovery onward, the directions to be taken in gaining a better understanding of the phage life cycle were clear. These have provided ever-increasing insights, thanks to the use of physico-chemical, genetic, serological, radiological, electron microscopic, and other tools. Thus the early vision of Max Delbrück of phage multiplication as "so simple a phenomenon that the answers cannot be hard to find" (Delbrück, 1946), gave way to a much more complex picture.

REFERENCES

Anderson, T. F., S. Boggs, and B. C. Winters. 1948. The relative sensitivities of bacterial viruses to intense sonic vibration. Science, *108*: 18.

Anderson, T. F., and A. H. Doermann. 1952. The intracellular growth of bacteriophages II. The growth of T3 studied by sonic disintegration and by T6-cyanide lysis of infected cells. J. Gen. Physiol., *35*: 657–667.

Bentzon, M. Weis, O. Maaløe, and G. Rasch. 1952. An analysis of the mode of increase in number of intracellular phage particles at different temperatures. Acta Path. et Microbiol. Scand., *30*: 243–270.

DELBRÜCK, M. 1940. The growth of bacteriophage and lysis of the host. J. Gen. Physiol., *23*: 643–660.

———1946. Experiments with bacterial viruses (bacteriophages). Harvey Lectures Series *XLI*: 161–187.

DELBRÜCK, M., and W. T. BAILEY, Jr. 1946. Induced mutations in bacterial viruses. Cold Spring Harbor Symp. Quant. Biol., *11*: 33–37.

DELBRÜCK, M., and S. E. LURIA. 1942. Interference between bacterial viruses I. Interference between two bacterial viruses acting upon the same host, and the mechanism of virus growth. Arch. Biochem., *1*: 111–141.

DEMARS, R. I. 1955. The production of phage related materials when bacteriophage development is interrupted by proflavine. Virology, *1*: 83–99.

DOERMANN, A. H. 1946. In Ann. Rpt. Biol. Lab., Long Island Biol. Assn., p. 22.

———1948. Lysis and lysis inhibition with *Escherichia coli* bacteriophage. J. Bacteriol., *55*: 257–276.

———1951. Intracellular phage growth as studied by premature lysis. Federation Proc., *10*: 591–594.

———1952. The intracellular growth of bacteriophages I. Liberation of intracellular bacteriophage T4 by premature lysis with another phage or with cyanide. J. Gen. Physiol., *35*: 645–656.

———1953. The vegetative state in the life cycle of bacteriophage: Evidence for its occurrence and its genetic characterization. Cold Spring Harbor Symp. Quant. Biol., *18*: 3–11.

DOERMANN, A. H., and C. DISSOSWAY. 1949. Intracellular growth and genetics of bacteriophage. Carnegie Inst. Wash. Year Book, *48*: 170–176.

FRASER, D. 1951. Bursting bacteria by release of gas pressure. Nature, *167*: 33.

HERSHEY, A. D. 1946. Mutation of bacteriophage with respect to type of plaque. Genetics, *31*: 620–640.

HERSHEY, A. D., and M. CHASE. 1952. Independent functions of viral protein and nucleic acid in growth of bacteriophage. J. Gen. Physiol., *36*: 39–56.

HERSHEY, A. D., and R. ROTMAN. 1949. Genetic recombination between host-range and plaque-type mutants of bacteriophage in single bacterial cells. Genetics, *34*: 44–71.

KAY, D. 1952. The intracellular multiplication of coli bacteriophage T5st. Brit. Jour. Exptl. Pathol., *33*: 236–243.

LATARJET, R. 1948. Intracellular growth of bacteriophage studied by roentgen irradiation. J. Gen. Physiol., *31*: 529–546.

LEVINTHAL, C., and H. FISHER. 1952. The structural development of a bacterial virus. Biochem. et Biophys. Acta, *9*: 419–429.

LURIA, S. E. 1945. Mutations of bacterial viruses affecting their host range. Genetics, *30*: 84–99.

———1947. Reactivation of irradiated bacteriophage by transfer of self-reproducing units. Proc. Natl. Acad. Sci., *33*: 253–264.

LURIA, S. E., and R. LATARJET. 1947. Ultraviolet irradiation of bacteriophage during intracellular growth. J. Bacteriol., *53*: 149–163.

PIRIE, N. W. 1946. The state of viruses in the infected cell. Cold Spring Harbor Symp. Quant. Biol., *11*: 184–192.

SÉCHAUD, J., and E. KELLENBERGER. 1956. Lyse précoce, provoquée par le chloroforme, chez les bactéries infectées par du bactériophage. Ann. Inst. Pasteur, *90*: 102–106.

UNDERWOOD, N., and A. H. DOERMANN. 1947. A photo-electric nephelometer. Rev. Sci. Instr., *18*: 665–669.

WOLLMAN, E., and E. WOLLMAN. 1937. Les "phases" des bactériophages (facteurs lysogènes). Compt. Rend. Soc. Biol., *124*: 931–934.

ANDRÉ LWOFF
Service de Physiologie Microbienne, Institut Pasteur, Paris, France

The Prophage and I

It is a long time since we have not seen Dr. Lwoff.
—ANTON CHEKOV, *Uncle Vanya*

The triumvirate responsible for this volume has asked the presumptive authors to "describe some significant contribution they had made to molecular biology." A scientist should never attempt to judge the value of his own achievements, whether significant or not, but especially when not. This is the Golden Rule of intellectual hygiene. And I would have been most embarrassed if the triumvirs had not made their own choice and decided that I had to deal with the prophage. Prophage is a remarkable entity indeed, a molecule I should say, for I have to be molecular. Who is not? The prophage is the basis of lysogeny, its essence, its very flesh; and it was whilst working on lysogeny, that I became pregnant with the molecule-prophage. How did this happen? Such is the question posed by the triumvirate. Of course the three gentlemen expressed themselves in a noble academic style: "What were the heuristic circumstances and ideas?" Yes.

The circumstances are shrouded in the hazes of a remote past. The ideas have been buried long ago. Let us assume, nevertheless, that Max Delbrück will appreciate as a connoisseur the subtle smell of exhumed remembrances, just as the divinities of the Olympus liked to suck in the fumes of grilled meats and fats of the sacrificed animals.

Today, I have to sacrifice myself. This paper is therefore a sort of self-combustion.

A few years ago, this was in 1921, I entered the Pasteur Institute. I rapidly became acquainted with Eugène Wollman, whom I visited from time to time. Eugène Wollman, who was working with bacteriophage and lysogeny, liked to show his experiments and was eager to discuss his ideas. Thus, especially between 1930–1939, I was introduced to *Bacillus megatherium.* I know: this is not the right spelling. But in the good old days the species did possess an *h*, which it lost only later in a taxonomic battle. Just the usual sort of thing. At the time, I was working on the nutrition of protozoa and bacteria and especially on growth factors. I should have been impressed by bacteriophage. Shall I confess that I was not? At least not deeply enough.

Later on, I did some work on the genetics of *Moraxella*. As a consequence, I was invited to the 1946 Cold Spring Harbor Symposium. I thus discovered America, Max Delbrück, and the powerful "phage church." The atmosphere was stimulating, and I swallowed everything with enthusiasm: genetics, as well as red beans seasoned with peanut butter, a most remarkable object. Catherine Fowler, who served the meals, was especially kind with the hungry Frenchmen. They were very hungry indeed, and perhaps, or I should say certainly, I ate more than my ration of cream. Anyhow, I did not attend the phage course, for I still considered viruses with some suspicion. I was in love with other creatures.

And now I am trying intensely to remember why and how, around 1949, I started working on lysogeny. I mean, of course, the heuristic and molecular circumstances. A digression concerning Jacques Monod is of necessity here. It is not so much a digression, for Jacques Monod (a) is a friend of Max Delbrück, (b) has played directly and indirectly an important role in the history of bacteriophage. At the time, Jacques Monod was working in the Department of Zoology of the Sorbonne on the growth of ciliates. As I was then classified as a specialist of the nutrition of ciliates, Monod was coming to me for discussions from time to time. Since I knew that Jacques Monod was more interested in growth than in ciliates, and since the nutrition and growth of a population of ciliates is something rather complex, I suggested that he work on a bacterium, such as *Escherichia coli*, for example. "Is it pathogenic?" asked Jacques Monod. The answer being satisfactory, Jacques Monod started working on *E. coli*. A few months later, he showed me beautiful growth curves exhibiting the diauxy phenomenon. "What could it mean?" he asked. I said this could have something to do with adaptive enzymes. The answer I received takes now its full fragrance in the light of Jacques Monod's achievements. This answer was: "adaptive enzymes, what is that?" Of course, he said this in French, because we used to speak French.

So Jacques Monod was working with bacteria and became deeply interested in bacteriophage and in lysogeny, and he even wrote a theoretical paper on the subject. This subtle analysis, which never saw the light of day, was unfortunately lost. But we had often discussed lysogeny. Why not attack the problem? Because of den Dooren de Jong—one of the classics in lysogeny, and because of Eugène Wollman, I selected *Bacillus megaterium* as material. Moreover, it is a large, that is, relatively large bacterium. This was of utmost importance, for I had decided to study single bacteria. The reason for this decision is simple. I dislike mathematics, for which I am not gifted, and I wanted to avoid formulas, statistical analysis and, more generally, calculations as much as possible.

However large *Bacillus megaterium* might be, a micromanipulator was needed. Moreover, the microscope had to be in an incubator. A plastic incubator was built, and it turned out that to maintain a constant temperature throughout the box was not so easy as it seemed. But the technical problems were solved; moreover I became an expert in handling paraffin oil, an oil chamber, microdrops, micropipettes and, of course, individual bacteria.

This experience turned out to be very useful in Pasadena in 1954, when Renato Dulbecco and I decided to attack the problem of the kinetics of release of *Poliovirus* by single animal cells. The California Institute of Technology, poor as many American institutions often are, could not afford to buy a microforge. The Chairman of the Department of Biology, George Beadle, kindly persuaded us that we should build one ourselves. This was done, thanks partly to the generous gift of a very, very old microscope—which was reluctantly sacrificed for this occasion—and thanks also to the remarkable mechanical skills of Renato Dulbecco, who—I wonder why—is better known as a virologist. The home-made microforge turned out to be functional. This digression has lasted long enough. We had better return to bacteria.

Thus, I was able to play with single bacteria. A lysogenic strain of *Bacillus megaterium* was put to work. Individual bacteria were inoculated into individual microdrops. What they were expected to do, what they had to do, was to grow and to divide. They grew, but the daughter bacteria did not separate. As a consequence, filaments were produced. Because of the size of the drop, spirals were formed. Long, beautiful, desperate spirals. The medium was held responsible. Its composition was changed, and thereon the bacteria behaved normally; that is, they grew, divided and separated. With the help of a micropipette, we managed to pump micro-samples of liquid out of the microdrop. The samples were transferred to petri dishes seeded with an indicator strain of *Bacillus megaterium* sensitive to the bacteriophage carried by the lysogenic strain. Thus it could be seen that lysogenic bacteria can grow and divide without liberating virions. Moreover, a lysogenic bacterium which had divided nineteen times without liberating virions, proved to be still lysogenic when seeded on an indicator plate. The experiments clearly showed that, contrary to the then current hypotheses, virions are not secreted by the lysogenic bacterium. They showed later on that virions are in fact liberated by the lysis of the bacterium.

Since I am trying to act as an historian, I must produce here a remark which, despite its non-molecular nature, is nevertheless part of the heuristic picture. During the course of my work, I received an unusual number of visits. Many of my colleagues were deeply interested in lysogeny. Perhaps

I should mention—although this is irrelevant—that my young technician, despite a remarkably high efficiency, was exceptionally beautiful and charming. Yet, as my visitors came in order to discuss lysogeny—of course—, and tried—God knows why—, to be brilliant, their intrusions were, on the whole, profitable. This remark might seem cynical. I have to confess that I never felt guilty. Alas! Perhaps I am cynical after all. Digging too deeply into the past is always dangerous, and I better come back to the real subject, in order to avoid more unexpected painful discoveries.

Thus, a lysogenic *Bacillus megaterium* can grow and divide without liberating virions. When lysogenic bacteria are lysed with lysozyme, which does not destroy virions, no virions are liberated. Thus lysogenic bacteria do not contain virions. Yet virions are always present in a culture of lysogenic bacteria. Why and how does a lysogenic bacterium manufacture and liberate virions? Single bacteria introduced into microdrops generally grow and multiply. But it happens, from time to time, that the totality of the micropopulation, say four to eight bacteria, lyses, i.e., disappears within an interval of a few minutes. When this happens, many virions are liberated, about a hundred per lysed bacterium.

Thus virions are liberated by bacterial lysis. The production of bacteriophage is a lethal event. The hereditary power to produce bacteriophage can be perpetuated only as a potential property. The expression of this property is lethal. A lysogenic bacterium perpetuates a non-infectious entity able to give rise to bacteriophage. The entity was christened ***prophage***. The term could be criticized, of course, for "prophage" means more or less "before the meal." Strictly speaking, prophage would be a sort of "hors d'oeuvre." But since it is customary to speak of "phage" instead of bacteriophage and since "probacteriophage" is a long word, it would inevitably have wound up as prophage anyway. So, hors d'oeuvre— I mean prophage—was proposed at the outset. The world was obviously eagerly awaiting the coming of the prophage, for, despite its French origin, the Greek term was rapidly and unanimously adopted (Lwoff and Gutman, 1949b, 1950a and b).

I had worked on lysogeny without ulterior motivation. Yet since prophage is a molecule—of nucleic acid—and since I was studying its biology, I later on became a molecular biologist. A redoubtable position, though round about 1950 the future molecular biologists did not think of themselves as such. The incommensurate virtue of this magic label was discovered only much later. Perhaps I should add that I am unable to decide to what extent I feel molecular, if I am molecular at all.

Let us now return to the production of bacteriophage by the lysogenic bacterium. Sometimes, as already stated, the bacteria introduced into the microdrop grew and divided, and nothing strange was seen in the population. Sometimes all the daughter bacteria lysed and liberated bacteriophage. Things happened as if the production of virions had been triggered, as if the production of bacteriophage by the bacteria had been induced by some environmental factor. We were so convinced of the validity of this conclusion that it was mentioned as an hypothesis in a paper dealing with lysogeny (Lwoff and Gutmann, 1949a).

Now the working hypothesis had to be put to work. What in the medium is responsible for bacteriophage development? Microdrops did not lend themselves easily to the type of experiments we contemplated. Experiments involving a few milliters were of necessity. For the handling of these mass cultures and for the solution of the problem we were fortunate enough to secure the cooperation of Louis Siminovitch and Niels Kjeldgaard.

Bacteria were seeded in various media. The kinetics of bacterial growth and of bacteriophage production were followed. It turned out that the frequency curve of bacteriophage production showed a peak toward the end of the exponential bacterial growth. If one of the kinetics of phage production is represented on linear, and the kinetics of bacterial growth on semilogarithmic coordinates, the phenomenon becomes perfectly clear. I know what you are going to say. Calculation showed that only a relatively small fraction of the bacteria, about 15%, are producers of bacteriophages. Nevertheless, it seemed clear, anyhow, this is what we concluded, that the culture medium becomes altered as a result of bacterial metabolism, and that the altered medium is responsible for the "induction." This reasoning proved to be correct, but only much later, after long and painful efforts.

So, the medium modified by bacterial metabolism was held responsible for the induction, and we decided to identify the responsible substance or factor: for instance, amino-acids, pH, redox potential, oxygen tension, and so on.

Our aim was to persuade the totality of the bacterial population to produce bacteriophage. All our attempts—a large number of attempts it was—were without result. Louis Siminovitch and Niels Kjeldgaard became very depressed, and even repressed—the concept of repression could have been discovered on this occasion. Yet I had decided that extrinsic factors must induce the formation of bacteriophage. Moreover, the hypothesis had been published already, and when one publishes an hypothesis, one is sentenced to hard labor. Finally, I am stubborn. So the experiments went on, day after day, and they continued to be desperately negative. A digression is necessary here.

Have I already told the story, and perhaps even published it? But I am not going to read all that I have written in order to find out; to read his own papers is punishment for any scientist. Our experiments consisted in inoculating exponentially growing bacteria into a given medium and following bacterial growth by measuring optical density. Samples were taken every fifteen minutes, and the technicians reported the results. They (the technicians, that is) were so involved that they had identified themselves with the bacteria, or with the growth curves, and they used to say, for example: "I am exponential," or "I am slightly flattened." Technicians and bacteria were consubstantial.

So negative experiments piled up, until after months and months of despair, it was decided to irradiate the bacteria with ultra-violet light. This was not rational at all, for ultra-violet radiations kill bacteria and bacteriophages, and on a strictly logical basis the idea still looks illogical in retrospect. Anyhow, a suspension of lysogenic bacilli was put under the UV lamp for a few seconds.

The Service de Physiologie Microbienne is located in an attic, just under the roof of the Pasteur Institute, with no proper insulation. The thermometer sometimes rises in a manner that leaves no conclusion other than that the temperature is high. It was a very hot summer day and the thermometer was unusually high. After irradiation, I collapsed in an armchair, in sweat, despair, and hope. Fifteen minutes later, Evelyne Ritz, my technician, entered the room and said: "Sir, I am growing normally." After another quarter of an hour, she came again and reported simply that she was normal. After fifteen more minutes, she was still growing. It was very hot and more desperate than ever. Now sixty minutes had elapsed since irradiation; Evelyne entered the room again and said very quietly, in her soft voice: "Sir, I am entirely lysed." So she was: the bacteria had disappeared! As far as I can remember, this was the greatest thrill—molecular thrill—of my scientific career. The identification of a few growth factors for protozoa and bacteria and the elucidation of their mode of action was the result of rational, logical work. Even the unexpected discovery of unknown, extraordinary ciliates did not produce such an emotional stress. With induction we had been in complete darkness.

The morale of Siminovitch and of Kjeldgaard suddenly rose to an alarming level. Fearing the worst, I had to remind them—as well as myself—that *Bacillus megaterium*, just as so many bacilli, often lyses without liberating any bacteriophage. Dilutions were made and samples plated on the indicator strain. The night was strange, and we all raced to the laboratory at an unusually early hour. The winner rushed to the incubator and picked up the Petri dishes. The concentration of bacteriophage was fantastic. Each lysed bacteria had liberated some hundred virions on the average. Induction was discovered.

May we be forgiven, but the first thing we did was to write a paper (Lwoff, Siminovitch and Kjeldgaard, 1950a). But we waited before sending it to press for three days, until the experiment had been twice repeated. Siminovitch and Kjeldgaard now became completely derepressed, and we started working on the physiology of induced bacteria and on the biology of prophage—this biology being molecular, of course (Lwoff, Siminovitch and Kjeldgaard, 1950b and c; Lwoff, 1951).

Another digression is again necessary. This one concerns radiations. I dislike radiations, whatever their wave-length. And I have to confess that my attitude towards radiations in general, and ultraviolet light in particular, has not been modified by the discovery of induction. This might look like black ingratitude, but it is so. "Black ingratitude" is the literal translation of the French "noire ingratitude." For a Frenchman *black* is so much better than *sheer* and, moreover, it sounds Shakespearian. Black ingratitude! Black, molecular ingratitude!

Thus irradiation of lysogenic bacteria with UV light can induce the vegetative phase of the bacteriophage. May I remind you that the first indication of induction came from the observation of single bacteria in microdrops, that we held the medium responsible, and that we had tried in vain to identify the responsible factor. Induction with ultra-violet light triggered a new series of experiments. It was finally found that thioglycolic acid and a number of other reducing substances are powerful inducers (Lwoff and Siminovitch, 1951a and 1952). Our experiments had been performed in a medium containing yeast extract. And everything went on beautifully until the sample of yeast extract was exhausted. The media prepared with another batch, and all other media, were devoid of inducing activity after addition of thioglycolic acid. A number of yeast extracts were prepared and one of them turned out to be "active." Thus we started investigating why certain media were "active" and others not. After numerous experiments it was found that the active medium lost its activity by treatment with 8-hydroxyquinolein, and the responsible factor was identified as copper (Lwoff, 1952). Inactive media could be activated by the addition of copper. It is known that the oxidation of sulfhydryl compounds by copper yields hydrogen peroxide. As a matter of fact, addition of hydrogen peroxide to an organic medium induces phage development. Organic peroxides are inducers, as are a number of mutagenic/carcinogenic agents (Lwoff and Jacob, 1952). Thus an explanation of the apparently spontaneous production of bacteriophage could be provided. Bacteria produce reducing substances. If the medium contains copper, the reducing substances are oxidized and hydrogen peroxide is formed. Hydrogen peroxide reacts with organic substances. Organic peroxides are formed, which produce the alteration of the bacterial metabolism responsible for induction. Of course, this alteration could as well be the result of rare mutations of critical bacterial genes.

Today, we are able to propose a coherent picture of the mechanism of induction. The prophage produces a messenger, which is responsible for the synthesis of a repressor, which is probably a protein. The repressor is responsible for the specific immunity of lysogenic bacteria. It blocks the expression of the other genes of the phage, and especially the genes responsible for the autonomous multiplication of the genetic material. Inducing agents alter the bacterial metabolism in such a way that a substance is produced which turns an active repressor into an inactive aporepressor. This substance could be the precursor of DNA (Goldthwait and Jacob, 1964). (For reasons which will become obvious later on, no reference is given for this paper.) Thus the problem of the mechanism of induction is now close to its solution.

Things are clear now, but have not always been. In the years 1950–1952, the concept of repression and derepression had not yet come to light. Nevertheless, Jacques Monod used to say that induction of enzyme synthesis and of phage development are the expression of one and the same phenomenon. This statement looked paradoxical, but was, paradoxically, a remarkable intuition.

During this phase of the work, I had numerous discussions with geneticists. Most of them considered that induction of bacteriophage development in a lysogenic bacterium is analogous to induction of a mutation. They were impressed by the fact that inducing agents are mutagens. My feeling was that induction and mutation must be completely different phenomena, for phage development is induced with a probability close to one. Today, induction of phage development is known to be a derepression. It does not involve an alteration of the base sequence of the nucleic acid; that is, it is not a hereditary modification of the genetic material. Yet if it should turn out that induction involves a detachment of the prophage from the bacterial chromosome, my opponents would not have been all that wrong. I would, nevertheless, have been right.

Thus a picture of lysogeny emerged. The lysogenic bacterium perpetuates the genetic material of the bacteriophage, the prophage. The prophage is an innocuous molecule. But when the prophage is induced, the vegetative phase of the bacteriophage is started, which ends in bacterial lysis. The prophage is a potentially lethal structure, and lysogeny can only be the perpetuation of the potentiality to produce virions. The prophage is a potentially lethal factor. However, some lysogenic bacteria perpetuate an abnormal prophage. Induction triggers the vegetative phase: the bacterium lyses, but no virions are produced. An altered genetic material, a defective prophage, is responsible for a fatal disease of the bacterium (Lwoff and Siminovitch, 1951b).

The data concerning lysogeny and its whereabouts were summarized in a review (Lwoff, 1953) and in a lecture (Lwoff, 1954).

Perhaps I should here make an attempt at a retroductive molecular and heuristic survey. It was the observation of the behavior of single bacteria in microdrops which led to the concept of induction. Once the concept was seeded, it grew, developed, and performed its task: it led to the discovery of induction. Today we know that some prophages are inducible, whereas others are not. We know that some inducible prophages are easy to induce, others are not. We know that some media develop inducing properties as a result of bacterial growth, others do not. Hence we had picked up the improbable: the right bacterium, the right prophage, and the right medium; and the conditions prevailing in the microdrops were such that induction took place only from time to time. Of course, sooner or later, someone would have irradiated an inducible lysogenic bacterium, either by accident or as the consequence of an acute ultra-violet mania. Induction would have been discovered anyhow.

As we are considering the history of lysogeny and of prophage, I have to do a horrible thing: to quote myself, namely a whole paragraph of a review (Lwoff, 1953).

Danger of hypothetical secretions. Twenty-five years elapsed between the discovery of lysogeny and its definition. Among the factors responsible for this long delay was the secretion theory which created the suspicion in which lysogeny itself has been held for a time. This theory was probably unconsciously reached by the following reasoning: lysogenic bacteria live and multiply, lysogeny is not lethal; lysogeny being phage production, phage production is not lethal. This view was certainly substantiated by the fact that lysogenic bacteria adsorb homologous phage without being killed: phage is not lethal. Thus, in lysogenic bacteria, prophage is not lethal, perpetuation of prophage is not lethal, phage is not lethal. How should a normally innocuous particle be lethal only when producing another similar innocuous particle? No example of this type of phenomenon was known. Phage must be secreted. Phage is secreted. Phage was secreted.

In early publications the definition of lysogenic bacteria was treated by preterition. But it is obvious that they were generally considered as phage-secreting bacteria, or, at least, as bacteria able to secrete phage. A scientifically minded bacteriophagist interested in lysogeny soon discovered that the secretion theory was devoid of any experimental basis. No evidence of phage secretion being available, he then logically decided that phage secretion was doubtful, that it did not exist. Therefore, phage-secreting bacteria could not exist and lysogeny did not exist. A wrong definition of lysogeny had led to the condemnation, not of the definition, but of lysogeny itself. And whilst lysogenic bacteria were utilized as instruments for the identification, or typing of bacteria, lysogeny itself had ceased to exist in the thinking of a number of prominent scientists.

Problems do not exist in nature. Nature only knows solutions. The solution, the lysogenic bacterium, was enslaved as a typing tool. The ghost of the problem, like those wisps of cloud that a breath of wind dispels, was blown away from the temple of science and a smell of sulfur was left floating in the air. For many years, many eminent scientists have considered lysogeny as a heresy.

This story was in a way symbolic. It is only after its publication that I learned from Max Delbrück himself why he had doubted the very existence of lysogeny. Max had asked X . . . What is the percentage of lysogenic bacteria in a lysogenic culture? The answer was: thirty per cent. This was the end . . . until Elie Wollman convinced the disciple of St. Thomas that such a thing as lysogeny does exist.

Now I am faced with a difficult problem. Contributors have been asked to give a "retrospective appreciation of the significance of their work in the light of later developments."

Cruel triumvirs! Unfortunate contributors! Dramatic option! Am I really expected to say that my own work has played a determining role in the development of virology, genetics, molecular biology and the like? It is fortunate that so many eminent scientists have worked on lysogeny before and after me. Thus, I shall turn the difficulty and consider the impact, not of my own contribution, but of lysogeny itself. Lysogeny has revealed a new type of cell/virus interrelation: the attachment of the genetic material of a virus onto a specific receptor of the host's chromosome. It has provided the first example of repression of viral functions. It has led to the recognition that host and viral genes can undergo recombination. The data and theoretical concepts concerning chemical inducers have provided a rational and efficient clue in the search for anticancerous drugs. Lysogeny has also led to the definition of viruses, to the very concept of virus (Lwoff, 1953). Moreover, lysogeny has played an important role in the development of our understanding of bacterial recombination and in the birth of the concept of episomes. But in reality, lysogeny is one of the manifold aspects of the biology of bacteriophage, and it is not so much lysogeny as bacteriophage which paved the way for the development of virology and played such an important role in the blooming of molecular biology. That is why I, like so many others, feel indebted to Max Delbrück.

The triumvirs have asked that the style of the article be "highly personal" and this is, of course, appropriate for a personal tribute. Yet one might fear this recommendation to be useless, for one writes like this or like that because one is like this or like that. I am precisely like this. May I call the attention of the reader to the fact that only my own papers are quoted in the bibliography? This is, I do hope, personal enough.

Whilst writing this article, I have experienced serious psychological difficulties, for the contributors are considered by the triumvirs as part of "the human panorama of the landscape of molecular biology." Good Lord!

Never did I think myself as panorama, or as part of a panorama, whether the panorama be human or not. No, never. But molecular biology nowadays invades everything, and if it is metamorphosing into a landscape, what can we do? Paint. This is my intimate, heuristic, molecular, and human conclusion.

Very important heuristic appendix

A few weeks after the manuscript was sent to the triumvirate it was read by my wife, and Marguerite told me what she remembered of the discovery of induction. As she knows and loves Max, her remarks are not out of place here.

On the induction day, as soon as I entered her room she noticed that I did not look normal. She immediately diagnosed an acute crisis of excitement. I had explained what was happening to the bacteria and added that, for the first time in my life, I had the feeling of having discovered something. This was a most unfortunate remark. For, instead of the congratulations I could have expected, I got a wigging: "induction was all right of course, but I should stop talking nonsense for I had already discovered a few things in the past." I am not going to comment on this particular point but simply give this as an example of how misunderstandings can arise in a couple.

Molecular appendix

At the end of October 1965, François Jacob also read the manuscript. Owing to the circumstances, he obviously felt more secure and dared to confess something which he had kept hidden for fifteen years. In June 1950, François Jacob visited me and expressed the wish to work in the attic (so-called Service de Physiologie Microbienne). Since 1940, his academic career had been rather strange: England, Africa with the division Leclerc, Normandy. But something compelled me to say "yes." And I added: "you know we have the induction" and François Jacob went, pleased, and saying to himself: "Induction, what the hell could that mean?"

Human appendix

And after all, why should I not also, just as Marguerite and François, reconsider induction in the light of October. I remember having long ago, this was in 1953, quoted Francis Bacon who, in the year one thousand six hundred and twenty, wrote this prophetic sentence: our only hope therefore lies in a true induction.

(*This is the fourth and last appendix*: "Je le jure."—André Lwoff)

REFERENCES

LWOFF, A. 1951. Conditions de l'efficacité inductrice du rayonnement ultraviolet chez une bactérie lysogène. Ann. Inst. Pasteur, *81*: 370–388.

———1952. Rôle des cations bivalents dans l'induction du développement du prophage par les agents réducteurs. C. R. Acad. Sci., *234*: 366–368.

———1953. Lysogeny. Bact. Rev., *17*: 269–337.

———1954. Control and interrelations of metabolic and viral diseases of bacteria. The Harvey Lectures, series L (1954–1955), p. 92–111, Academic Press, New York.

Lwoff, A., and A. Gutmann. 1949a. Production discontinue de bactériophages par une souche lysogène de *Bacillus megatherium.* C. R. Acad. Sci., *229*: 679–682.
———, ———1949b. La perpétuation endomicrobienne du bactériophage chez un *Bacillus megatherium* lysogène. C. R. Acad. Sci., *229*: 789–791.
———, ———1950a. Recherches sur un *Bacillus megatherium* lysogène. Ann. Inst. Pasteur, *78*: 711–739.
———, ———1950b. La libération de bactériophages par la lyse d'une bactérie lysogène. C. R. Acad. Sci., *230*: 154–156.
Lwoff, A., and F. Jacob. 1952. Induction de la production de bactériophages et d'une colicine par les peroxydes, les éthyléneimines et les halogénoalcoylamines. C. R. Acad. Sci., *234*: 2308–2310.
Lwoff, A., and L. Siminovitch. 1951a. Induction par des substances réductrices de la production de bactériophages chez une bactérie lysogène. C. R. Acad. Sci., *232*: 1146–1147.
———, ———1951b. Induction de la lyse d'une bactérie lysogène sans production de bactériophage. C. R. Acad. Sci., *233*: 1397–1399.
———, ———1952. L'induction du développement du prophage par les substances réductrices. Ann. Inst. Pasteur, *82*: 676–690.
Lwoff, A., L. Siminovitch, and N. Kjeldgaard. 1950a. Induction de la production de bactériophages chez une bactérie lysogène. Ann. Inst. Pasteur, *79*: 815–858.
———, ———, ———1950b. Sur les conditions de production du bactériophage chez une bactérie lysogène. C. R. Acad. Sci., *230*: 1219–1221.
———, ———, ———1950c. Induction de la lyse bactériophagique de la totalité d'une population microbienne lysogène. C. R. Acad. Sci., *231*: 190–191.

A. D. HERSHEY
Carnegie Institution of Washington, Genetics Research Unit
Cold Spring Harbor, New York

The Injection of DNA into Cells by Phage

My assignment in this book is to recall the circumstances surrounding the blender experiment of Hershey and Chase (1952), which showed that a particle of phage T2 infects a bacterium by injecting its DNA through the bacterial cell wall. To carry out that assignment, I shall try to evoke the period conveniently bounded by two symposia, both organized by Max Delbrück, that are listed in the bibliography of this essay as "Viruses 1950" and "Viruses 1953." During that period major consensuses were reached concerning the life cycle of phages, the central genetic role of DNA, and DNA structure. By 1953, phage workers had for the first time what Delbrück used to call a party line. For the first time, too, biologists were in possession of a chemical structure that said something about function, but that is another story.

Certain ideas concerning the structure and reproduction of phage particles had been excluded by 1950. The particles were not composed of some sort of primordial gene-stuff (nucleoprotein) whose major function was replication. The electron micrographs published by Luria and Anderson (1942), showing a complex, species-specific, phage-particle structure, disposed of that idea (and subjected me to the first of a series of mental shocks of a kind to which biologists nowadays, no doubt, acquire immunity in the cradle). To the same purpose, Anderson (1949) showed that the particles possess an osmotic membrane, and Novick and Szilard (1951), pursuing one of several germinal observations made by Delbrück and Bailey (1946), demonstrated the independence of phage-particle genotype and phenotype.

Neither could the phages be regarded simply as very small intracellular parasites, though d'Herelle favored the analogy and Burnet (1934) effectively argued its biological aptness. (I shouldn't oversimplify; the points of view of d'Herelle and Burnet were poles apart.) The weaknesses of the analogy were that it failed to account for the metabolic inertness of the particles and for the human hope that unique information could be got by studying them.

The admissible view of the time was presented by Luria (1950). The phage particle is a dual structure (mainly protein and DNA) in which differentiation of nuclear and cytoplasmic functions is vaguely discernible. After phage infection, the bacterial metabolic system falls under the control of the nuclear apparatus of the phage, producing mainly phage-specific materials (Cohen, 1949). The infecting phage particle does not itself survive the infection, losing at least its ability to infect another bacterium (Doermann, 1948). Thus intracellular phage growth is characterized by a phase of eclipse, lasting about half the life of the infected bacterium, during which the cell does not contain demonstrable phage particles. During the phase of eclipse, genetic recombination takes place, apparently preceded by replication of genetic determinants of the phage. Phage growth therefore consists of two stages: replication of some noninfective form of virus, followed by conversion of the products of replication back into finished phage particles which do not themselves participate in reproduction.

In truth, the facts available in 1950 did not demand any fundamental distinction between growth of phage and (as I remarked at the "Viruses 1950" meeting) growth of malaria parasites, which also multiply in eclipse. The conviction of phage workers about the uniqueness of phages was, for the time being, an article of faith, the origin of which seems obscure in retrospect (but see Muller, 1922). By way of example, I like to recall an incident fixed in my mind by the occasion of a visit to my laboratory by Max Delbrück. The time was 1946, when I was about to make the first quantitative cross with mutants of T2. I remarked to Delbrück that I rather expected to see a random assortment of markers. "Do you think phage particles dissolve?" he asked. To make it a wager I answered "Yes," and owe him a dollar to this day, feeling, I suppose, that the decision against me was only technical. In fact, the issue is complex and still undecided. Whereas today I am firmly convinced that the unit of T2 replication is the phage chromosome, if not something larger, Delbrück is now sometimes in favor of genetic subunits.

Luria (1947, 1950, 1953) must be credited with the idea of subunits of genetic replication and, indeed, with the first logically developed scheme of phage growth. His scheme was provocative, useful, and basically correct insofar as it emphasized the difference between growth of phage and multiplication of nuclei. It seems worthwhile to recall it briefly even though some important details proved to be wrong.

(1) To explain the high efficiency of "multiplicity reactivation" after infection with phage particles inactivated by ultraviolet light, Luria postulated that the phage genetic material comes apart into subunits that multiply independently (except subunits damaged by radiation), and then reassemble to form the nuclei of phage particles. The reassembly fails only if undamaged subunits of any specified kind are lacking.

The subunit hypothesis also explained the high frequency (in T2) of genetic recombination. There is no equally plausible way to explain either phenomenon today.

(2) When Cohen and Arbogast (1950) found that irradiated phage particles could not initiate DNA synthesis in the absence of multiplicity reactivation, postulate (1) demanded a phage genetic material containing little DNA. The DNA content of the particles would then belong to a cytoplasm acquired in the final "baking" (Luria, 1950). The current interpretation, of course, is that postulate (1) is incorrect.

(3) The increase in resistance to ultraviolet light of the phage-producing ability of T2-infected bacteria that occurs before replication starts (Luria and Latarjet, 1947) suggested to Benzer (1952) that some photosensitive structure becomes dispensable after performing its task. This interpretation agreed nicely with postulate (2) (Luria, personal communication, March, 1952). The current interpretation in terms of DNA lesions that interfere with gene function but not directly with genetic replication (Symonds and McCloy, 1958; Krieg, 1959), perhaps because they are subject to repair processes of various kinds, is less elegant if nearer the truth.

(4) Finally, study of proflavine lysates in Luria's laboratory (De Mars et al., 1953) and of premature lysates by Levinthal and Fisher (1952) had revealed DNA-poor formed elements that at first appeared to be precursors of phage particles, possibly nuclei that had not yet got their DNA cytoplasm. These subsequently proved to be unfinished phage particles from which the DNA had leaked out (Anderson et al., 1953; Kellenberger, 1959).

The foregoing adequately explains why few people in 1950 were primarily interested in phage DNA. Some were, however, for rather different reasons.

Northrop (1951) wrote, concerning phage growth, "The nucleic acid may be the essential, autocatalytic part of the molecule, as in the case of the transforming principle of the pneumococcus (Avery et al., 1944), and the protein portion may be necessary only to allow entrance to the host cell."

A more specific version of this thought was expressed by Roger Herriott in a letter dated November 16, 1951, from which I quote: "I've been thinking—and perhaps you have, too—that the virus may act like a little hypodermic needle full of transforming principles; that the virus as such never enters the cell; that only the tail contacts the host and perhaps enzymatically cuts a small hole through the outer membrane and then the nucleic acid of the virus head flows into the cell."

Presumably Herriott's (1951) analysis of the phenomenon of osmotic shock was the immediate stimulus to his idea. His letter went on to say that if the notion was correct, phage ghosts ought to be found in the cell debris after lysis. But by that time the prediction had already been confirmed (see below).

DNA Injection

Probably A. F. Graham and R. C. French were the first people to start thinking along the proper experimental lines. Lesley et al. (1950) had shown that the DNA of superinfecting T2 phage particles (unlike the DNA of the first particle to infect) quickly appears in the culture medium, mostly in acid-soluble form. The "superinfection breakdown" suggested to them that attachment of phage particles to bacteria is followed by disruption of the particles, which exposes the DNA to potential enzymatic cleavage. They confirmed their thought by allowing phage particles to attach to heat-killed bacteria in the presence of pancreatic deoxyribonuclease (Graham, 1953). They told me about their experiment in 1951, when Chase and I were groping toward the idea of the blender experiment. We readily confirmed their results.

The work of Hershey and Chase had taken an entirely different ideological line, starting from the demonstration by Putnam and Kozloff (1950) that DNA phosphorus is transferred from parental to offspring phage. Their discovery seemed to offer a very direct method for analyzing the material basis of heredity, a possibility that was explored by numerous investigators in the years following and is still being exploited in several ways.

The interpretation of such experiments was complicated enormously when it was found that DNA phosphorus not of parental origin (notably that of bacterial DNA present in the cells before infection) could be transferred to phage particles. In fact, Kozloff (1952a,b) and Watson and Maaløe (personal communication, summer of 1951) suspected that all transfer of both protein and nucleic acid labels proceeded through an indirect metabolic route. Hershey et al. (1951) reported equal and low (35 per cent) transfer of P^{32} and S^{35} from parents to progeny, but refused to conclude that the transfer was indirect. First, they realized that equal transfer of the two major components of the phage particle would have to reflect some sort of coincidence dependent on experimental conditions. Second, if the original protein coats should survive the infection, they might reappear as contaminants among the phage progeny and vitiate analysis. Third, Watson and Maaløe (experiments reported at the Cold Spring Harbor phage meeting, summer of 1951) had noticed that much of the labeled parental protein in the lysates remained attached to the cellular debris, and had observed a correspondingly low apparent transfer of label to progeny (cf. the experiment proposed by Herriott).

Hershey et al. (1951) had used in their experiments a culture medium poor in electrolytes, which, as it turned out, encouraged the detachment of the protein coats of the parental phage from the bacterial cell walls during lysis. By switching to a medium containing more salt, Hershey and Chase (1952) confirmed the results of Watson and Maaløe and showed that little or no protein label was in fact transferred from parents to progeny. Everyone agreed, on the other hand, that DNA phosphorus was authentically transferred, though adequate proof of transfer exclusively in the form of large polynucleotides came much later (Meselson, 1960; Kellenberger et al., 1961; Kozinski, 1961). The results of the isotope transfer experiments, together with Anderson's and Herriott's descriptions of the structure and mode of attachment of phage particles, led directly to the notion of the blender experiment.

Anderson (1949) had found that stirring the cell suspension in a Waring Blendor prevented attachment of phage particles to bacteria, and perhaps Chase and I should have thought of using that machine first. Instead, we tried various grinding arrangements, with results that weren't very encouraging. But when Margaret McDonald loaned us her blender the experiments promptly succeeded.

The blender experiment has been described incorrectly in print several times, and in the hope of preserving its essential simplicity I shall recall it here. A chilled suspension of bacterial cells recently infected with phage T2 is spun for a few minutes in a Waring Blendor and afterwards centrifuged briefly at a speed sufficient to throw the bacterial cells to the bottom of the tube. One thus obtains two fractions: a pellet containing the infected bacteria, and a supernatant fluid containing any particles smaller than bacteria. Each of these fractions is analyzed for the radiophosphorus in DNA or radiosulfur in protein with which (in different experiments) the original phage particles have been labeled. The results are:

(1) Most of the phage DNA remains with the bacterial cells.

(2) Most of the phage protein is found in the supernatant fluid.

(3) Most of the initially infected bacteria remain competent to produce phage.

(4) If the mechanical stirring is omitted, both protein and DNA sediment with the bacteria.

(5) The phage protein removed from the cells by stirring consists of more-or-less intact, empty phage coats, which may therefore be thought of as passive vehicles for the transport of DNA from cell to cell, and which, having performed that task, play no further role in phage growth.

The particular way by which phage T2 and all or most other phages infect bacterial cells probably represents a specific adaptation to the structural peculiarities of bacteria. Many other viruses apparently penetrate bodily into their host cells. The manner of penetration seen with phage nevertheless served to indicate two general principles: that the first step in viral growth is the liberation of nucleic acid from a proteinaceous coat that is functionless thereafter, and that nucleic acid alone should suffice to infect. The latter inference, also predicted in Herriott's letter of 1951, was proved in due course (Fraenkel-Conrat, 1956; Gierer and Schramm, 1956; "Viruses 1962"). These two principles may, in fact, be universal and exclusive viral attributes, unless sperm cells share them too.

Experimental infection of cells with free viral DNA may succeed or fail for unknown reasons. Infection of *Escherichia coli* with T2 DNA, for example, has not proved possible. When experimental infection succeeds it is not, apparently, by accident. Ottolenghi and Hotchkiss (1960) showed that direct transfer of DNA fragments from pneumococcal cell to cell is probably a natural phenomenon, and Tomasz and Hotchkiss (1964) have demonstrated a complex mechanism ensuring that pneumococci will be receptive to DNA only when others of their kind are in the vicinity.

Green's (1964) work shows, as might have been anticipated, that there are sometimes enzymic barriers to entry of DNA into cells. *Bacillus subtilis* can be infected with DNA extracted from phage SP82, but several molecules per cell are required and these get fragmented on the way in, afterwards recombining in a sort of multiplicity reactivation. In natural infection this does not happen, either because the organ designed for the purpose injects DNA past the enzymic barrier or, as Green suggests, because natural infection occurs quickly, permitting viral genes to start functioning in concert before conditions are set up conducive, perhaps, to something like superinfection breakdown in T2.

In natural infection with T5, injection is characteristically slow, apparently because it requires ordered, sequential processes (Lanni, 1965). Here the first segment of the DNA molecule to enter the cell sets the stage for the remainder. Evidently the DNA molecule in the T5 phage particle is in effect one-ended.

If phage growth is initiated by DNA only, and the protein coat is needed only for extracellular functions, replication of phage DNA could be in important respects autonomous. This principle also was confirmed (Burton, 1955; Hershey and Melechen, 1957; Tomizawa, 1958; Sinsheimer et al., 1962), and tended to justify the venerable thought that growth of viruses differs in important ways from that of other intracellular parasites. The difference, insofar as it has been clarified, is that phage chromosomes multiply without benefit of a proper cytoplasm and nuclear membrane. The critical factor may well be the way in which chromosomes are distributed. In bacteria, which, like phages, have a single chromosome, the cell membrane probably functions as a rudimentary mitotic apparatus (Jacob et al., 1963). In phages, the particle is apparently constructed around a single chromosome, or at any rate couldn't possibly contain two, and a distribution problem of the numerical sort never arises.

These remarks about comparative biology are tentative because we know so little about the process by which phage particles are put together.

In addition it should be recalled that we are almost equally ignorant of the manner in which DNA replicates, and precisely what degree of autonomy DNA synthesis enjoys under various circumstances. Up till now, quite properly, uninfected bacteria (Maaløe, 1961) and cell-free systems (Lehman et al., 1958) have claimed the center of attention.

Where do these historical notes belong in a larger history of molecular biology? It is clear that the discoveries of bacterial transformation, of the manner of natural infection by viruses, of infective viral nucleic acids, of DNA polymerase, and of the base-pairing rules of Watson and Crick, each contributed in a different way to focus attention on the biological role of nucleic acids. It is useless to speculate, of course, what the impact of any one would have been without the support of the others. Surely, though, the base-pairing rules plus any one of the others could have generated very much the same molecular genetics that we pursue today. But not so quickly, because the record itself since Avery, MacLeod, and McCarty (1944) shows that some redundancy of evidence was needed to be convincing and that diversity of experimental materials was often crucial to discovery.

REFERENCES

Anderson, T. F. 1949. The reactions of bacterial viruses with their host cells. Botan. Rev., *15*: 464–505.

Anderson, T. F., C. Rappaport, and N. A. Muscatine. 1953. On the structure and osmotic properties of phage particles. Ann. Inst. Pasteur, *84*: 5–14.

Avery, O. T., C. M. MacLeod, and M. McCarty. 1944. Studies on the chemical nature of the substance inducing transformation of pneumococcal types. Induction of transformation by a desoxyribonucleic acid fraction isolated from pneumococcus Type III. J. Exptl. Med., *79*: 137–158.

Benzer, S. 1952. Resistance to ultraviolet light as an index to the reproduction of bacteriophage. J. Bacteriol., *63*: 59–72.

Burnet, F. M. 1934. The bacteriophages. Biol. Rev. Cambridge Phil. Soc., *9*: 332–350.

Burton, K. 1955. The relation between the synthesis of deoxyribonucleic acid and the synthesis of protein in the multiplication of bacteriophage T2. Biochem. J., *61*: 473–483.

Cohen, S. S. 1949. Growth requirements of bacterial viruses. Bacteriol. Rev., *13*: 1–24.

Cohen, S. S., and R. Arbogast. 1950. Chemical studies in host-virus interactions. VIII. The mutual reactivation of $T2r^+$ virus inactivated by ultraviolet light and the synthesis of desoxyribose nucleic acid. J. Exptl. Med., *91*: 637–650.

Delbrück, M., and W. T. Bailey, Jr. 1946. Induced mutations in bacterial viruses. Cold Spring Harbor Symp. Quant. Biol., *11*: 33–37.

De Mars, R. I., S. E. Luria, H. Fisher, and C. Levinthal. 1953. The production of incomplete bacteriophage particles by the action of proflavine and the properties of the incomplete particles. Ann. Inst. Pasteur, *84*: 113–128.

Doermann, A. H. 1948. Intracellular growth of bacteriophage. Carnegie Inst. Wash. Yr. Bk., *47*: 176–182.

Fraenkel-Conrat, H. 1956. The role of the nucleic acid in the reconstitution of active tobacco mosaic virus. J. Am. Chem. Soc., *78*: 882–883.

Gierer, A., and G. Schramm. 1956. Infectivity of ribonucleic acid from tobacco mosaic virus. Nature, *177*: 702–703.

GRAHAM, A. F. 1953. The fate of the infecting phage particle. Ann. Inst. Pasteur, *84*: 90–98.
GREEN, D. M. 1964. Infectivity of DNA isolated from *Bacillus subtilis* bacteriophage, SP82. J. Mol. Biol., *10*: 438–451.
HERRIOTT, R. M. 1951. Nucleic-acid-free T2 virus "ghosts" with specific biologic action. J. Bacteriol., *61*: 752–754.
HERSHEY, A. D., and M. CHASE. 1952. Independent functions of viral protein and nucleic acid in growth of bacteriophage. J. Gen. Physiol., *36*: 39–56.
HERSHEY, A. D., and N. E. MELECHEN. 1957. Synthesis of phage-precursor nucleic acid in the presence of chloramphenicol. Virology, *3*: 207–236.
HERSHEY, A. D., C. ROESEL, M. CHASE, and S. FORMAN. 1951. Growth and inheritance in bacteriophage. Carnegie Inst. Wash. Yr. Bk., *50*: 195–200.
JACOB, F., S. BRENNER, and F. CUZIN. 1963. On the regulation of DNA synthesis in bacteria. Cold Spring Harbor Symp. Quant. Biol., *28*: 329–347.
KELLENBERGER, E. 1959. Growth of bacteriophage, p. 11–33. *In* A. Isaacs and B. W. Lacey [Ed.] Virus Growth and Variation, Ninth Symp. Soc. Gen. Microbiol., University Press, Cambridge.
KELLENBERGER, G., M. L. ZICHICHI, and J. J. WEIGLE. 1961. Exchange of DNA in the recombination of bacteriophage λ. Proc. Natl. Acad. Sci., *47*: 869–878.
KOZINSKI, A. 1961. Fragmentary transfer of P^{32}-labeled parental DNA to progeny phage. Virology, *13*: 124–134.
KOZLOFF, L. M. 1952a. Biochemical studies of virus reproduction. VI. The breakdown of bacteriophage $T6r^{+}$. J. Biol. Chem., *194*: 83–93.
———1952b. Biochemical studies of virus reproduction. VII. The appearance of parent nitrogen and phosphorus in the progeny. J. Biol. Chem., *194*: 95–108.
KRIEG, D. R. 1959. A study of gene action in ultraviolet-irradiated bacteriophage T4. Virology, *8*: 80–98.
LANNI, Y. T. 1965. DNA transfer from phage T5 to host cells: dependence on intercurrent protein synthesis. Proc. Natl. Acad. Sci., *53*: 969–973.
LEHMAN, I. R., S. B. ZIMMERMAN, J. ADLER, M. J. BESSMAN, E. S. SIMMS, and A. KORNBERG. 1958. Enzymatic synthesis of deoxyribonucleic acid. V. Chemical composition of enzymatically synthesized deoxyribonucleic acid. Proc. Natl. Acad. Sci., *44*: 1191–1196.
LESLEY, S. M., R. C. FRENCH, and A. F. GRAHAM. 1950. Breakdown of infecting coliphage by the host cell. Arch. Biochem., *28*: 149–150.
LEVINTHAL, C., and H. FISHER. 1952. The structural development of a bacterial virus. Biochem. Biophys. Acta, *9*: 419–429.
LURIA, S. E. 1947. Reactivation of irradiated bacteriophage by transfer of self-reproducing units. Proc. Natl. Acad. Sci., *33*: 253–264.
———1950. Bacteriophage: an essay on virus reproduction. Science, *111*: 507–511. (Also in Viruses 1950.)
———1953. An analysis of bacteriophage multiplication, p. 99–113. *In* Paul Fildes and W. E. van Heyningen [Ed.] The Nature of Virus Multiplication, Second Symp. Soc. Gen. Microbiol., University Press, Cambridge.
LURIA, S. E., and T. F. ANDERSON. 1942. The identification and characterization of bacteriophages with the electron microscope. Proc. Natl. Acad. Sci., *28*: 127–130.
LURIA, S. E., and R. LATARJET. 1947. Ultraviolet irradiation of bacteriophage during intracellular growth. J. Bacteriol., *53*: 149–163.
MAALØE, O. 1961. The control of normal DNA replication in bacteria. Cold Spring Harbor Symp. Quant. Biol., *26*: 45–52.
MESELSON, M. 1960. The deoxyribonucleic acid of coliphage T7 and its transfer from parental to progeny phages, p. 240–245. *In* J. S. Mitchell [Ed.] The Cell Nucleus. Academic Press, New York.
MULLER, H. J. 1922. Variation due to change in the individual gene. Am. Naturalist, *56*: 32–50.

NORTHROP, J. H. 1951. Growth and phage production of lysogenic *B. megatherium*. J. Gen. Physiol., *34*: 715–735.

NOVICK, A., and L. SZILARD. 1951. Virus strains of identical phenotype but different genotype. Science, *113*: 34–35.

OTTOLENGHI, E., and R. D. HOTCHKISS. 1960. Appearance of genetic transforming activity in pneumococcal cultures. Science, *132*: 1257–1258.

PUTNAM, F. W., and L. M. KOZLOFF. 1950. Biochemical studies of virus reproduction. IV. The fate of the infecting virus particle. J. Biol. Chem., *182*: 243–250.

SINSHEIMER, R. L., B. STARMAN, C. NAGLER, and S. GUTHRIE. 1962. The process of infection with bacteriophage ΦX174. I. Evidence for a "replicative form." J. Mol. Biol., *4*: 142–160.

SYMONDS, N., and E. W. McCLOY. 1958. The irradiation of phage-infected bacteria: its bearing on the relationship between functional and genetic radiation damage. Virology, *6*: 649–668.

TOMASZ, A., and R. D. HOTCHKISS. 1964. Regulation of the transformability of pneumococcal cultures by macromolecular cell products. Proc. Natl. Acad. Sci., *51*: 480–487.

TOMIZAWA, J. 1958. Sensitivity of phage-precursor nucleic acid, synthesized in the presence of chloramphenicol, to ultraviolet irradiation. Virology, *6*: 55–80.

Viruses 1950. Ed., M. Delbrück. California Institute of Technology, Pasadena, 1950.

Viruses 1953. Cold Spring Harbor Symp. Quant. Biol., Vol. 18.

Viruses 1962. Basic mechanisms in animal virus biology. Cold Spring Harbor Symp. Quant. Biol., Vol. 27.

LLOYD M. KOZLOFF
Department of Microbiology
University of Colorado Medical Center, Denver, Colorado

Transfer of Parental Material to Progeny

INTRODUCTION

Two biochemists from the University of Chicago, Earl Evans and Birgit Vennesland, took the second phage course at Cold Spring Harbor under Max Delbrück in the summer of 1946. I had elected to work on a doctoral thesis with Earl Evans and, in September of 1946, Evans suggested to me that *E. coli* infected with a virulent phage might offer considerable promise for the study of the mechanism of protein synthesis.

Some of the chemical and physical properties of purified T2 phage were reported that fall by the group working with Beard (Hook *et al.*, 1946) and we learned that Seymour Cohen (1947) had already begun a study of the biochemistry of phage-infected bacteria. Our initial efforts consisted mainly of straightforward chemical analysis of the virus, and shortly after we had started this work, Frank Putnam, an experienced protein chemist, joined us.

By 1947 we were able to prepare fairly large quantities of purified T2 phage and found that the virus particle consisted of roughly 50% DNA and 50% protein by weight. But we had no clear concepts of the organization of these viral components, and did not suspect that they played separate and distinct roles in viral multiplication. Although protein synthesis was our main interest, we did not attempt an amino acid analysis of what was obviously a multicomponent system, since that seemed unrewarding. For less compelling reasons, we also did not attempt to analyze for the nature of the purine and pyrimidine bases in the phage DNA, and thus missed out on the important discovery that these viruses contain, instead of cytosine, a unique phage DNA pyrimidine that was identified, in 1953, by Wyatt and Cohen as hydroxymethyl cytosine (Wyatt and Cohen, 1953).

Because of the earlier training they had received at Columbia University in the use of isotopic tracer techniques in biological systems, many of the staff of the Biochemistry Department at Chicago were investigating biosynthetic pathways in intermediate metabolism, using both radioactive and stable isotopes. In this environment it was natural for a biochemist interested in phage growth to ask where the material used for viral synthesis comes from. Three sources of this material were obvious possibilities: (a) the host cell, (b) the growth medium, and (c) the parent phage particle. Since in one phage growth cycle there was an about 100-fold increase in viral material, the fraction contributed to the phage progeny by the parent had to be small.

In 1948, Seymour Cohen (Cohen, 1948) reported that most of the phosphorus used for viral DNA synthesis in T2 and T4 infected bacteria was assimilated from the medium only after infection. We repeated these experiments using T6 phage, and were impressed by the observation that a significant fraction, some 20–30% of the viral phosphorus, was contributed by the host cell (Kozloff and Putnam, 1950), that is, had been assimilated before infection. Further experiments with P^{32} and N^{15} tracers (Kozloff et al., 1951) and with labeled purine bases (Koch et al., 1952) then showed that the host cell DNA was being used as a material source for the synthesis of viral nucleic acid. In the case of T7, a phage smaller than T2, T4, and T6, all the host cell DNA was converted into viral DNA (Putnam et al., 1952). Although later experiments showed that, contrary to an earlier notion, the host cell DNA contributes no genetic information to the viral progeny, two facts were established by this work which were pertinent to later studies: (1) Intracellular growth of the T-phages destroys biosynthetic control by the host cell nucleus, and (2) there are vigorous nucleases present in the infected bacteria, whose presence presaged complexities for the interpretation of control mechanisms and the fate of the infecting parent virus nucleic acid (Kozloff, 1953). These experiments also ruled out the remote though sometimes promulgated possibility that virus reproduction consists of the mere conversion of rather specific host precursors into new virus particles.

THE INITIAL TRANSFER EXPERIMENT

In 1948, S. E. Luria and some other phage workers urged us to measure the extent of the parent contribution to the material of the progeny viruses. Only a few biological laboratories were then using isotopes, and there was hope that these experiments would help us understand viral reproduction, and in particular the replication of the viral nucleic acid. We spent several months assuring ourselves that phage did not have the capacity to exchange its phosphorus with that of the medium, and determining that the levels of radioactivity used were not grossly injurious to the phage.

The first transfer experiments were performed in 1948, using P^{32}-labeled T6 phage (Putnam and Kozloff, 1950). The experiments were simple in design, being largely based on the one-step growth experiment of Ellis and Delbrück (1939). Purified, radioactive parent T6 virus, grown on a medium containing $HP^{32}O_4^{=}$, was allowed to infect host bacteria grown and maintained in the absence of the isotope. Generally, enough parent virus was added to the bacterial culture so that all host cells were infected at the outset, and only a single cycle of phage growth could occur. The culture was incubated until lysis had occurred, after which the distribution of the P^{32} in the various fractions of the lysate was measured. In this way, it was found that about 40% of the parental phosphorus was transferred to the phage progeny; however, significant amounts were also found attached to bacterial cell debris and in the "soluble fraction," both as relatively large pieces of DNA and as lower molecular weight compounds.

One difficulty in determining the amount of material actually transferred to the progeny was, as Delbrück had pointed out to us, that in our cultures considerable adsorption of progeny phage was probably occurring to infected but unlysed bacteria and to cell wall debris of lysed bacteria. Thus it was possible that the amount of parental phosphorus transferred was really greater than that actually measured. In later transfer experiments, Watson and Maaløe (1953) allowed lysis to occur in salt-free broth, thus preventing readsorption, and found that a maximum of 50% of parental phosphorus was transferred to the progeny.

FURTHER TRANSFER EXPERIMENTS

These initial observations made in 1950 of transfer of parental phosphorus to phage progeny were immediately confirmed and extended in several laboratories. They promised to offer a new method of studying viral replication (French et al., 1952; Watson and Maaløe, 1953; Mackal and Kozloff, 1954). All the T-phages were shown to transfer their nucleic acid phosphorus from parent to progeny with similar efficiencies, and the transfer of nucleic acid components was confirmed in studies on purine (Watson and Maaløe, 1953) and thymine transfer (Kozloff, 1953). In contrast to the extensive transfer of parental nucleic acid, little parental protein was found to be transferred to the progeny (Hershey and Chase, 1952; Mackal and Kozloff, 1954). These results were in agreement with the conclusions of the Hershey and Chase experiment that the DNA is the viral germinal substance.

Most of the subsequent transfer experiments were devoted to determining why only a fraction of the DNA was transferred, and how it was transferred. It was realized that there might be a direct connection between the transmission of genetic information and the transmission of parental material. But all these early experiments failed to show any such connection. Thus, in the presence of undamaged helper phage, labeled parental phage genetically damaged by UV or X-rays appeared to transfer its DNA material almost as efficiently as undamaged phage. Furthermore, small amounts of parental atoms from the excluded phage T6 appeared in the progeny of the excluding phage T7 when bacteria were simultaneously infected with both phage types (Kozloff, 1952). It was even possible to detect a small randomization of the nitrogen of the DNA bases and the phosphorus atoms during parental transfer. It was concluded (Kozloff, 1953) that, if there is an essential connection between transfer of parental DNA and its replication, this connection is obscured by unspecific reactions proceeding in the infected cell.

Within a few years, however, it was shown that the transfer of atoms from parent to progeny phage did possess some genetic specificity. Experiments using a combination of genetic markers with the isotopic technique revealed that parental atoms did, by and large, stay with parental genetic material (Hershey and Burgi, 1956). Atoms from parent phage traced by both isotopes and radiation lesions appeared together in the viral progeny. Further evidence that complete breakdown of the parental DNA to the nucleoside level was not occurring was provided by showing that the addition of competitive nucleic acid components such as nucleosides did not influence the transfer of parental material.

The transfer of only a fraction of the parental atoms led to the hypothesis that the parental DNA was bipartite: one part being transferred and one degraded. The first test of this hypothesis was suggested by Leo Szilard and carried out by Maaløe and Watson (1951). Progeny containing 50% of the original parental P^{32} were carried through a second generation (and eventually through a third generation [Stent et al., 1956]). In each generation approximately half the parental phosphorus was transferred and half was lost. This argued against any special part of the parent DNA having special value and being conserved. At this time it seemed possible that the 50% loss was some physiological feature of the whole infection sequence, and not necessarily just of the replication process.

Although these experiments did not support any simple concept of the replication mechanism, the details of the transfer were examined further. One critical consideration was the distribution of the parental material among the progeny. None of the previous methods were sensitive enough to determine whether a fairly large piece of viral DNA might be contributed to a single progeny phage. Two ingenious methods for examining the content of individual phage particles for parental atoms were invented. Levinthal (1956) directly examined progeny phage by embedding them in a sensitive photographic emulsion and counting the tracks left by the decay of individual P^{32} atoms. Stent and Jerne (1955) determined how parental P^{32} was transferred to the individual progeny phage by examining the rate at which they died due to the lethal effects of disintegration of their P^{32} atoms. Both types of experiment showed that, whereas fairly large pieces (some 10–20%) of parental DNA were transferred intact through successive generations, there was also considerable dispersion of the bulk of parental P among many viral progeny.

TRANSFER AND REPLICATION MECHANISMS

The dispersive nature of phage DNA transfer shown by these earlier experiments did not lead to the firm conclusion that DNA replication must involve breakdown of the phage genome. Even in 1953 (Kozloff, 1953) a dispersive mechanism (although not called by that name) was rejected partly because it was unesthetic, and partly because it rested on negative results. In 1957 a new technique using density gradient centrifugation was developed by Meselson, Stahl and Vinograd (1957) which could measure the isotopic content of DNA molecules containing stable heavy isotopes such as N^{15}. Meselson and Stahl (1958) then showed that the DNA of *E. coli* was replicated semi-conservatively as predicted from the Watson-Crick structure, by following the distribution of parental N^{15} atoms over the progeny DNA molecules.

Using these newer techniques for analysis, Kozinski (1961) later confirmed the earlier results that transfer of the parental T-even phage DNA was apparently dispersive. But more importantly, he was able to fragment the progeny phage DNA into small portions and examine these fragments for their content of both parental and progeny material. This analysis showed that those fragments of progeny phage DNA that carry the transferred parental atoms are half parental material and half newly synthesized. These results clearly supported the conclusion that the replication mechanism of phage DNA, like that of its host bacterium, was also semi-conservative.

IMPLICATIONS AND RETROSPECTIONS

It should be emphasized that many of these experiments were carried out before the Watson-Crick structure for DNA was advanced. Although phage transfer experiments failed to give the first clue to DNA replication, they did lead to additional concepts and experiments which established current ideas on replication. Once a likely structure of DNA was proposed, possible mechanisms for DNA replication, including the relationship of transferred parental material to progeny, were outlined by Delbrück and Stent (1957). These theoretical considerations greatly aided the planning and analysis of the phage transfer experiments. Since then transfer experiments with other phages such as single-stranded DNA phages and the RNA phages have been carried out to examine various other replication mechanisms, and it has been established that simple semiconservative replication does not fit these two situations. These mechanisms are discussed elsewhere in this book.

In retrospect it appears that the early transfer experiments using T-even phages were the right kind of experiment in the wrong viral system. The DNA of the T-even viruses is apparently too large and too susceptible to breakage for any simple analysis; but more importantly still, the many recombinational events which take place in T-even infected cells prevent any simple analysis of the replication process. Recombination probably involves breakage and reunion, including the formation and removal of any unequal segments of the different DNA strands. Probably, there occurs also significant enzymatic degradation of portions of single DNA strands. This is undoubtedly one of the factors responsible for the 50% efficiency of transfer of parental material. In addition, the finding of Hershey that the parental phage DNA enters a pool from which molecules are later withdrawn at random to be converted into finished phage particles also would account for a transfer efficiency of less than 100%.

If some of the components of degraded nucleic acid enter the pool of potentially usable compounds for later synthesis, their reuse (in addition to recombination of strands of original parent with the daughter replicas) would account for the apparent dispersive transfer and partial randomization of the parental nucleic acid, as well as for the possibility of transfer of parental atoms without transfer of genetic information. Some of this dispersal might have been predicted from the known facts of genetic recombination, and perhaps could have been foreseen from the nuclease activities observed in the infected cell. But the mere survival of the viral genome in this disruptive milieu implied the existence of protective or repair mechanisms, whose nature was then obscure and is still largely unknown. The transfer experiments thus led indirectly to the question of the function of nucleases and their relationship to DNA polymerase and repair enzymes, problems that now, 18 years later, are of some considerable concern.

REFERENCES

COHEN, S. S. 1947. Synthesis of bacterial viruses in infected cells. Cold Spring Harbor Symp. Quant. Biol., *12*: 25.

———1948. Synthesis of bacterial viruses; synthesis of nucleic acid and protein in *Escherichia coli* infected with T2r$^+$ bacteriophage. J. Biol. Chem., *174*: 295.

DELBRÜCK, M., and G. S. STENT. 1957. On the mechanism of DNA replication, p. 699. *In* W. D. McElroy and B. Glass [Eds.], The Chemical Basis of Heredity. The Johns Hopkins Press, Baltimore.

ELLIS, E. L. and M. DELBRÜCK. 1939. The growth of bacteriophage. J. Gen. Physiol., *22*: 365.

FRENCH, R. C., A. F. GRAHAM, S. M. LESLEY, and C. D. VAN ROOYEN. 1952. The contribution of phosphorus from T2r$^+$ bacteriophage to progeny. J. Bacteriol., *64*: 597.

HERSHEY, A. D. and E. BURGI. 1956. Genetic significance of the transfer of nucleic acid from parental to offspring phage. Cold Spring Harbor Symp. Quant. Biol., *21*: 91.

HERSHEY, A. D., and M. CHASE. 1952. Independent functions of viral protein and nucleic acid in growth of bacteriophage. J. Gen. Physiol., *36*: 39.

HOOK, A. E., D. BEARD, A. R. TAYLOR, D. G. SHARP, and J. W. BEARD. 1946. Isolation and characterization of the T2 bacteriophage of *Escherichia coli*. J. Biol. Chem., *165*: 241.

KOCH, A. L., F. W. PUTNAM, and E. A. EVANS, Jr. 1952. Biochemical studies of virus production. VIII. Purine metabolism. J. Biol. Chem., *197*: 113.

KOZINSKI, A. W. 1961. Fragmentary transfer of P^{32}-labeled parental DNA to progeny phage. Virology, *13*: 124.

KOZLOFF, L. M. 1952. Biochemical studies of virus reproduction. VII. The appearance of parent nitrogen and phosphorus in the progeny. J. Biol. Chem., *194*: 95.

———1953. Origin and fate of bacteriophage material. Cold Spring Harbor Symp. Quant. Biol., *18*: 209.

KOZLOFF, L. M., K. KNOWLTON, F. W. PUTNAM, and E. A. EVANS, Jr. 1951. Biochemical studies on virus reproduction. V. The origin of bacteriophage nitrogen. J. Biol. Chem., *188*: 101.

KOZLOFF, L. M., and F. W. PUTNAM. 1950. Biochemical studies of virus reproduction. III. The origin of virus phosphorus in the *Escherichia coli* T6 bacteriophage system. J. Biol. Chem., *182*: 229.

LEVINTHAL, C. 1956. The mechanism of DNA replication and genetic recombination in phage. Proc. Natl. Acad. Sci., *42*: 394.

MAALØE, O., and J. D. WATSON. 1951. The transfer of radioactive phosphorus from parental to progeny phage. Proc. Natl. Acad. Sci., *37*: 507.

MACKAL, R. P., and L. M. KOZLOFF. 1954. Biochemical studies of virus reproduction. XII. The fate of bacteriophage T7. J. Biol. Chem., *209*: 83.

MESELSON, M., and F. W. STAHL. 1958. The replication of DNA in *Escherichia coli*. Proc. Natl. Acad. Sci., *44*: 671.

MESELSON, M., F. W. STAHL, and J. VINOGRAD. 1957. Equilibrium sedimentation of macromolecules in density gradients. Proc. Natl. Acad. Sci., *43*: 581.

PUTNAM, F. W., and L. M. KOZLOFF. 1950. Biochemical studies of virus reproduction. IV. The fate of the infecting virus particle. J. Biol. Chem., *182*: 243.

PUTNAM, F. W., D. MILLER, L. PALM, and E. A. EVANS, Jr. 1952. Biochemical studies of virus reproduction. X. Precursors of bacteriophage T7. J. Biol. Chem., *199*: 177.

STENT, G. S., and N. K. JERNE. 1955. The distribution of parental phosphorus atoms among bacteriophage progeny. Proc. Natl. Acad. Sci., *41*: 704.

STENT, G. S., G. H. SATO, and N. K. JERNE. 1956. Dispersal of the parental nucleic acid of bacteriophage T4 among its progeny. J. Mol. Biol., *1*: 134.

WATSON, J. D., and O. MAALØE. 1953. Nucleic acid transfer from parental to progeny bacteriophage. Biochim. Biophys. Acta, *10*: 432.

WYATT, G. R., and S. S. COHEN. 1953. The bases of the nucleic acids of some bacterial and animal viruses: The occurrence of 5-hydroxymethylcytosine. Biochem. J., *55*: 774.

EDWARD KELLENBERGER
Institut de Biologie Moléculaire, Université de Genève, Switzerland
and Kansas State University, Manhattan, Kansas

Electron Microscopy of Developing Bacteriophage

I. INTRODUCTION

One morning in 1949 Jean Weigle came down the narrow helical staircase in the old University of Geneva Physics Institute (which he then still headed) with a bunch of reprints and journals. Downstairs we were, a few young enthusiasts gathered around an electron microscope, always eager to hear news from Weigle. This particular morning, he talked to us about genetic recombination and multiplicity reactivation of phage. I remember, as if it were yesterday, his schematic drawing of a possible explanation of these phenomena in terms of the reassembly of chromosomal subunits. This was certainly the decisive day for me, for I then decided that electron microscopy ought to become useful in the understanding of the physiological genetics of bacteriophage.

I can no longer date this day, though I am sure of the names that were mentioned: Delbrück, Luria, and Hershey. But it must have been shortly after Weigle and I decided that we had now done all that a University Physics Institute should and could do in collaboration with a firm producing an electron microscope (among other things, we had designed a new type of electrostatic lens), and that we had better turn our future efforts toward application of this instrument. Our personal feelings oriented our interests toward biology, and in 1950 we embarked on the application of electron microscopy to the question of how the bacteriophage components are synthesized and assembled into complete, intact virus particles. In this work, I benefited continuously and greatly from the help of my wife, Grete, who often contributed to experiments without appearing later as an author of the relevant papers.

As soon as we tried to look at phage-infected cells, we discovered that the cytology of uninfected bacteria was barely known. It was necessary, therefore, to investigate first the normal cell before attacking the problem we had set out to solve. This turned out to be a long project in itself (E. Kellenberger and A. Ryter, 1964). It is fair to say that only during the last few years has electron microscopy become a really useful tool in molecular biology. But recent progress in instrumentation and technique has now made electron microscopy quite indispensable.

II. TECHNICAL AND PSYCHOLOGICAL ASPECTS

Technical problems are generally boring, but most fundamental advances are attributable to the application of a new method, a new gadget, or a new instrument. The student nowadays is tempted to ignore technical problems and is, in fact, encouraged to do so by the boredom that most of us show as soon as techniques are being discussed.

Electron microscopy is certainly one of the most difficult techniques presently used in molecular biology; without devoting a personal effort to its inherent problems one cannot hope for any real success. No wonder that only few young people are attracted to the electron microscope.

The electron microscope has clearly two roles: it can be the instrument for the discovery of new structures, or it can serve as an accessory in an integrated research project. In the former role it had certainly been very rewarding in the beginning, when everything about the fine structure of cells, viruses, and ribosomes had yet to be discovered. Nowadays, new structures are discovered less readily, and so the use of electron microscopy as a mere research accessory has become the more important role. But this lacks the excitement which one feels when hunting in new lands; it involves hard work, tedious repetitions, and boring controls.

The data provided by most scientific instruments can be translated into numbers, because they involve only few measurable parameters. This is not so for a micrograph: Its general and particular aspects, though visually apparent, are not commensurable. Furthermore, under the electron microscope the same structure may appear with quite different morphology, depending on the preparative methods used and the controllable and uncontrollable conditions they entail. Which of several morphologies is significant? How does one collect statistics when one has no measurable parameters? Some people have the necessary psychological equipment for this line of work: They have an acute sense of observation and a well developed (subconscious) memory for optical sensations that allows them to make correct observations *before* any wishful thinking has influenced their conscious analysis of the visual image. Such people can develop a feeling for what is seen repeatedly and can, therefore, distinguish intuitively between adventitious and inherent aspects. The other requirement is a capacity to integrate this observational sense with a critical mind and an experimental attitude. A good electron microscopist will try to get under control the conditions which influence the final aspect he sees; that is, he will study what we call "sources of artifacts." He will design methods which either avoid these sources or at least hold them constant.

When an electron microscopist reaches this point, he will become interested in quantitative methods. When studying particulate material he will realize that the field he observes is not always truly representative of the original material. Indeed, many of the most widely used methods are based on the adhesion of particles to the surface of the specimen-supporting film. He will find out that only a few of the particles put into contact with the film adsorb to it and, what is worse, that from a heterogeneous population one or another component may be selectively adsorbed. Such selective adsorption is obviously very serious in biological systems where impurities of the order of a few per cent are quite usual and may even be present in higher amounts than the particles at which we think we are looking.

Several methods have been devised in order to overcome such errors of selection. Unfortunately, they are all rather delicate and cumbersome, so that they are not in general use. As an illustration, let us consider the procedure of agar-filtration. This method is an extension of that devised by Hillier, Knaysi, and Baker (1948), in which bacteria are deposited on a collodion film spread over nutrient agar. The bacteria are then grown on the surface of this film, through which nutrients diffuse. The film carrying the cells is later placed on the specimen grids and observed. In the course of such studies my wife Grete discovered, quite accidentally, a way to make permeable collodion films whose pores are small enough to retain particles on their surface and yet allow filtration of the suspending liquid through the film into an underlying agar gel (E. Kellenberger and G. Kellenberger, 1954). This filtration allowed development of a quantitative method (E. Kellenberger and W. Arber, 1957) in which polystyrene latex particles of known concentrations are used as a reference. Such reference particles were first used by Williams and Backus (1949) for electron micrographic counting of particles in small droplets deposited from a spray.

One of the most important advances in the electron microscopic observation of particles was the introduction of negative staining (Brenner and Horne, 1959). Electron microscopy of biological objects is limited mainly by one's ability to achieve sufficient contrast, rather than by the resolving power of the microscope. What is the use of having a resolution of 4 A if organic particles of diameter less than 20–30 A do not provide enough contrast to be directly visible? The negative staining has pushed this limit down by approximately a factor of two. But negative staining has also increased the possibility of misinterpretation: Whereas in an unstained but shadowed specimen any deformation introduced by drying and other influences becomes obvious and clearly visible in relief, specimens prepared with the negative stain phosphotungstate do not reveal such deformations, which are often taken as biologically significant features (Bradley, 1962).

Cytological fixation is a mystical matter to many people, in that it represents to them a supernatural force that preserves a "natural state." The truth is, however, that by fixation one deliberately modifies the specimen so it is less sensitive to the effects of dehydration. One modifies the specimen often by causing aggregation to such a degree that the sum of the changes introduced by the dehydration and by fixation is smaller than either change alone. The resolution of the light microscope is too low to see aggregates of diameter less than 1000 A. Since aggregates are useful in retaining stains, and since light microscopists choose their fixation in order to have good staining, it is only natural that their recommendations are quite unsuitable for electron microscopy, which shows up such aggregates.

III. PHAGE DEVELOPMENT STUDIED BY THE OBSERVATIONS OF LYSATES AND THIN SECTIONS

In the late 1940's and early 1950's many papers appeared showing wonderful electron micrographs of phage infected cells. But, since in much of that work infection was made under uncontrolled conditions (the multiplicity of infection being usually much too high, for instance), it was impossible to assess the relevance of the pictures to phage growth. To render electron microscopy useful for phage research, it was necessary to resort to the use of genetic, physiological, and chemical data. Four different methods are currently used in exploring phage growth and assembly:

(1) Examination of artificial lysates of phage-infected cells at different times after infection by the agar filtration method, in order to determine the kinds and amounts of different structures produced.

(2) Study of thin sections of phage-infected cells taken from the cultures prior to lysis. This method avoids some of the artifacts that are caused by lysis, such as emptying or destruction of the very fragile phage precursor particles.

(3) Purification and enrichment, under control of electron microscopy, of the materials liberated by lysis. The fine structure of these can be studied by negative staining.

(4) Observation of the products of single cell bursts in order to gain insight into the events happening in one cell, for comparison with the average results obtained by observations of the whole population of phage-infected cells.

SOME OF THE RESULTS ACHIEVED

(1) *Observation of phage-related particles*

Much can be learned about the assembly process of phages by study of genetic variants which carry one or more hereditary defects that cause the production of only incomplete or immature particles. One such class of variants comprise the defective-lysogenic strains which produce phage-related materials (Jacob and Wollman, 1956) readily observable in the electron microscope (W. Arber and G. Kellenberger, 1958). Study of defective lysogens is, however, very time-consuming because it is impossible to assign such mutants into complementation groups without the help of other (e.g., conditional lethal) mutants.

Another useful application of electron microscopy is the diagnosis of lysogeny by means of the visual detection of phage-like particles (Fig. 1) in bacterial lysates for which no sensitive indicator strains are known (G. Kellenberger and E. Kellenberger, 1952).

Bacteriocinogenic strains had long been suspected to be defective lysogens. In agreement with this notion, morphologically defined as well as active phages were found in many of the studied strains—in particular in inducible colicinogenic strains (G. Kellenberger and E. Kellenberger, 1956). But, since lysogenic strains can be supposed to be very common, demonstrating the presence of phages in a colicinogenic strain is not *a priori* proof of the identity of the phage with the colicine. For instance, in the colicinogenic *E. coli* strain ML we detected at least two morphologically different and hitherto unknown phage types (G. Kellenberger and E. Kellenberger, 1956). But, in addition to these phages, another bacteriocidal particle could be inferred to be present which could not be identified morphologically and which proved to be quite unrelated to the two phages. The activity of this additional, or "real" colicine of strain ML showed a different degree of sensitivity to chloroform and heat than the two phages. It is quite possible that some colicines are not related to phages at all, but are (or derive from) bacterial regulatory substances; in this case, the morphological identification of such substances in a crude lysate would be a hopeless enterprise. Only quantitative electron microscopy combined with methods of purification could provide the necessary experimental approach.

(2) *Single Cell Lysis*

A special adaptation of the agar filtration method allowed observation of the products of a single lysed cell *in situ* (Fig. 3, 4) (E. Kellenberger and G. Kellenberger, 1954, 1957). A first control experiment showed that the number of morphologically countable phages liberated per lysed cell has a distribution which is very similar to the distribution of infective phage produced in single burst experiments. With this new method it could be seen that, in the lysate of a single cell, both intact and unfinished particles appear. This indicates that in a single cell phage maturation is not a synchronous process. In these single cell lysates we also noticed the small globular particles that had been seen in normal lysates and misinterpreted as phage precursors by many workers. Since we found these particles also in penicillin-induced bacterial lysates, it seemed unlikely that they were phage precursors (Fig. 2). In fact, they turned out to be ribonucleoproteins

(Schachman et al., 1952), and are now called ribosomes. Looking now at these micrographs with hindsight one can easily discern clusters of ribosomes: the polyribosomes (Fig. 4).

(3) *The Life Cycle of T-even Phage*

About 1950 we started to study the growth and development of phage T2 and λ, a project that received a big boost when Janine Séchaud and Antoinette Ryter joined us. In retrospect, the effort was very great compared to the results achieved. Some people had tried to make us understand this beforehand, but others egged us on, especially Seymour Benzer and Gunther Stent, who gave us moral support in the very early days by their positive attitude. Anyway, the ultra-thin sections of phage-infected bacteria showed—for the first time convincingly—the pool of vegetative phage. This pool was predicted by Visconti and Delbrück on genetic grounds, and now it could be seen to be similar in aspect to the nucleus of the normal bacterium, i.e., as an area of finely fibrillar DNA. After this pool is formed in the infected cell, "black bodies" appear, which are phages, or their precursors (Fig. 5). At still later stages of phage growth, the cell becomes crowded with progeny phages, which frequently form crystal-like arrangements. In order to study the process of phage particle maturation, we counted the number of "black bodies" per cell seen on thin sections and the number of protein capsids, either full or empty, seen in parallel artificial lysates. In this way we found that the number of "black bodies" is higher than the corresponding number of capsids. This indicated that some of the "black bodies" are phage precursor particles which disintegrate upon lysis without leaving an empty capsid. Since, furthermore, empty capsids were never found in thin sections of well preserved specimens, we concluded that empty capsids are not produced first and only later filled with DNA. We called these morphogenetic precursor particles, devoid of a stable capsid, *phage condensates*. We postulated that they are composed of the phage DNA in a rather compact form produced through the action of a postulated condensing principle. This condensing principle was thought to be a protein, because condensation does not occur in the presence of chloramphenicol, a condition under which phage DNA synthesis continues to proceed (Tomizawa and Sunakawa, 1956; Hershey and Melechen, 1957). But as soon as chloramphenicol is removed, after a large pool of vegetative DNA has been accumulated in its presence, maturation of phage particles starts immediately. Under these conditions, the number of condensates is more than ten times greater than the number of capsids. These experiments provided further evidence against the notion that phage capsids are made first, and only later filled with phage DNA, because they showed that the appearance of complete, stable capsids follows that of the condensates. They provided, however, no insight into the nature of the condensation process, and they did not rule out the possibility that a thin, unstable envelope is produced simultaneously with, or as part of, phage DNA condensation. Indeed, recent unpublished experiments made with M. and F. Eiserling seem to support that idea.

(4) *The Production of Empty Capsids*

T2-infected cells treated with proflavine produce a high proportion of empty capsids, observable in the lysate (De Mars, et al., 1953). Are these empty capsids produced as such, or are they the break-down products of a precursor particle? In the case of proflavine lysates, there was some hope that the study of ultra-thin sections of infected bacteria could provide the answer. When observing well-prepared thin sections of proflavine-treated T2-infected bacteria, one finds that most of the capsids are full. Their contents do not, however, give a completely normal appearance, being frequently slightly retracted from the capsid. Although this retraction is certainly an artifact, it nevertheless indicates some abnormal property of the proflavine-phages. It explains also why some apparently empty capsids can be found in sections, since the empty capsids are probably seen in sections which pass through the emptied part of the particle. These experiments indicate that, even in the presence of proflavine, empty capsids are not produced as such. In the case of lysates of phage λ, we discovered that about half of the particles are smaller than λ (Karamata et al., 1962). The capsid of these smaller variants is serologically indistinguishable from that of the normal λ particles. But the content of these small particles is neither RNA or DNA and its nature has not been determined.

IV. MORPHOPOIESIS

As the preceding sections have shown, we had not progressed very far in the solution of the problem we set out to solve: the mechanism of assembly of a virus particle. Later, a new experimental system became available in which successive steps of phage maturation can be specifically inhibited, namely the conditional lethal mutants of T4 introduced by Edgar, Steinberg, and Epstein (Epstein et al., 1963). They already had a collection of such mutants when I joined them in winter 1960/61, so that we could start together on a physiological survey of the residual functional capacities of the different mutants. For this purpose, the formation of phage tail fibers was determined by serology, and of phage DNA by chemical tests; the occurrence of nuclear break-down of the infected host cell by phase contrast microscopy, and the production of phage-related materials by electron microscopy. This preliminary survey helped to classify the main physiological mutant groups, and led to the study of the morphopoiesis of

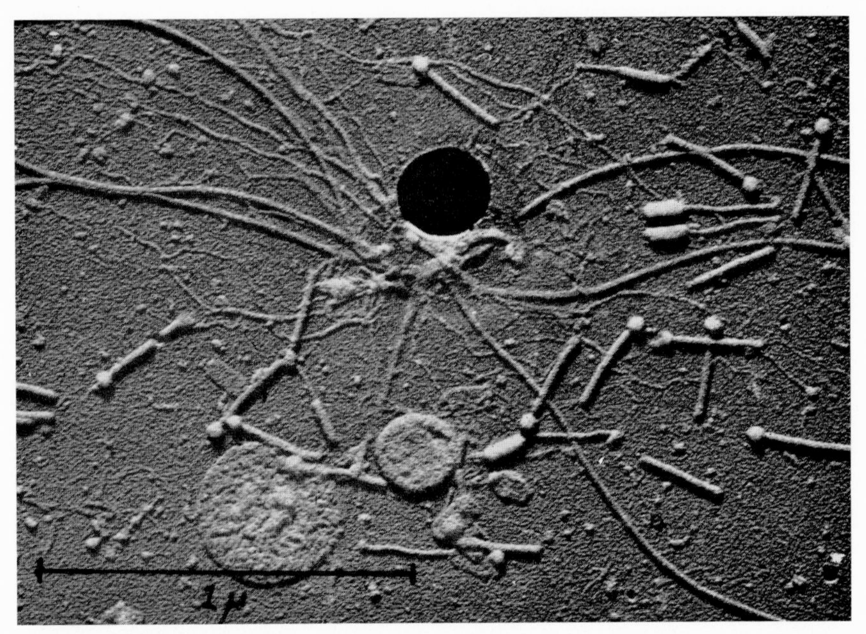

FIGURE 1. Lysate of strain Caron of *B. subtilis* (or *cereus*) after induction with UV-light. Two types of phages can be seen, free tails with extended and contracted sheaths as well as flagella. No indicator strain has been found for this phage. (Micrograph No. 13563, January 1954)

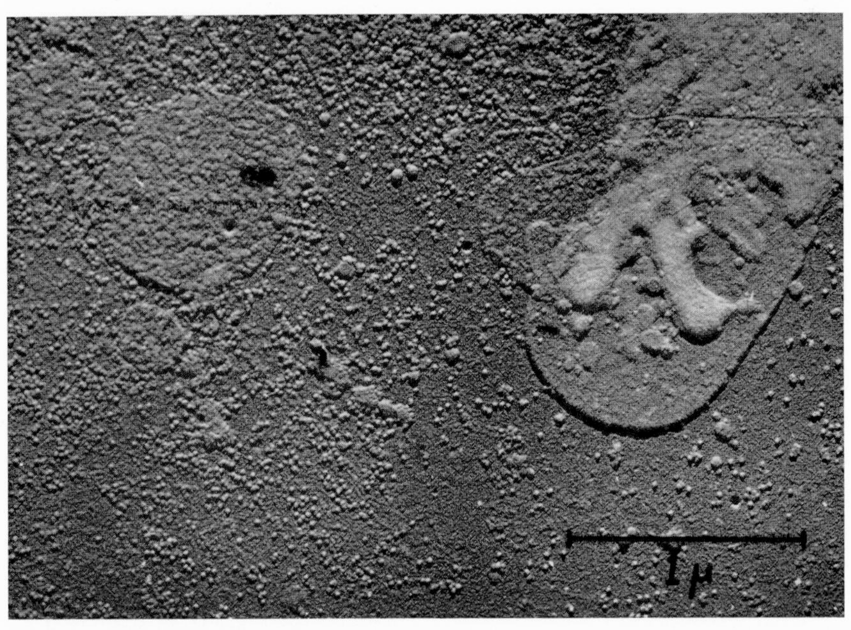

FIGURE 2. Coli cells lysed by penicillin and having released a great number of ribosomes. (Micrograph No. 16738, April 1955, in collaboration with G. Kellenberger)

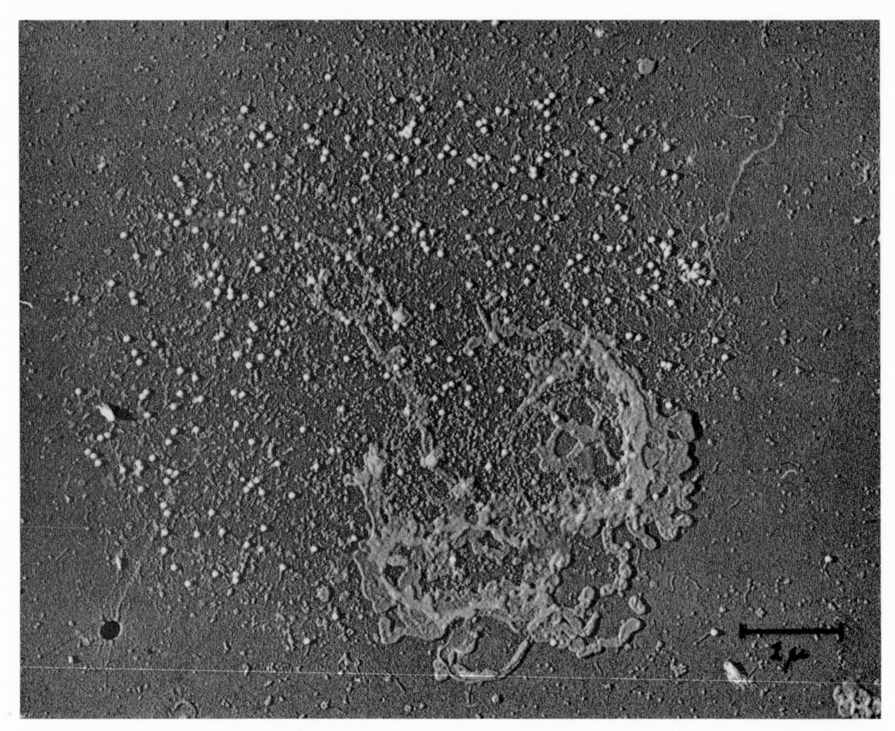

FIGURE 3. An area showing the lytic products of a single phage-λ-infected cell. Phages free tails, some empty capsids, and cellular debris can be seen. The ribosomes are clearly visible. (Micrograph No. 15764, January 1955, in collaboration with G. Kellenberger)

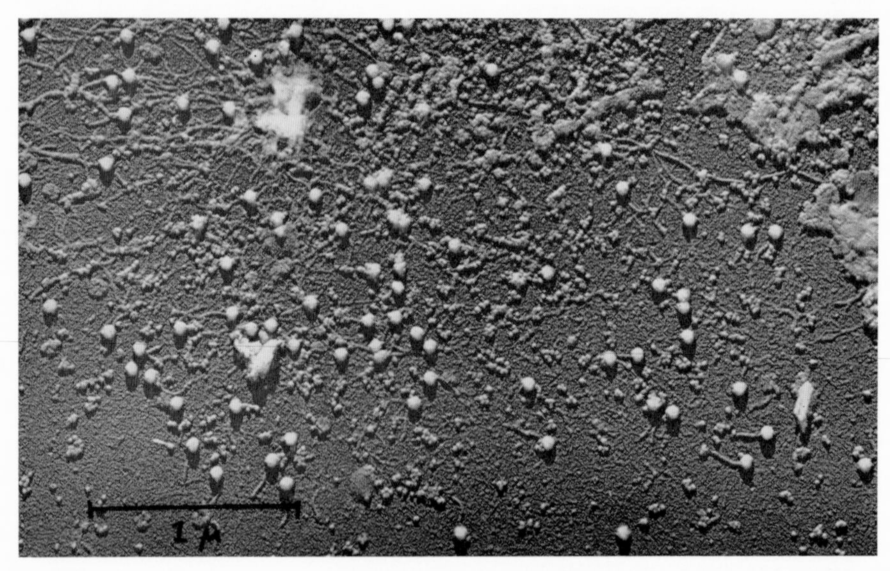

FIGURE 4. A further enlargement of part of an area of a single burst similar to that of Figure 3. (Micrograph No. 15769, January 1955, in collaboration with G. Kellenberger)

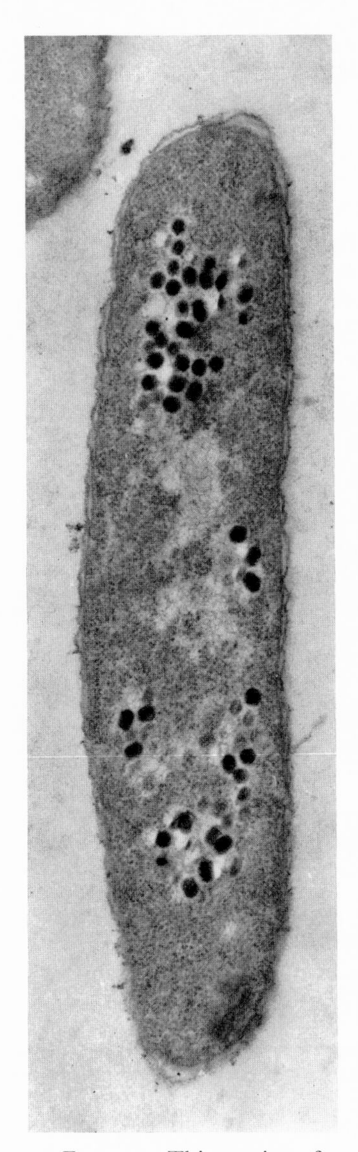

FIGURE 5. Thin section of a cell infected with normal T4. Note the pool of vegetative DNA and "black bodies," which can be either finished or unfinished phage particles. The less dark particles are not necessarily maturating phages, but are more likely partial sections of "black bodies." (Micrograph No. 26103, May 1961, in collaboration with J. Bron)

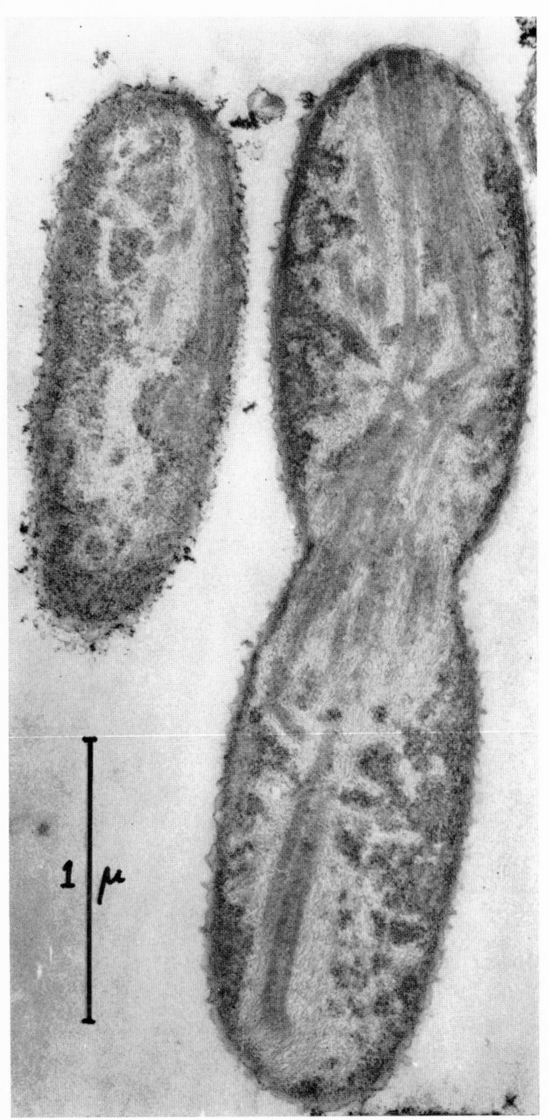

FIGURE 6. A thin section of a cell infected with a conditional lethal amber mutant in gene 20. Instead of phages, tubular structures can be seen. The DNA pool is very much enlarged, indicating that these polyheads are not filled with DNA. (Micrograph No. 31470, August 1965, in collaboration with E. Boy de la Tour and R. Favre)

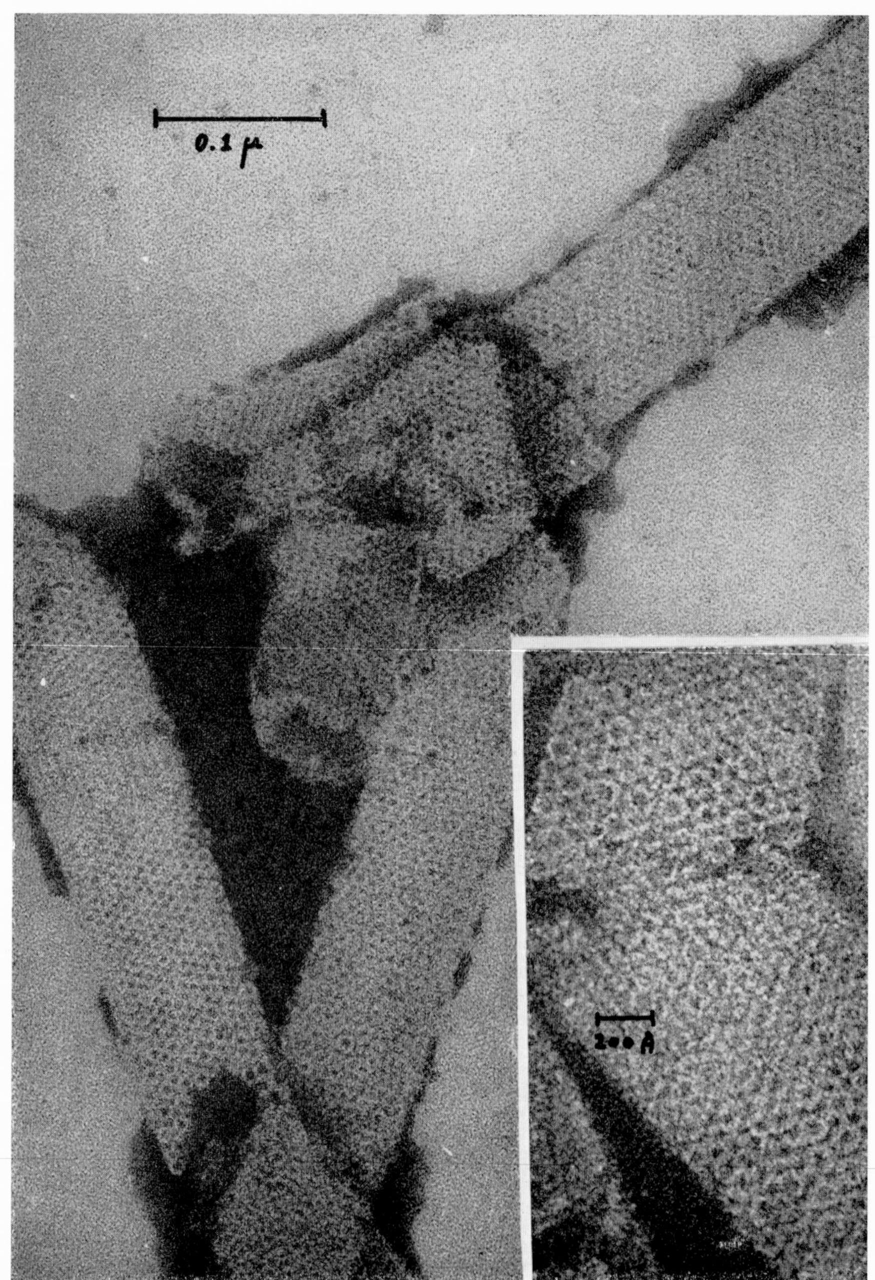

FIGURE 7. "Polyheads" produced by infection with a T4 amber mutant in gene 20. Specimen prepared in phosphotungstate. One can discern the "capsomers," which are shown also at a higher magnification in the insert. Some capsomers can be distinguished which show a substructure of 6-fold symmetry. On the polyheads a moiré pattern can be observed, which occurs by superposition of lower and upper layers of the completely flattened tube. (Micrographs 30168 and 30172, 1964, by E. Boy de la Tour)

bacteriophage (E. Kellenberger, 1965). We have defined morphopoiesis as follows:

The process leading from a pool of subunits to a morphologically characterized biological entity (like an organelle or a virus). Within one system, only a determined number of different types of subunits participate. Substances contributing specifically to the specific morphology are called 'morphopoietic factors or principles.' They can, but need not be a part of the final product.

A morphopoiesis is of the *first order* if only one type of subunits is used to build a structure by self-assembly.

In morphopoiesis of *higher orders*, supplementary bits of morphopoietic information are contributed through morphopoietic factors.

Morphopoietic regulation appears to act at the level of the gene products through phenomena of sequential triggering.

Table I

GENES IMPLICATED IN THE FORMATION OF THE PHAGE HEAD

Gene No.	*Result of mutation in this gene*	*Function of the gene*	*Reference*
20	Polyheads	Morphopoietic principle of unknown function	Epstein et al., 1963; Favre et al., 1965
21	Abnormal heads, adhering to the envelopes	?	Epstein et al., 1963
22	No heads	Minority subunit of the capsid?	Levinthal and Hosoda, personal commun.
23	No heads	Majority subunit of the capsid	Sarabhai et al., 1964
66	Increase of short-headed variants	Morphopoietic principle of unknown function	Geiduschek, Eiserling and Epstein, personal commun.
24	No heads	?	
31	Aggregation of product of gene 23 into lumps	Morphopoietic principle or solubilizer of product of gene 23?	Kellenberger et al., unpublished

Listed in Table I are the genes which are implicated in this process. It can be seen that gene 23 is responsible for what we call the "majority protein" of the capsid. As for the other genes, we only know the phenomena which occur when their products are inactive or absent. Most noteworthy are the following effects: Mutations in gene 20 result in the production of long tubular structures, which are built of the protein product of gene 23 (Figs. 6 and 7) (Favre et al., 1965). These *polyheads* provide an interesting material for the study of surface lattices (Finch et al., 1964; E. Kellenberger and Boy de la Tour, 1965). Polyheads are thought to represent morphopoiesis of low order: in addition to the information contained in the protein subunit encoded in gene 23, only a few, or no, supplementary bits of information are contributed by other morphopoietic factors. It is most intriguing, furthermore, that polyheads—when observed inside the cell—are filled. Whether their content is a specific substance, acting as a morphopoietic core, or only a selective sample of cellular sap, is not known. We only know that polyheads contain practically no nucleic acid. This situation seems comparable to that already described for the small variant of phage λ. The hypothesis of a so-called "morphopoietic core" which has more or less the shape of the final particle, and on which the subunits settle in order to form the capsid, is a tempting working hypothesis. Mutations in gene 31 result in the aggregation of the protein coded by gene 23 into lumps, as though the protein product of gene 23 is soluble only when in contact with, or in the presence of the product of gene 31; this product seems to act as if it were a solubility factor.

It seems obvious to us that this system of conditional lethal mutants of T4 will provide very good material for study of the mechanism of morphopoietic regulation. It should become possible to understand the way in which irregular polyhedral capsids are produced, and it should also become possible to understand the dynamic aspects of phage assembly. As we have seen, within any infected cell the assembly of phage particles is not synchronous; all stages exist simultaneously. Nevertheless, it seems likely that the initiation of the assembly process is not merely the consequence of an accidental encounter of several different types of subunits. Assembly must be sequential. Hence the regulation of events must be based on a sequential triggering of reactions induced by the different morphopoietic factors, acting successively on the subunits that are to form the intact phage progeny particle.

REFERENCES

Arber, W., and G. Kellenberger. 1958. Study of the properties of seven defective-lysogenic strains derived from *E. coli* K12. Virology *5*: 458–475.

Bradley, D. E. 1962. A study of the negative staining process. J. Gen. Microbiol., *29*: 503–516.

Brenner, S., and R. W. Horne. 1959. A negative staining method for high resolution electron microscopy of viruses. Biophys. Biochim. Acta, *34*: 103–110.

DeMars, R. I., S. E. Luria, H. Fisher, and C. Levinthal. 1953. The production of incomplete bacteriophage particles by the action of proflavine and the properties of incomplete particles. Ann. Inst. Pasteur, *84*: 113–128.

Epstein, R. H., A. Bolle, Ch. Steinberg, E. Kellenberger, E. Boy de la Tour, R. Chevalley, R. S. Edgar, M. Susman, G. H. Denhardt, and A. Lielausis. 1963. Physiological studies of conditional lethal mutants of bacteriophage T4D. Cold Spring Harbor Symp. Quant. Biol., *28*: 375–394.

FAVRE, R., E. BOY DE LA TOUR, N. SEGRÉ, and E. KELLENBERGER. 1965. Studies on the morphopoiesis of the head of phage T-even. I. Morphological, immunological and genetic characterization of polyheads. J. Ultrastruct. Res., *13*: 318–342.

FINCH, J. T., A. KLUG, and A. O. W. STRETTON. 1964. The structure of the "polyheads" of T4 bacteriophage. J. Mol. Biol., *10*: 570–575.

HERSHEY, A. D., and N. E. MELECHEN. 1957. Synthesis of phage precursor nucleic acid in the presence of chloramphenicol. Virology, *3*: 207–236.

HILLIER, J., G. KNAYSI, and R. F. BAKER. 1948. New preparation techniques for the electron microscopy of bacteria. J. Bacteriol., *56*: 569–576.

JACOB, F., and E. L. WOLLMAN. 1956. Recherches sur les bactéries lysogènes défectives. I. Déterminisme génétique de la morphogenèse chez un bactériophage tempéré. Ann. Inst. Pasteur, *90*: 282–302.

KARAMATA, D., E. KELLENBERGER, G. KELLENBERGER, and M. TERZI. 1962. Sur une particule accompagnant le développement du coliphage λ. Pathol. Microbiol., *25*: 575–585.

KELLENBERGER, E. 1965. Control mechanisms of bacteriophage morphopoiesis. Ciba Found. Symp., Principles of Biomolecular Organization, Wolstenholme [Ed.], Churchill, London, in press.

KELLENBERGER, E., and W. ARBER. 1957. Electron microscopical studies of phage multiplication. I. A method for quantitative anlysis of particle suspensions. Virology *3*: 245–255.

KELLENBERGER, E., and E. BOY DE LA TOUR. 1965. Studies on the morphopoiesis of the head of phage T-even. II. Observations on the fine structure of polyheads. J. Ultrastruct. Res., *13*: 343–358.

KELLENBERGER, E., and G. KELLENBERGER. 1954. The process of filtration on agar used for electron microscopic studies of bacteriophages in lysates and in single bacterial bursts. Proc. Intern. Conf. Electron Microscopy, London, 1954, p. 268–270.

KELLENBERGER, E., and A. RYTER. 1964. Bacteriology, p. 335–393. *In* B. M. Siegel [Ed.], Modern Developments in Electron Microscopy. Academic Press, New York.

KELLENBERGER, G., and E. KELLENBERGER. 1952. La lysogénie d'une souche Bacillus Cereus, mise en évidence par microscopie électronique. Schweiz. Z. Allgem. Pathol. Bakteriol., *15*: 225–233.

———, ——— 1956. Etude de souches colicinogènes au microscope électronique. Schweiz. Z. Allgem. Pathol. Bakteriol., *19*: 582–597.

———, ——— 1957. Electron microscopical studies of phage multiplication. III. Observation of single cell lysis. Virology, *3*: 275–285.

SARABHAI, A. S., A. O. STRETTON, S. BRENNER, and A. BOLLE. 1964. Co-linearity of the gene with the poly peptide chain [Bacteriophage T4D]. Nature, *201* (4914): 13–17.

SCHACHMAN, H. K., A. B. PARDEE, and R. Y. STANIER. 1952. Studies on the macromolecular organization of microbial cells. Arch. Biochem. Biophys., *38*: 245–260.

TOMIZAWA, I., and S. SUNAKAWA. 1956. The effect of chloramphenicol on deoxyribonucleic acid synthesis and the development of resistance to ultraviolet irradiation in *E. coli* infected with bacteriophage T2. J. Gen. Physiol., *39*: 553–565.

WILLIAMS, R. C., and R. C. BACKUS. 1949. Macromolecular weights determined by direct particle counting. I. The weight of bushy stunt virus particle. J. Amer. Chem. Soc., *71*: 4052–4057.

III. *Phage Genetics*

L. Szilard and A. D. Hershey, Cold Spring Harbor, 1951

AARON NOVICK
Institute of Molecular Biology, University of Oregon, Eugene, Oregon

Phenotypic Mixing

The discovery of phenotypic mixing between related bacterial viruses was a wonderful introduction to biology for Leo Szilard and me. What could be more appropriate for a physicist and a chemist who wanted to become biologists than to find that a virus can change its spots?

I had been trained as a physical organic chemist, having given up childhood dreams of becoming an astronomer. I turned to chemistry largely in response to the stimulation provided by Frank Westheimer, with whom I studied, rather than because of an intense interest in chemistry. In Westheimer's laboratory at the University of Chicago, I had my first glimmering of the possible excitement of studying biology at the molecular level, when he proposed an experiment in 1941 that might have provided information about the sequence of amino acids in the protein, silk fibroin. The results of that experiment were not promising, and because of the pressures of the war I turned to other things to complete my training. In March of 1943 I joined the Atomic Energy Project at the University of Chicago and was abruptly introduced to the developments which even then held awesome portents for mankind.

Leo Szilard was a legendary figure even among the many legendary figures then associated with the "Project." He had provided not only the principal initiative to get it started, but also contributed to its momentum by appearing wherever there was a block to its progress and intervening with sound good sense and ingenious suggestions.

Szilard preferred the company of the younger men to that of his own age group and never succumbed to the stuffiness that often comes with middle age. No doubt he seemed eccentric to many, certainly so to the generals, because he declined to confine his work to his office, or his thinking to prescribed subjects. Conversations with him could occur anywhere, on any subject. One might run into him on a street corner, in a restaurant, or wherever he could sit in a sunny and quiet spot. He was often to be found near the tennis courts at the Quadrangle Club, sitting with his face toward the sun, asking questions and listening with half-closed eyes in a conversation that might be on child-rearing or aging, physics or sociology, medicine or law. He was fascinated by the laws of man as well as by the laws of nature. Thinking was his greatest pleasure, and he enjoyed thinking about an incredibly wide range of subjects. It was in such contexts that I got to know him. He enjoyed advising the young and was excellent at it. Whether the problem was personal or scientific, one found oneself editing the nonsense out of one's thoughts when talking to him.

At the end of the war Szilard turned his energy and ingenuity toward helping mankind in its moral and political adjustments to the technology of atomic energy. During this period I became active politically, along with many other of the younger scientists. One spring evening in 1947, as we were leaving a meeting of the Atomic Scientists of Chicago, Szilard approached me and asked whether I would care to join him in an adventure into biology. Despite his caution to think his proposition over carefully, I accepted immediately.

Apparently he had been considering a move to biology for some time, in part because he saw that the era of excitement in nuclear physics, where one man could contribute significantly, had ended and in part because he sensed that biology was on the threshold of an era much like that of physics prior to World War II.

But at that time it was unheard of for biology departments to appoint physicists, or even chemists, to their staffs. Fortunately, Robert Hutchins, Chancellor of the University of Chicago, appreciated Szilard's greatness and had the courage to appoint him as Professor of Biology and of Sociology, with specific departmental affiliations to be worked out later. Szilard was placed administratively in the new Institute of Radiobiology and Biophysics, one of three institutes set up at Chicago to adapt to post-war scientific developments. Szilard tried without success to get an appointment for me in that Institute. But I was so excited about the prospects of working with him in biology that I was happy to join him even in a position where I would be dependent upon his research grants for my support. He was never pleased with this arrangement and, because he felt I was being underpaid, insisted on supplementing my salary from his own pocket.

Szilard proposed that we get started in biology by taking the Cold Spring Harbor phage course that had been recently started by Max Delbrück. So we enrolled for the summer session of 1947, which was to have been taught by Mark Adams. But because Adams became ill, Delbrück delayed his first post-war return to Europe and stayed on at Cold Spring Harbor to give the course. We were as delighted as the many others introduced by Delbrück to a biology that had been made comfortable for people with backgrounds in the physical sciences. In that three-week course we were given a set of clear definitions, a set of experimental techniques and the spirit of trying to clarify and understand. It seemed to us that Delbrück had created, almost single-handedly, an area in which we could work, and after the three-weeks course we felt ready to embark on our own without further preparation.

It was evident to me that Szilard regarded Delbrück highly. Usually Szilard listened to people only as long as they had something to say that interested him and made sense. This meant that he sometimes turned away in the middle of a conversation. But whenever Delbrück was talking, he stayed to listen.

We had hardly returned to Chicago that summer, where I was trying to complete my work in nuclear physics, when Szilard persuaded me to return with him to Cold Spring Harbor to try there some genetic crosses between differently marked phages. These primitive experiments had to be abandoned for lack of laboratory facilities when we returned to Chicago that fall. We arranged to start work in our own laboratory in January of 1948. The Institute of Radiobiology and Biophysics was located in a former synagogue of a Jewish orphanage whose buildings had been taken over by the University, and our laboratory was a room in the basement with a very low ceiling laced with pipes.

We decided to study mutation in bacteria and tried a number of schemes to find a simple and accurate way of measuring mutation rates. We also looked for cases where we could study forward and reverse mutation at the same genetic locus.

We were both impressed by Joshua Lederberg and the implications of his by then two-year-old discovery of genetic recombination in bacteria. Delbrück, however, had reservations about Lederberg's interpretations. Because of our very high regard for Delbrück's opinions we tried to design an experiment that, we hoped, would test Lederberg's conclusions and convince Delbrück. For this purpose we compared the results of a bacterial cross with those of the reciprocal cross and found clear evidence of genetic recombination, independent of the particular markers used. Szilard wrote to Delbrück and to Luria reporting the results and offered to eat his hat (which he never wore anyway) if this was not genetic recombination. Delbrück replied that recombination or not, one still did not understand the basic nature of the phenomenon and he urged us to continue our studies. But we were by then being attracted by other problems and had little incentive to continue the work when we learned that in essence our experiment had been done already and reported as part of a table in a paper by Lederberg (Lederberg, 1947).

As part of our schooling the preceding summer at Cold Spring Harbor we had studied attentively the 1946 Cold Spring Harbor Symposium volume which contained the papers that all the early phage workers considered basic to the new field, later to be called molecular biology. We were particularly intrigued by the Delbrück and Bailey paper (Delbrück and Bailey, 1946), which reported the discovery that a single bacterium can be infected simultaneously with two genetically distinct phages and that new phage genotypes issue among the progeny of this mixed infection, a discovery that made phage genetics possible. A paradox presented in this paper quickly caught our attention. Like many theoretical physicists, Szilard loved paradoxes—they are certain indications of error in understanding and they provide solid food for thought.

The paradox was this: When bacteria, mixedly infected with phages T2 and T4, are plated before lysis, those yielding T2 are scored as plaque yielders on plates seeded with B/4, those yielding T4 as plaques on B/2, and those yielding both as clear plaque formers on B/2+B/4; those yielding T2 or T4 alone give turbid plaques on B/2+B/4. One would expect that the number of plaques on B/4 should be *greater* than the number of clear plaques on B/2+B/4, since the latter plate indicates only those T2 yielders which also yield T4. But, as Delbrück and Bailey pointed out, the number of plaques on B/4 is generally much *less* than the number of clear plaques on B/2+B/4.

Our interest in this paradox was revived by a talk with Luria who had come to Chicago to give a memorial lecture for Louis Slotin, a physicist who had been killed in a laboratory accident at Los Alamos. He told us that the discrepancy was much greater for UV-irradiated T2.

It is difficult to recall the line of our reasoning, which must have followed from the inference drawn by Delbrück and Bailey that the presence of B/2 on the mixed indicator plate raises the plating efficiency of the liberated T2 particles. Apparently, we must have hit upon the idea that the effect was on a sub-class of the liberated T2 particles and that the critical factor was B/2 itself rather than some chemical it made. By studying the process in liquid suspension rather than on a plate, we found that the proportion of T2 type phages capable of plating on B/4 was strikingly increased by one step of growth of the mixed phage lysate on strain B/2, and we were able to show that strain B worked as well for increasing the proportion of such phages.

Reproduced on pages 138 and 139 are some of the first notebook entries on our work of that May in 1948. Szilard, who made these records, always used the right-hand page first. This allowed him to examine simultaneously a page, and either the page which preceded it or the page which followed it, without having to turn a page. He sometimes reminded himself that he was following this convention by putting a capital H (for Hebrew) in the upper right-hand corner.

As can be seen from these notebook entries, we soon realized that we were dealing with phenotypic mixing. That is, the resolution of the paradox was that, among the phage progeny of a bacterium simultaneously infected with T2 and T4, there appear phage particles having the T2 genotype but carrying the T4 phenotype. In further experiments, which were never published, we were able to demonstrate that these T2 particles with T4 host range are far more sensitive to neutralization by anti-T4 serum than normal T2, showing that the T4 phenotype was probably the consequence of a T4 protein in these particles. We seem to have had the correct insight into the nature of these particles. We explained to our friends that they arose in the following way: At some point in the development of the phage in the host bacterium, the phage particles put on coats, much like customers leaving a restaurant. For some unknown reason there seemed to be an excess of T4 coats.

In retrospect, it is clear that we did not appreciate sufficiently the importance of the clue provided by this discovery. We were even uncertain at first whether to publish it and reported it only in the "Phage Information Service," which circulated privately among the members of the Phage Group. Apparently we were not alone in our lack of appreciation, because Delbrück subsequently advised us to forget about publishing it. Finally, we did publish our results (Novick and Szilard, 1951), though whether through appreciation of their importance or through some kind of vested interest I no longer recall.

Fortunately, A. D. Hershey did take the result seriously. He wrote to us describing his finding of phenotypic mixing between T2 wild type and T2h, its host range mutant. I like to believe that this effect led him to search for the physical basis of phenotypic mixing which culminated in his imaginative and courageous use of the Waring Blendor to separate genotype from phenotype, an experiment which provided the ultimate proof that the genotype of a phage is established by its DNA (Hershey and Chase, 1952).

Of course, another reason why we did not pursue the implications of our findings on phenotypic mixing is that we became interested in other things. Szilard had been intrigued by Monod's discovery that bacteria fed a mixture of glucose and lactose at first consume only the glucose and metabolize the lactose only after they have depleted the glucose (Monod, 1942). Szilard argued that if the glucose concentration of the medium were lowered sufficiently, a point should be reached below which the bacteria would use both sugars at the same time. This point could be found, he reasoned, if the bacteria were grown in a continuous-culture apparatus in which both sugars were fed simultaneously at less than saturating concentrations.

This reasoning led him to think about how a continuous-culture apparatus might work, and it was not long before he discovered the principle

Filtrate	B/2	B/4	Mix
Assay (Titer 1.5 × 10^8 T_2 / Titer 1.4×10^9 T_4)	137	15	161 no clear
A — ads on B/2, centr. wash (50 to 100 T_4 per B/2)			66 clear 444 total
B — ads on B/2 centr. wash, high ratio bacteria to phage			2 clear 213 total
B — ads on B/2 no centrif	391	151	8 clear 480 total
after lysis	confl.	1540	
B — on Broth	—	154.5	
ads on B/2	442	163	
B* — 100 times larger ratio B/2 to phage, adsorp. B/2	458	182	clear pl < 4
after lysis	56 ×1000	2365	
after second lysis	48 ×1000 ×10	from 264×10 to 287×10	
B — on broth	393	144	
ads on B/2	366	172	
after lysis	27000	1540	
after second lysis	31000 ×10	157 ×10	
A — ads on B/2 centr. wash as above			77 clear 628 tot

Controls T_2, $T_2 + T_4$	B/2	B/4	Mix
control $T_2 + T_4$	396	34	253
control $T_2 + T_4$	298	85	
after lysis	confl.	46	
T_2 as control → broth	215		
→ ads on B/2	189		

FIGURE 1

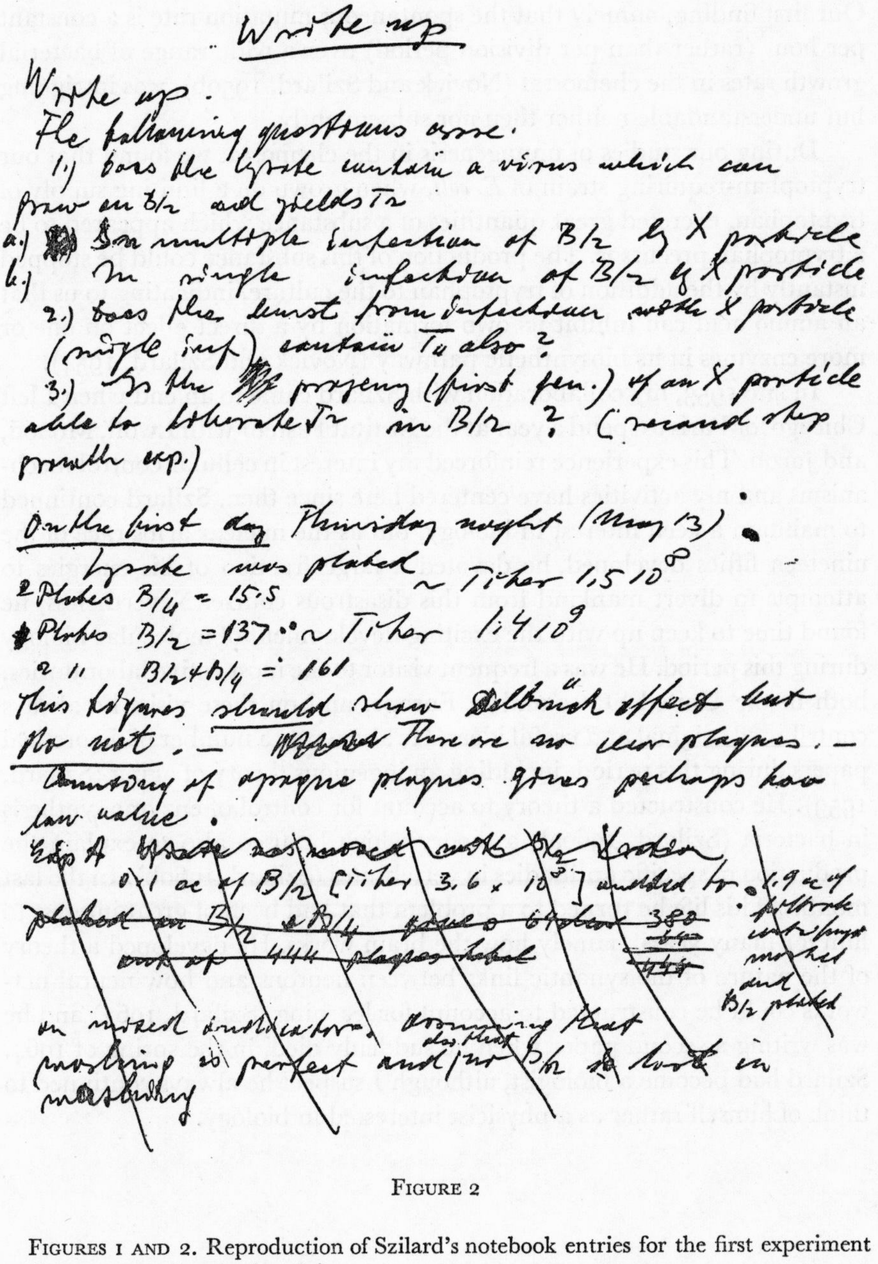

Write up

Write up.

The following questions arose:

1.) Does the lysate contain a virus which can grow on B/2 and yields T2

a.) In multiple infection of B/2 by X particle

b.) In single infection of B/2 by X particle

2.) Does the burst from infection with X particle (single inf.) contain T4 also?

3.) Is the progeny (first gen.) of an X particle able to tolerate T2 in B/2? (second step growth exp.)

On the first day Thursday night (May 13)

Lysate was plated ~ Titer $1,5\ 10^8$ —

2 Plates B/4 = 15.5 ~ Titer $1,5\ 10^8$

2 Plates B/2 = 137 ~ Titer $1.4\ 10^9$

2 " B/2 + B/4 = 161

this lysate should show Delbrück effect but

Do not. — There are no clear plaques. —

Counting of opaque plaques gives perhaps too low value.

Exp A Lysate was mixed with B/2 [illegible]

0.1 cc of B/2 [Titer 3.6×10^8 added to 1.9 cc

plated on B/2 + B/4 [illegible] 380 / 444

out of 444 plaques total

in mixed indicator. Assuming that mixing is perfect and for B/2 distinct [illegible] in marking

[illegible] B/2 plated

FIGURE 2

FIGURES 1 AND 2. Reproduction of Szilard's notebook entries for the first experiment to demonstrate phenotypic mixing.

of the chemostat. With an appropriate apparatus we were able to demonstrate that the principle worked (Novick and Szilard, 1950a). We realized that the chemostat could be used not only for the physiological studies for which it was invented but also for an accurate measure of mutation rate. Our first finding, namely that the spontaneous mutation rate is a constant per hour (rather than per division period) over a wide range of bacterial growth rates in the chemostat (Novick and Szilard, 1950b), was intriguing but understandable neither then nor subsequently.

During our studies of mutagenesis in the chemostat we found that our tryptophan-requiring strain of *E. coli*, when grown on a limiting supply of tryptophan, excreted great quantities of a substance which appeared to be a tryptophan precursor. The production of this substance could be stopped instantly by the addition of tryptophan to the culture, indicating to us that an amino acid can inhibit its own formation by a direct effect on one or more enzymes in its biosynthetic pathway (Novick and Szilard, 1954).

In late 1953, my collaboration with Szilard came to an end when I left Chicago for Paris to spend a year at the Institut Pasteur with Lwoff, Monod, and Jacob. This experience reinforced my interest in cellular control mechanisms and my activities have centered here since then. Szilard continued to maintain a keen interest in biology, but as the nuclear arms race of the nineteen fifties developed, he devoted a large fraction of his energies to attempts to divert mankind from this disastrous course. Nevertheless, he found time to keep up with the exciting development of molecular biology during this period. He was a frequent visitor to the most active laboratories, both in the United States and in Europe, and on these visits he always contributed original and useful ideas. He also wrote a number of theoretical papers during this period, including an ingenious theory of aging (Szilard, 1959). He constructed a theory to account for control of enzyme synthesis in bacteria (Szilard, 1960a), a theory which he used also to explain the production of specific antibodies in vertebrates (Szilard, 1960b). In the last months of his life he turned to a problem that had been of great interest to him for many years, namely how the brain works. He developed a theory of the nature of the synaptic links between neurons and how neural networks could be constructed to account for learning (Szilard, 1964) and he was writing a second paper when he suddenly died, in the spring of 1964. Szilard had become a biologist, although I suspect he always continued to think of himself rather as a physicist interested in biology.

REFERENCES

Delbrück, M., and W. T. Bailey, Jr. 1946. Induced mutations in bacterial viruses. Cold Spring Harbor Symp. Quant. Biol., *11*: 33–37.

Hershey, A. D., and M. Chase. 1952. Independent functions of viral protein and nucleic acid in growth of bacteriophage. J. Gen. Physiol., *36*: 39–56.

Lederberg, J. 1947. Gene recombination and linked segregation in *Escherichia coli*. Genetics, *32*: 505–525.

Monod, J. 1942. Recherches sur la Croissance des Cultures Bactériennes. Paris: Hermann et Cie.

Novick, A., and L. Szilard. 1950a. Description of the chemostat. Science, *112*: 715–716.

———, ———1950b. Experiments with the chemostat on spontaneous mutations of bacteria. Proc. Natl. Acad. Sci., *36*: 708–719.

———, ———1951. Virus strains of identical phenotype but different genotype. Science, *113*: 34–35.

———, ———1954. Experiments with the chemostat on the rates of amino acid synthesis in bacteria, p. 21. *In* Dynamics of Growth Processes. Princeton University Press.

Szilard, L. 1959. On the nature of the ageing process. Proc. Natl. Acad. Sci., *45*: 30–45.

———1960a. The control of the formation of specific proteins in bacteria and in animal cells. Proc. Natl. Acad. Sci., *46*: 277–292.

———1960b. The molecular basis of antibody formation. Proc. Natl. Acad. Sci., *46*: 293–302.

———1964. On memory and recall. Proc. Natl. Acad. Sci., *51*: 1092–1099.

N. VISCONTI
Istituto di Genetica Medica di Torino, Italy

Mating Theory

When I was invited to contribute to the Delbrück Festschrift volume with an article on the "Mating Theory," I felt like an apostate who had been asked to talk of his early religious vocation. It is a rather embarrassing subject, particularly for one who has "left the order"; thus, my first reaction was to refuse. However, I was induced to make my contribution for two reasons: first, my great devotion to Max Delbrück, whose inspiring personality acted very strongly on my intellectual formation; second, the thought that episodes of my work with Delbrück might be of some interest to all those who, on the premises of Delbrück's work and teaching, have contributed to the development of molecular biology.

Although my work on the subject of this article came to a close in 1953 when I abandoned science after six years of intense work, and although over ten years have passed since then and my recollections may be a bit rusty, I hope to manage to portray faithfully the thrill we sensed at Cold Spring Harbor of working on the brink of new discoveries in a practically new science in the early 1950's.

For somebody interested in testing simple ideas in science, working at the Carnegie Institution in Cold Spring Harbor between the years 1950 and 1953 was a unique opportunity. Life was easy-going and very informal, work schedules completely free, under the benevolent supervision and constant encouragement of M. Demerec. One could obtain technical help or check on a literature reference in the library at any time of day or night. The field of phage genetics was developed enough so that all sorts of experiments could be planned and discussed on the beach with people of different backgrounds, and then performed in the laboratory in the following days. It was enough to expound your ideas to Alfred Hershey or Evelyn Witkin to get a quick impression such as "It sounds good" or "It might be feasible," and have all materials prepared for the following day, along with some advice on the methods to follow. Collaboration was complete and the atmosphere very friendly.

It was in this particular environment that I met Max Delbrück in the summer of 1950. I was taking the course on phage with great enthusiasm and with some of the fervency of a neophyte. Experiments that I would never have dreamed or dared to perform before were run as routine exercises for the students. Mark Adams had taken up the teaching of the course, and it was extraordinary to notice the disparity of his pupils—some were full professors, and some were youngsters still waiting to get an M.A. degree. Mark Adams was the first person to give me some idea of the personality of Max Delbrück, who was already presented in the course as the Venerable Founder of the Science. When I learned that Delbrück was arriving in Cold Spring Harbor, I was in quite a dither over meeting him. I had the impression I was off to a good start with him when I explained the programs I had in mind for my work in Cold Spring Harbor, and I had the feeling he might be interested in some of the things I was going to do.

Then our relation suffered an abrupt collapse, and likewise all my hopes, when one evening I felt the sting of his criticism which left burning scars for some time to come. We often had seminars in the evening and I remember once giving a brief lecture on some work I had done back in Italy on the action of nitrogen mustard on spores of *Aspergillus niger.* The experimental data were uninspiring and the theory built on them wrong. On previous occasions, and on that particular one, no one had paid enough attention to my exposition to discover where the false step of my mistaken theory lay. Delbrück saw it right away and pointed it out with great accuracy and some violence. That was the end of my relation with Delbrück for the first season.

The next summer Delbrück came back to Cold Spring Harbor and my hopes reflowered. He got interested in some experiments I was doing on three factor crosses in phage T2. Using marked strains from Hershey's collection I was trying all sorts of crosses using linked and unlinked markers and infecting bacteria with an excess of one parent type. Up to that moment no complete theory had been drawn on phage recombination and, although the idea of successive rounds of mating between vegetative forms was vaguely implied by several workers, nobody had made a specific effort to build a complete theory which could predict all possible results on the basis of pure calculations.

After looking at my results Delbrück thought that such an attempt could be made. We had several discussions on the data I already had and on new ones I was accumulating every day. While I was preoccupied at the moment with statistical methods for getting the best information out of my data and for checking simple assumptions of equality between the numbers of complementary recombinants, Delbrück scorned such an effort and started speculating about the assumptions for the new theory. Having had some training as a classical geneticist, I was inclined to put a lot of effort into obtaining the best estimates of recombination values, and I was attracted by the use of elaborate methods such as those designed by R. A. Fisher, J. B. S. Haldane, K. Mather, and other statistical geneticists. Delbrück, on the other hand, showed very little interest in this path; he thought the data were all right, and that what was needed was a theory that would first assume the basic elements for a Mendelian recombination and should therefore predict results for future experiments. His feeling was that future attempts to disprove the theory would bring new ideas. For the moment he was interested in: (a) formulating the theory; (b) checking if the data we had would match. Accurate fitting of the data bored him slightly.

All the facts underlying the recombination theory in phage had been described in the fundamental work of Hershey and Rotman (1949). We can briefly summarize them here as follows:

(1) Upon infecting the bacterium the phage particle undergoes a modification in a state called vegetative stage. The vegetative phage is non-infecting.

(2) The mature phage starts appearing in the bacterium after a certain length of time and grows linearly with time from then on.

(3) Phage can undergo mutation. Mutations are generally stable. We can suppose genetical continuity between the infecting phage and phage yielded upon lysis. The first phage we call parental phage, the second offspring phage. The yield of a burst can be compared to a culture of phage originating from a single infecting particle. The notion of a vegetative phage particle first multiplying in the bacterial cell and maturing afterwards can account for these facts.

(4) If the bacterium is multiply infected with several genetically marked phage particles, the progeny generally contains particles which show a combination of markers derived from the parental types.

(5) If different ratios of marked phage particles are used for infection, the same ratios for the single markers are found in the progeny. This means there is no selection for these markers either during growth or maturation of the vegetative phage.

(6) The proportion of recombinants in a two factor cross have different values depending on the markers. The same frequencies are obtained if crosses are made in coupling or repulsion. This means that loci are the segregating units. Values as high as 40% recombinants can be obtained. Markers showing this high recombination value are called unlinked.

(7) If only unlinked markers are used, the different recombinant frequencies in the yield are those which should be obtained if the population of infecting phage went almost, but not quite, to complete mixture or to complete statistical equilibrium through an infinite number of rounds of mating. This conclusion was reached in our work after analyzing the results of two factor, three factor, biparental, and triparental crosses with equal or

unequal multiplicities of the infecting phages. The notion of repeated matings among vegetative particles can account for all these results. In a random mating Mendelian population this approximation to equilibrium requires at least 4 to 5 rounds of mating.

In a three factor cross involving three unlinked markers, ABC × abc, if recombinants of the Ab type are selected, in this class of recombinants C and c will be distributed in the ratio 1:1. We interpret this result as meaning that by selecting only recombinants AB we make sure that the selected phage particles have undergone at least one mating with the opposite parent.

Many phage markers show some degree of linkage, which means a frequency of recombinants lower than 40%. If we choose a group of linked markers we find that recombination frequencies are roughly additive, which is a strong indication that the arrangement of the genes on the linkage groups is linear (Hershey and Rotman, 1949; Doermann and Hill, 1953). On the other hand, we know that a mixed infection of a bacterium is not a straightforward analogue of a simple genetic cross. If we consider the phage particles in the yield as a sample of a Mendelian random breeding population, then the fact that linkage is detectable means that the number of matings is not infinite. For a given number of rounds of mating, the closer the linkage for a single mating, the smaller the frequency of recombinants in the final yields. Two very striking facts confirm the idea that we are dealing with a population of mating vegetative phages. The first appears from a comparison of the result of a two factor cross AB × ab (A and B linked) when the cross is made either with equal multiplicity or with very unequal multiplicity of the two infecting parental types. In the first case, speaking in a rather approximate way, AB has a 0.5 chance of mating with ab at every round of mating. This means that half of the mating cannot give recombinants Ab or aB. In contrast, in the case of very unequal multiplicity, let us suppose that AB is the minority parent, and that every bacterium has been infected by 1 AB and 20 ab. At every round of mating AB will mate with an ab, and in this case every round of mating will be useful in obtaining recombinants. If the frequency of recombinants as compared to the minority parental type (which in the first case is equal to the majority parental type) is p in the first case, it should be 2p in the second one. Experiments of this kind give the predicted results (Visconti and Delbrück, 1953).

Another striking fact of phage genetics can be indicated by the name of apparent negative interference (Hershey and Rotman, 1949; Doermann and Hill, 1953). Let us consider three linked factors ABC. Let the recombination frequency between A and B in a cross AB × ab be p. We can now repeat the cross using a third marker C closely linked to B. If we select recombinants Bc, and score the frequency of recombinants aB only within this class, then this value is greater than p. This means that recombination between B and C increases the probability of recombination between A and B. In conventional breeding tests, this result would be called negative interference. Visconti and Delbrück (1953) have shown that this result in population genetics is a consequence of any type of successive mating, if mating is not in synchronous rounds, but if its frequency has some kind of distribution per single particle. Supposing each particle has the same probability of mating per time unit: after a period of time, some particles will have mated more than others. The closer the linkage between B and C, the rarer will be the recombinants. If we choose these recombinants we select those particles which have undergone more mating.

(8) Although in a two factor cross the number of recombinants of the complementary type are equal in yields averaged over many bacteria, in a single burst there is no correlation between the frequencies of the two recombinants (Hershey and Rotman, 1949). This shows that formation of a mature recombinant particle is a process independent of the formation of the complementary one. This lack of correlation can be explained, at least in part, by assuming that vegetative phage particles are extracted at random from the pool of mating particles, and after becoming mature phage particles, they are accumulated in the bacterium.

(9) Rare recombinants do not appear in clones in the yields from single bacteria. Their number in different bursts forms a nearly random distribution (Hershey and Rotman, 1949). This shows that recombinants appear late in the course of multiplication. We can assume that vegetative phage undergoes a period of multiplication without or with very little mating. We also assume that the vegetative population multiplies by simple division until a high density is reached before mating becomes frequent.

A theory of phage genetics (Visconti and Delbrück, 1953) can now be expressed in the following terms:

The parental particles enter the bacterial cell and are transformed into vegetative phage particles. These multiply, and as they multiply the rate of mating becomes appreciable. At a certain moment, shortly after mating has begun, vegetative phage particles are randomly sorted out at a constant rate and then transformed into mature phage particles which accumulate in the bacterium. Maturation is irreversible; mature phage particles do not mate and do not grow.

The accumulation of mature phage goes on linearly with time up to the moment of lysis.

The progeny examined at the time of lysis consists of a mixture of samples withdrawn at various times from the mating pool.

This theory places the genetics of phage very much in line with classical Mendelian theory which has been worked out for higher organisms. The difference from higher organisms consists mainly in the fact that we cannot directly study the recombination from a single mating. Only by applying the methods of population genetics can we calculate the recombination values per single mating and thus build the linkage maps. Interference is either absent or very small in these linkage maps. These are, in brief, the essential features of the theory, the mathematical elaboration of which was based on all sorts of multifactor crosses using different parental inputs.

What I considered one of the best achievements of the theory was the exact prediction of what happened when lysis was inhibited. A few months later, still in Cold Spring Harbor working with Levinthal, we showed that the frequency of recombinants can be greatly increased by inhibiting the lysis.

During this prolonged intracellular phase both the proportion of recombinants among the mature particles and the burst size increase linearly with time (Levinthal and Visconti, 1953).

When I sent these results to Delbrück he was moderately interested. I had the feeling that Delbrück hoped for some new idea to come out of the theory. But from what I understand, it has not yielded any important development in subsequent years.

Going back to the summer of 1951, Delbrück was spending most of his time in front of blackboards. He could stand for hours in front of one, writing down and then erasing complicated matrices on which he worked before coming down to simpler methods. When he had settled a step in the theory, he would copy on paper the result of the calculations, erase the blackboards, and start on a new step. People would walk in and out, ask for an explanation, or make a suggestion. Delbrück was rarely disturbed by intruders, and could easily talk about what he was doing while doing it. He was also not disturbed by suggestions, even if they were wrong, as long as they followed some logical pattern. In a sense his teaching was Socratic. Neville Symonds and Dale Kaiser would often sit in, take bits of the problems away, and come back later with their own solutions.

All that summer Delbrück was very busy with the mathematics of the theory, and it was only a few months later in California that we were able to discuss our work in more general terms. In January 1952, I left Cold Spring Harbor for Pasadena, where I spent a month writing with Delbrück "The Mechanism of Genetic Recombination in Phage" (Visconti and Delbrück, 1953). During this period there was a feeling that dramatic developments in biology were in the offing.

The central problem in biology appeared to be the gene and its control of the metabolism of the cell. It was clear that duplication had to be interpreted as the way in which the genes controlled protein synthesis. All kinds of models to explain duplication of genetic structures were being discussed in those days. The fact that vegetative phage, one of the simplest of genetic structures, was duplicating and recombining in the bacterial cell tended to give way to all kinds of speculation as to the model of such a mechanism. In those days I was still hopeful that our work on recombination would give some clue to the structure of the gene. On this matter Delbrück was very cautious. Although the theory indicated the notion of rounds of mating between vegetative phage particles, we rarely discussed what the implication would be in mechanical terms.

James Watson, who was one of Luria's students, was very much attracted by this problem. Watson had been interested in my experiments on recombination in phage, not so much in the formal theory of the mating mechanism as by the idea of a template that should account for both duplication and recombination. (Playing around with the idea of a template was a favorite game of young biologists in those days.) Watson was almost obsessed with the idea of a structural model for the vegetative phage particle, and was starting to think in terms of a relation to a chemical structure. This turned out to be the right approach.

While we were writing our paper in Pasadena, Delbrück told me that he had advised Watson to go to England and study the X-ray diffraction pattern of crystalline DNA at the Cavendish Laboratory. Delbrück was diffident about chemists in general, and although he was hopeful about Watson's work he was certainly much more attracted to the theory of recombination as an approach to the study of genetical structures. Maybe there was in him some quirk which led him to distrust chemistry; but with his wry sense of humor he also used to say that this diffidence had already played him a bad trick when, years before, while working with Lise Meitner in Germany, he had been one of the first physicists to learn of the experiments on uranium fission. When, in the Chemistry Department of the Kaiser Wilhelm Institut it was found that, after neutron bombardment of uranium, traces of barium were present, Delbrück's conclusion was that either the chemists had messed up their analyses, or otherwise he was confronted with the greatest discovery in modern physics. Diffidence toward chemists, and probably toward great discoveries as well, had put him on the wrong track.

Still, it was the chemical investigation of the structure of DNA that, in a way, broke the demarcation of genetics from biochemistry, thus starting a new trend to which all the smart members of the group working on phage genetics adhered in the following years. Many in this group had either been Delbrück's students or had started out in phage work under Delbrück's inspiration.

Sensing the new trend, I was suffering from a feeling of great frustration. I had enjoyed first learning and then using the fairly simple techniques of phage experiments. Any idea could be rapidly tested and experiments succeeded one another like games in tennis. I remember having discussed this aspect of biology with Cyrus Levinthal while working with him on increase of phage T2 recombination with delayed lysis (Levinthal and Visconti, 1953). He was attracted to biology because he thought that in physics nobody could test his own theories any more. He thought it was dull to make experiments on other people's theories, or to try and convince other people to make experiments to test your own theories. But Levinthal was also very attracted by new techniques, while I felt scared and incapable of handling new materials and methods. This was one reason for my frustration; the other was a more general one: working in close contact with brilliant people gave me a sort of inferiority complex. For these reasons I was considering abandoning science, which in fact I did a few months later. I often opened my heart to Delbrück on this subject. I remember he once said to me, "You don't have the inspiration or the talent to be an artist; then what else do you want to do in life besides be a scientist?" For Max Delbrück it was as simple as that.

REFERENCES

Doermann, A. H., and M. B. Hill. 1953. Genetic structure of bacteriophage T4 as described by recombination studies of factors influencing plaque morphology. Genetics, *38*: 79.

Hershey, A. D., and R. Rotman. 1949. Genetic recombination between host range and plaque-type mutants of bacteriophage in single bacterial cells. Genetics, *34*: 44.

Levinthal, C., and N. Visconti. 1953. Growth and recombination in bacterial viruses, Genetics, *38*: 500.

Visconti, N., and M. Delbrück. 1953. Mechanism of Genetic Recombination in Phage. Genetics, *38*: 5.

A. D. KAISER
Department of Biochemistry, Stanford University School of Medicine
Palo Alto, California

On the Physical Basis of Genetic Structure in Bacteriophage

At Cold Spring Harbor in the summer of 1951 Max Delbrück became interested in Nicolo Visconti's experiments with phage T2. Visconti had been studying the effect on recombinant frequency of relative input of the two parental phage types in a cross. When he made a cross with nine particles per cell of one parent and one particle per cell of the other parent, Visconti obtained a striking and unexpected result: the burst contained more recombinant particles than minority parental particles. Delbrück realized that this result could be explained if each phage has several opportunities to recombine during a single growth cycle. This was the origin of the Visconti-Delbrück (1953) theory of phage genetics which Delbrück worked out in quantitative detail.

This was also the origin of my interest in phage genetics. Delbrück had invited me to accompany him to Cold Spring Harbor that summer; in this way I shared in the excitement of seeing a complex body of experimental facts fall into line as the logical conclusions of a small number of assumptions. Consequently, when Jean Weigle suggested on my return to Caltech that I might map the mutants of phage λ which he had been isolating (Weigle, 1953), I eagerly accepted the opportunity. I have been working on the genetics of phage λ ever since.

INFECTIVITY OF DNA ISOLATED FROM PHAGE λ

By 1956, I had received my Ph.D. from Caltech, had spent two years at the Pasteur Institute, and was working at Washington University in St. Louis. The RNA of tobacco mosaic virus had just been shown to be infective (Gierer and Schramm, 1956; Fraenkel-Conrat, Singer and Williams, 1957). And so David Hogness, like myself a recent alumnus of the Pasteur Institute, and I decided that in order to learn something about the relation of genetics to DNA structure it would be necessary to isolate genetically active DNA from a bacterium or bacterial virus with known genetics. The same thought had apparently occurred to Guthrie and Sinsheimer (1960) and to Meyer, Mackal, Tao and Evans (1961). We tried to obtain active DNA from *E. coli* and from λ phage; the λ DNA worked (Kaiser and Hogness, 1960). It is a favorable material, because λ DNA can be isolated as a single molecular species, weighing about 30×10^6 daltons.

We started our experiments by using DNA from the defective λdg variant of λ carrying the bacterial *gal* genes concerned with conversion of galactose to glucose, and looked for a transformation of gal^- recipient bacteria to gal^+ induced by the addition of λdg DNA. In this first assay for biological activity of λ DNA we added intact, infective "helper" λ to the recipient bacteria, in addition to the λdg DNA, because Arber, Kellenberger, and Weigle (1957) had shown that the efficiency of transduction by λdg can be increased 20-fold by simultaneous infection with active phage.

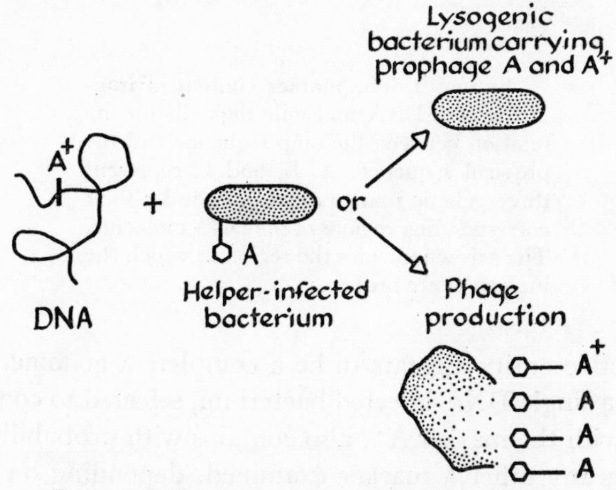

FIGURE 1. This assay system for λDNA utilizes whole bacteria recently infected with complete phage (helper). DNA and helper phage carry different genetic markers so that offspring of the infecting DNA can be recognized. Either a lysogenic or a productive response can be detected by appropriate plating.

After we found in this way that the *gal* genes can be transduced by λdg DNA, we used DNA from genetically marked infective λ phage and saw that also phage genes carried by the free viral DNA can appear either in the prophage resulting from the lysogenic response or in the infective progeny phage resulting from a productive response. In all of our experiments which employed *intact* recipient bacteria, addition of a "helper" phage proved necessary for infection by λ DNA, but Meyer et al. (1961) found that helper is not essential if bacterial protoplasts are used as recipients. The λ DNA assay we have used is depicted in Fig. 1. This assay gives 4×10^7 plaques per μg λ DNA. Expressed another way, the assay detects about 0.5% of the added DNA molecules. The presence of helper makes it necessary to distinguish the DNA genome from the helper genome by differences in genetic markers, as indicated in Fig. 1 by the symbols A and A^+.

The infective entity in a λ DNA extract is a typical DNA molecule: it is sensitive to pancreatic DNase, resistant to anti-λ serum, and it has the same buoyant density, 1.71 gm cm^{-3}, as the average DNA molecule in the extract. This density is also predicted by the base composition of λ DNA.

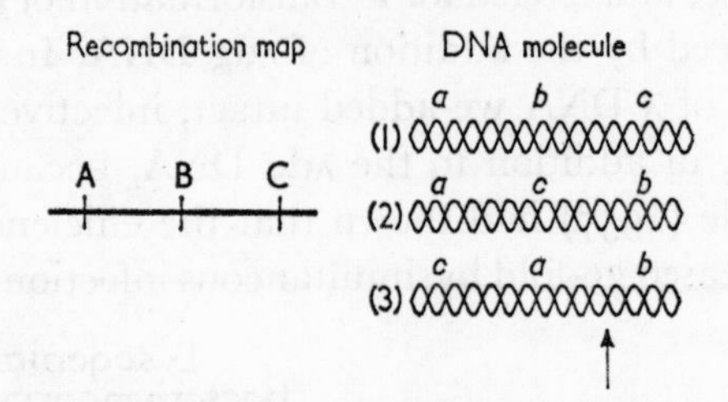

FIGURE 2. The marker content of fragments of a DNA molecule depends on the relation between the map sequence and the physical sequence. A, B, and C represent three genetic markers and *a*, *b*, and *c* their corresponding regions of the DNA molecule. The arrow indicates the region in which the molecules are broken.

The infective entity appears to be a complete λ genome. The phage produced by a single DNA-infected bacterium, selected to contain at least one particle with the marker A^+, also contains with probability 0.6 to 0.9 particles with any other λ marker examined, depending on the marker (Kaiser, 1962). Since this happens under conditions when the number of infected bacteria is directly proportional to the DNA concentration, all markers must be on the same DNA molecule.

EQUIVALENCE OF MAP SEQUENCE AND PHYSICAL SEQUENCE

A genetic map reflects both the structure of a DNA molecule and the mechanism of recombination. Is the recombination map a faithful representation of the structure of DNA, in the sense that the sequence of genetic markers on the recombination map is the same as the sequence of the nucleotides, or blocks of nucleotides, which correspond to the genetic markers? The idea for an experiment to test the equivalence of map and physical sequence grew out of Hershey and Burgi's (1960) paper on shear breakage of T2 DNA. Their most relevant finding for our purpose was that application of the proper amount of hydrodynamic shear breaks DNA molecules at or near their midpoints.

Consider three markers genetically linked in the sequence A, B, C. The physical sequence of the segments of the polynucleotide chain corresponding to each of the markers could be any one of the three possibilities shown in Fig. 2. If the DNA molecule were broken in the region indicated by an arrow, then the marker content of the two fragments would depend on the physical sequence. The marker content would be AB and C if the order were (1) in the figure, or AC and B if the order were (2) or (3). Order (1) is equivalent to the map sequence and is unique among the three possibilities in that both of its fragments are continuous segments of the recombination map.

Our experimental plan was, therefore, to detect the activity of genetic markers on DNA fragments, in order to determine which markers became unlinked and which remained linked after breakage. We found that the sets of genes carried by half-molecules are indeed halves of the recombination map, see Fig. 3, the break being located between the loci of *h* and *s* (Kaiser, 1962; Radding and Kaiser, 1963). The first results were obtained at a DNA concentration so low that each bacterium would have taken up no more than one half-molecule. Hogness and Simmons (1964) were later able to separate the two half-molecules of λ dg DNA from each other by column chromatography, since the two half-molecules differ in their guanine-cytosine content and hence in their binding to methylated albumin. In this way, different genetic markers were found in different fractions, leaving no doubt that some genes occur only on one physical half of the phage chromosome and others only on the other half. Therefore, the chromosome of the intact λ phage particle is a single DNA molecule having a unique nucleotide sequence.

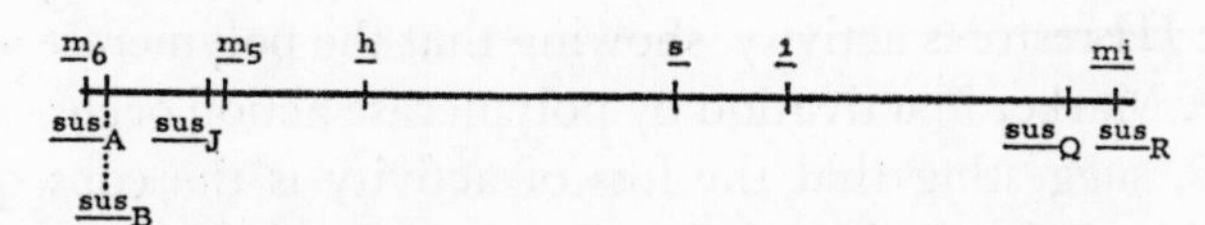

FIGURE 3. A linkage map of bacteriophage λ, based on crosses in the vegetative phase.

Although breakage into half-molecules now permitted the assignment of markers on the left half of the map to one half-molecule (the half with higher guanine-cytosine content) and the markers on the right half of the map to the other half-molecule, nothing could be inferred from those experiments about the physical sequence of markers on the same half-molecule. More breaks were necessary to do this, and hence we analyzed one-sixth molecules (Kaiser and Inman, 1965). Again, the results were in agreement with those predicted by the recombination map of Fig. 3: *i* is unlinked from sus_Q but sus_Q and sus_R remain linked to each other. Curiously, the marker *i*, located about one-third of the way from the right end of the map, is not active in the sixth-molecules. The reason for this was not apparent until it was learned that a cohesive end is required for λ DNA infection of helper-infected bacteria.

STRUCTURE OF THE COHESIVE ENDS

Although the ideas to be described in this section are a digression from the main line of argument, they arose from an attempt to test the equivalence of map sequence and physical sequence by another method. Richardson, Lehman, and Kornberg (1964) had isolated a DNase from *E. coli*, called exonuclease III, which degrades double-stranded DNA stepwise starting from the 3′-hydroxyl terminus of each nucleotide strand. We expected that the enzyme would destroy the activity of markers near the ends of the map before it destroyed those in the middle. When Hans Strack carried out this experiment he found instead that the marker *i*λ, which is about ⅓ of the way in from the right end of the map, lost activity at the same rate as the marker pair $sus_A sus_J$ which are at the left end of the map. At first we suspected that the exonuclease III preparation was contaminated with an endonuclease. To rule out this possibility Charles Richardson suggested trying to repair exonuclease damage with *E. coli* DNA polymerase as he, Inman, and Kornberg (1961) had done with phage T7 DNA. Instead of repair, Strack found a further loss of infectivity. (Later, Strack was able to demonstrate a transient repair by polymerase acting after exonuclease III.)

As a control experiment Strack examined the effect of DNA polymerase on native λ DNA. To our surprise he found that polymerase rapidly inactivates λ DNA, destroying markers in the middle of the map at the same rate as those near the end. Treatment of the polymerase product with exonuclease III restores activity, showing that the polymerase action is not degradative. Marker inactivation by polymerase action occurs at 15°C and even at 0°C, suggesting that the loss of activity is the consequence of a polynucleotide chain continuation rather than of a chain initiation reaction. The specificities of polymerase and exonuclease III led us to the following view (Strack and Kaiser, 1965) of the structure of one end of a molecule of λ DNA.

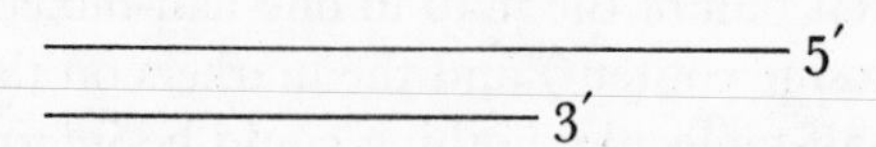

Hershey, Burgi, and Ingraham had reported their discovery of the specific cohesiveness of λ DNA in 1963. Ris and Chandler (1963) and MacHattie and Thomas (1964) then proved that the cohesive sites were at the ends of the molecule by comparison of the contour length of open with closed (circular) molecules. The presence of protruding 5′-terminated single strands at both ends of a molecule of λ DNA provides a natural explanation of cohesion. If the two strands have complementary base sequences then the formation of a helix between the two single strands would cause the ends to cohere.

MAPPING THE COHESIVE ENDS

Concurrently with the breakage and the enzyme experiments described above, Ross Inman and I were studying the effect of cohesion on the infectivity of λ DNA. We found that closed molecules are less infective than open molecules and that cohered half-molecules (which form a kind of inverted whole molecule) are less infective than free half-molecules. These observations suggested that a DNA molecule must have a free cohesive end to be infective, or at least to have normal infectivity, toward helper-infected recipient bacteria. This idea helps to explain why the product of DNA polymerase action on λ DNA is not infective, as these molecules would also lack cohesive ends.

The cohesive ends are linked to markers at the ends of the recombination map. The infectivity of the marker pair $sus_A sus_B$ and of $sus_Q sus_R$ falls when a mixture of sixth-molecules cohere to themselves. Their infectivity rises when the cohered fragments disjoin (Kaiser and Inman, 1965). Therefore the sixth-molecules which carry $sus_A sus_B$ and $sus_Q sus_R$ also carry cohesive ends. Reference to the recombination map, Fig. 3, will show that sus_A and sus_B are at the left end of the map and sus_Q and sus_R at the right end. Moreover, the marker *i* is more than one-sixth of the total length of the map from either end. The marker *i* is not infective in sixth-molecules, we believe, because it is carried by a fragment which lacks a cohesive end. Thus the ends of the map are at the ends of the molecule and the map and physical sequence are equivalent.

SUMMING-UP

The experiments described here have involved genetic markers which are separated from each other by a thousand nucleotide pairs or more. The support these experiments furnish for the assumption that genetic and physical maps are equivalent is valid only at that level of resolution. To extend the argument to the level of individual nucleotides would require a more refined technique for breaking DNA or the certainty that the mechanism of recombination for genetic markers separated by a single nucleotide is the same as that for markers separated by thousands of nucleotides.

It is difficult to determine the nucleotide sequence of a very long polynucleotide. And hence localization of a single base alteration in DNA by chemical means is not yet practical, although rapid advances are being made in the necessary techniques. But genetic tests can give us already the map position of base alterations with high precision. The studies with λ DNA have furnished evidence that in trying to characterize the precise structure of a genetically active DNA macromolecule, map sequence and nucleotide sequence may be equated.

REFERENCES

ARBER, W., G. KELLENBERGER, and J. J. WEIGLE. 1957. La deféctuosité du phage lambda transducteur. Schweiz. Z. Path. Bakt., *20*: 659.

FRAENKEL-CONRAT, H., B. SINGER, and R. C. WILLIAMS. 1957. Infectivity of viral nucleic acid. Biochim. Biophys. Acta, *25*: 87.

GIERER, A., and G. SCHRAMM. 1956. Infectivity of ribonucleic acid from tobacco mosaic virus. Nature, *177*: 702.

GUTHRIE, G. D., and R. L. SINSHEIMER. 1960. Infection of protoplasts of *Escherichia coli* by subviral particles of bacteriophage ΦX 174. J. Mol. Biol., *2*: 297.

HERSHEY, A. D., and E. BURGI. 1960. Molecular homogeneity of the deoxyribonucleic acid of phage T2. J. Mol. Biol., *2*: 143.

HERSHEY, A. D., E. BURGI, and L. INGRAHAM. 1963. Cohesion of DNA molecules isolated from phage lambda. Proc. Natl. Acad. Sci., *49*: 748.

HOGNESS, D. S., and J. R. SIMMONS. 1964. Breakage of λdg DNA: chemical and genetic characterization of each isolated half-molecule. J. Mol. Biol., *9*: 411.

KAISER, A. D. 1962. The production of phage chromosome fragments and their capacity for genetic transfer. J. Mol. Biol., *4*: 275.

KAISER, A. D., and D. S. HOGNESS. 1960. The transformation of *Escherichia coli* with deoxyribonucleic acid isolated from bacteriophage λdg. J. Mol. Biol., *2*: 392.

KAISER, A. D., and R. B. INMAN. 1965. Cohesion and the biological activity of bacteriophage lambda DNA. J. Mol. Biol., *13*: 78.

MACHATTIE, L. A., and C. A. THOMAS, Jr. 1964. DNA from bacteriophage lambda: molecular length and conformation. Science, *144*: 1142.

MEYER, F., R. P. MACKAL, M. TAO, and E. A. EVANS, Jr. 1961. Infectious deoxyribonucleic acid from λ bacteriophage. J. Biol. Chem., *236*: 1141.

RADDING, C. M., and A. D. KAISER. 1963. Gene transfer by broken molecules of λDNA: activity of the left half-molecule. J. Mol. Biol., *7*: 225.

RICHARDSON, C. C., R. B. INMAN, and A. KORNBERG. 1964. Enzymic synthesis of deoxyribonucleic acid. XVIII. The repair of partially single-stranded DNA templates by DNA polymerase. J. Mol. Biol., *9*: 46.

RICHARDSON, C. C., I. R. LEHMAN, and A. KRONBERG. 1964. A deoxyribonucleic acid phosphatase-exonuclease from *Escherichia coli*. II. Characterization of the exonuclease activity. J. Biol. Chem., *239*: 251.

RIS, H., and B. L. CHANDLER. 1963. The ultrastructure of genetic systems in prokaryotes and eukaryotes. Cold Spring Harbor Symp. Quant. Biol., *28*: 1.

STRACK, H. B., and A. D. KAISER. 1965. On the structure of the ends of lambda DNA. J. Mol. Biol., *12*: 36.

VISCONTI, N., and M. DELBRÜCK. 1953. The mechanism of genetic recombination in phage. Genetics, *38*: 5.

WEIGLE, J. J. 1953. Induction of mutations in a bacterial virus. Proc. Natl. Acad. Sci., *39*: 628.

SEYMOUR BENZER
*Department of Biological Sciences, Purdue University, Lafayette, Indiana**

Adventures in the rII Region

Writing a paper about genetic fine structure means overcoming two powerful blocks imposed upon me by Max Delbrück. One was laid down in 1955, when Delbrück said that the problem would keep me occupied for ten years. He was right: in 1965 my interest suddenly turned off post-hypnotically, and it is now more than I can do even to think about the subject. The second block was laid down more recently, when several papers of mine happened to appear at about the same time. To a letter from his wife to mine, Delbrück appended a footnote: "Dear Dotty, please tell Seymour to stop writing so many papers. If I gave them the attention his papers *used* to deserve, they would take all my time. If he *must* continue, tell him to do what Ernst Mayr asked his mother to do in her long daily letters, namely, *underline what is important*." It is very difficult for me now to think of anything worthy of being underlined.

Delbrück first entered my life in the form of the chapter heading "Delbrück's Model" in Schrödinger's book, "What is Life?" I read that book at an impressionable age, while still a graduate student in pre-transistor solid state physics at Purdue University. Not long afterward, at a meeting of the American Physical Society at Bloomington, Indiana, a friend took me to visit the home of a former coed classmate of his. Her husband not only knew Delbrück personally, but even pulled a snapshot of him out of a drawer. I could not have been more impressed. The husband's name was Salvador Luria, and it was not long before he had persuaded me to enroll in the phage course at Cold Spring Harbor. Thus I was suddenly plunged into the biology business.

After spending an initial postdoctoral year in A. Hollaender's newly organized Biology Division at Oak Ridge, I had a choice of going to Luria's or to Delbrück's laboratory and asked Luria's student James Watson for advice. Watson thought that it depended on what I wanted. Luria, he said, would be likely to ask me every day what I had done, whereas I might not see Delbrück for a week at a time. I chose to join Delbrück at Caltech. That was sixteen years ago. I have just returned to Caltech and many of the old

**Presently on leave at Biology Division, California Institute of Technology, Pasadena, California*

memories seem quite fresh. But this time it is to work on neurobiology with Roger Sperry, who also has the virtue of disappearing for a week at a time.

At Caltech in 1949, Jean Weigle and I shared a room. Since he was a "lark," rising at 4 a.m., and I was an "owl," we could do round-the-clock phage experiments together. I would put the plates into the incubator before retiring and he would take them out a few hours later, record the results and do the next experiment. When I arrived, we could do a further experiment together, and then I would do one more experiment after he went to bed. This system broke down on Mondays, however. There was always a shortage of pipettes on that day, since the kitchen staff did not work on weekends. Complaints to Delbrück were in vain—he insisted that it was good to have one pipetteless day during which one was obliged to think.

I recall one experiment that Weigle and I tried. We thought that one might be able to transfer genes from one bacterium to another via a temperate phage. Delbrück said that the idea was crazy and bet us (50 milkshakes to one) that our experiment would not work. And he won his bet. But the experiment failed not because the idea was crazy, but because it failed to incorporate what Delbrück calls "the principle of limited sloppiness." To keep the background of spontaneous mutants low, we used as donor and recipient a pair of bacterial strains that differed in *two* genes, and scored, as Lederberg and Tatum had done in the discovery of bacterial conjugation, only cases in which *both* genes were transferred. Not long afterward, Zinder and Lederberg did discover phage-mediated transduction, but found, of course, that only genes that are very closely linked (the ones we used had not been) are transferred jointly.

Others working in Delbrück's group at Caltech at that time were Renato Dulbecco, Gunther Stent, Elie Wollman, Wolfhard Weidel, and, as graduate students, Armin D. Kaiser and George Bowen. The urge to do experiments was always so strong that we could not get ourselves to sit down and write up the results. Delbrück had a solution for this. He assembled all who had papers to write and whisked us off to Caltech's Marine Biology station at Corona del Mar. There, we were locked up for three days and ordered to write. Delbrück's wife, Manny, typed as rapidly as we could spew the stuff out; we mercilessly criticized each other's drafts, and in three days every one had a completed paper. That was how my paper on UV irradiation of intracellular phage came to be written.

Delbrück deprecated biochemistry, and this influenced some of us to avoid it. In fostering this attitude, he was assisted by Weidel, the only biochemist in the Caltech phage group at that time, who, when he became impatient with some of our purely formal discussions, would annoy us by saying things like, "Now let a biochemist tell you what this is *really* all about." Weidel, after some months of work, isolated the "phage-receptor substance" from *E. coli*, which seemed very impressive until he looked at his "molecules" in the electron microscope and saw that they were lovely, empty, complete cell walls. Though by means of these studies Weidel was ultimately to elucidate the structure of the bacterial cell wall, this dénouement simply confirmed our prejudice as to the despicability of biochemistry, a comforting prejudice to justify our ignorance. Time has changed that for most of us, however. Even Delbrück himself recently spent time studying *lipid* biochemistry in connection with his Phycomyces work.

Sometimes Delbrück would proclaim Wednesday and Thursday as a weekend, to avoid crowds and highway traffic on camping trips. The first camping trip in which I participated, into the Anza desert, was fairly typical. We just kept driving until the car got stuck in the sand, and that determined the campsite. Most of the following day was spent in digging the car out. Such visits to the desert were (and still are) the favorite means of entertaining visitors, although some people, like Luria, will not come to Caltech unless guaranteed immunity from camping. It was on one such trip in 1950 that André Lwoff invited me to spend a year in his laboratory at the Institut Pasteur in Paris.

In Lwoff's laboratory I shared a room with François Jacob. It was a lively year. In addition to the other regulars like Jacques Monod, Elie Wollman and Pierre Schaeffer, there was a fine crew of long- and short-term transients: Melvin Cohn, Louis Siminovitch, Annamaria Torriani, Germaine Cohen-Bazire, Roger Stanier and Gunther Stent.

I first discovered the *r*II phenomenon in Paris, but did not then recognize its significance. As Pasteur would say, "my mind was not prepared." My research project in Paris was inspired by a remark made by Roger Stanier in a review on "Enzymatic Adaptation," in which he had said that while it would be interesting to know whether all bacteria in a culture adapt simultaneously to fermentation of a new carbon source, that would be "almost impossible" to find out. That statement was a challenge for me, since the Luria-Latarjet experiment of UV irradiation of vegetative phage, on which I had been working at Caltech, made it possible to follow the progressive intracellular development of phage from minute to minute and also the distribution of rates of development among the various cells of a phage-infected population. In those experiments I had shown that in bacteria starved before infection, development of phage is arrested at a very early stage. I thought, therefore, that it should be possible to test the simultaneity of adaptation in a bacterial culture by making the metabolism of the cells, and hence intracellular phage development, dependent upon the presence of an inducible enzyme. For instance, in a bacterial population "half-adapted" to lactose, all cells might have half the maximal level of galactosidase, or half the cells might have the full enzyme complement. If starved cells were infected with phage and placed in a medium in which lactose was the only carbon source, only those cells having galactosidase could support phage growth. Heterogeneity in enzyme level in the bacterial population would then show up in the Luria-Latarjet experiment as dispersion in the sensitivity of infective centers to UV irradiation.

Once I had started to work in Paris, however, I hit on a more direct non-radiobiological approach to this problem. Monod and Wollman had shown three years earlier that in the bacterial strains they used galactosidase synthesis is arrested by phage infection and that the enzyme is released intact upon the eventual lysis of the infected cells. Thus, if prior to phage infection, a cell population were homogenous in its galactosidase content, a plot of cells lysed vs. enzyme released at various times after phage infection, in a medium containing lactose as only carbon-energy source, would give a straight line. But if the population were heterogenous such a plot would give a curve, its initial slope corresponding to the cells with the highest enzyme level, provided that the enzyme content of each cell were the limiting factor for the rate of phage development within it.

It seemed only natural for me to use one of the bacterial strains then under study in Lwoff's laboratory, a K12 derivative of *E. coli* K12 lysogenic for phage lambda, inducible for galactosidase formation, and sensitive to phage T2. To avoid complications arising from re-infection by early-released phages leading to "lysis inhibition" I chose an *r* (rapid lysis) mutant of T2. But I could not make the experiment work; for some reason the T2-infected cells did not lyse—even when they had been fully induced in lactose medium prior to infection. I checked the phage stock titer again, plating as usual for plaque formation on *E. coli* strain B, and there seemed to be nothing wrong with the stock. That was, of course, the first discovery of the *r*II phenomenon, namely the inability of *r*II mutants of T-even phages to grow on *E. coli* strains carrying the lambda prophage. Rather than trying to figure out at that point what was going on in the system I had chosen, I shifted to the bacterial strain and phage combination used by Monod and Wollman in the first place.This promptly worked and made it possible to demonstrate that, under certain conditions of adaptation of an *E. coli* culture, there is striking heterogeneity in the amount of induced enzyme appearing in individual cells.

That was in 1952, the year of the Hershey-Chase experiment showing the germinal role of the phage DNA. The next year brought the Watson-Crick model, and now DNA was really *in*. Upon my return to Purdue University, I was invited to give a genetics seminar and chose the topic "The Size of the Gene." It was largely based on a review that had recently been published by G. Pontecorvo in *Advances in Enzymology*. The article made the point that the various definitions of the gene were not necessarily equivalent and that high resolution genetic mapping would be required to distinguish them. High resolving power requires detection of small numbers of recombinants, so that to resolve details on the level of the size of a gene it would be necessary to apply selective techniques to large mating populations.

The second time I encountered the *r*II phenomenon my mind *was* prepared. I had started out to attempt the Hershey-Chase experiment with genetic markers, to show sequential injection of the various parts of the phage genome, as Jacob and Wollman had done with bacterial conjugation. For that experiment, a stock of an *r* mutant of phage T2 was needed. Now stocks of *r* mutants grown on strain B of *E. coli* usually have titers much lower than the wild type r^+ stocks, since r^+ phages induce lysis inhibition and hold the cells together for a longer period of intracellular phage multiplication. But I had just read in George Streisinger's thesis that on certain strains of *E. coli* other than strain B, *r* mutants of T2 do yield titers as high as r^+. Could that mean that the *r* mutant can produce lysis inhibition on those strains? To test this possibility, I plated out some T2*r* and T2r^+ on the strains I had on hand in my laboratory. If *r* produced lysis inhibition, it should make small, fuzzy-edged plaques similar to r^+, rather than the large, sharp-edged *r*-type plaques seen on strain B. On that day, I happened to be preparing an experiment on lysogeny for my phage class and was growing cultures of K12(λ) and its non-lysogenic derivative K12S (obtained via Luria from Esther Lederberg). Plating T2*r* and T2r^+ on those strains certainly gave different results from plating them on strain B. On K12S, *r* and r^+ both gave small, fuzzy plaques. On K12(λ), r^+ made small fuzzy plaques, but the plate to which *r* was supposed to have been added had no plaques at all. I was sure that in the rush to prepare for class, I had neglected to add phage to that plate. But repetition confirmed the result.

To me, the significance of this result was now obvious at once; here was a system with the features needed for high genetic resolution. Mutants could be detected by the plaque morphology using strain B. Good high-titer stocks of the *r* mutants could be grown using strain K12S. Strain K12 (λ) could be the selective host for detecting r^+ recombinants arising in crosses between *r* mutants. A quick computation showed that if the phage genome were assumed to be one long thread of DNA with uniform probability of recombination per unit length, the resolving power would be sufficient to resolve mutations even if they were located at adjacent nucleotide sites. In other words, here was a system in which one could, as Delbrück later put it, "run the genetic map into the ground." I dropped everything else and embarked on this project.

It soon became evident that this plating behavior was shown only by certain *r* mutants, namely those belonging to one of the map "clusters" found by Alfred Hershey. These mutants all mapped within a few per cent recombination of each other, but were located at a large number of distinct sites. A unique order of the sites could be established, and the map could be sharply divided into two contiguous segments or *cistrons*, as I later called them, such that any mutant located within one cistron would functionally complement any located within the other cistron. Certain mutants were anomalous in the sense that they would not yield recombinants when crossed with two or more other mutants that did give recombinants with each other. These "deletions" turned out to be especially useful for later work.

The state of thinking in genetics at that time can perhaps be judged by the following experience. In April, 1954, at a meeting at Oak Ridge, I asked a geneticist who was very familiar with Hershey's cluster of mutations and the *r*II region what he thought was their significance. He said he thought that all were really the same mutation, but located at different places in the heterochromatin.

In that summer of 1954 the Hersheys generously lent their Cold Spring Harbor house to my family. There I met Sydney Brenner. Brenner later described this encounter in a talk at a Brookhaven Symposium: "I was carrying around a book on sequence analysis in proteins and Seymour was carrying around a map of the *r*II region—consisting of two mutants that mapped in a straight line." The future seemed to us quite straightforward —isolate the *r*II protein from various mutants, then establish the colinearity of alterations in amino acid sequence with the locations of the mutations in the genetic map. If we could somehow identify the DNA bases, we could even solve the genetic code.

I wrote up the story of the *r*II mutants and showed the manuscript to Delbrück in Amsterdam later that summer. One of his typically succinct comments was: "Delusions of grandeur." Delbrück knocked the paper so badly that not until a visit to Caltech the following spring did I dare to approach him with another version. He submitted it to the *Proceedings of the National Academy of Sciences*, and it contained an appropriate acknowledgment of his "moderating influence."

Alan Garen came to Purdue and embarked on the isolation of "the *r*II protein." He began by finding that an extract from r^+ infected cells could stimulate development of *r*II mutants in K12(λ). To identify the active principle, he tried inactivating it with specific enzymes. But adding deoxyribonuclease to the principle greatly *enhanced* the effect, and he soon found that this was due to the Mg^{++} that had been added to activate the enzyme. From then on, his time at Purdue was spent in analyzing the Mg^{++} effect on growth of *r*II mutants in K12(λ). Since I had bet that the active principle was a protein, I paid off a bottle of champagne on that one. Strictly speaking, however, Garen never really established that the activity in the original extract was due to traces of Mg^{++}, and it is conceivable that the bottle of champagne may have to be paid back someday. The tradition established by Garen of not finding the elusive *r*II protein was later valiantly continued at Purdue by Masayasu Nomura, F. Robert Williams, and Mutsuo Sekiguchi, and elsewhere by others, using a variety of techniques. Some interesting facts, nonetheless, have emerged from these studies. In spite of the lack of success so far, there is reason to believe that an *r*II protein actually *does* exist because suppressors that act to produce structural changes in other proteins will also suppress some *r*II mutations.

Mapping the *r*II region was more rewarding. It was an example of what we called "Hershey Heaven." This expression comes from a reply that Alfred Hershey gave when Garen once asked him for his idea of scientific happiness: "To have one experiment that works, and keep doing it all the time." Making use of the properties of overlapping deletions, it was possible to divide the map into segments that fitted a strictly linear topology. The *r*II region also had striking differences in local topography, as could be shown by the different frequencies of occurrence of mutations at various "hot spots."

The results brought into focus the distinction between the various definitions of a gene and led me to propose the names *cistron*, *recon*, and *muton* for the units of function, recombination, and mutation.

Ernest Freese came to Purdue about that time. He had gone to Caltech from Germany, to do physics, but became interested in biology. Delbrück invoked his "moderating influence" to talk him out of going into biology, which was, of course, a very effective seduction technique. At Purdue, Freese followed up on the discovery by Rose Litman and Arthur Pardee that one can induce mutations in T2 phage with 5-bromouracil. This chemical mutagenesis had appeared to them as "nonspecific," because the pyrimidine analogue induced mutations in many different phage genes. By mapping the induced *r*II mutants, however, we found that the mutation rate at certain *r*II sites was raised 10,000-fold, while remaining unchanged at some other sites, giving a completely different site distribution for induced and spontaneous mutations. This finding opened the field of relating mutagenic specificity to DNA structure, which Freese and his associates pursued vigorously in the following years.

In 1957, I went to Cambridge for a year to join up, at last, with Sydney Brenner to finish that little problem of colinearity of gene and protein structure. George Streisinger and Sewell Champe joined in. By this time, several other laboratories were in the race. Instead of trying to find the *r*II protein, we tried to find a gene to fit a phage protein that could be isolated. We shifted our choice from phage heads to tails to tail sheaths and back again to heads. That was the heyday of fingerprinting, which had recently been developed by Vernon Ingram, and we had the privilege of working under the supervision of the Master. It was always easy to find structural differences between the corresponding proteins of T2 and T4, but in that year we never did find a difference that could be related to a single mutation. Renato Dulbecco, who was also spending that year at Cambridge, was a spoilsport. He maintained that the result would be interesting only if the genetic map of a cistron and its polypeptide product were *not* colinear. In the end, it was Charles Yanofsky, who, several years later, achieved the first demonstration of colinearity for the tryptophan synthetase of *E. coli.*

The year at Cambridge was hardly wasted, however. In addition to analyzing mutant hemoglobins with Vernon Ingram and building models with Francis Crick (in a very drafty tower room at the Cavendish), it was that year that Brenner, Barnett, and I mapped the *r*II mutations induced by proflavine, finding that its specificity of action was utterly different from 5-bromouracil. The significance of this observation was not evident at the time, but important developments grew out of it later, when it occurred to Crick that proflavine might cause insertions or deletions in the DNA that would shift the reading frame for codons in protein synthesis.

Returning to Purdue, I resumed the project of running the map into the ground with the aid of all the new mutagens that had been found. But I was jarred out of this by the discovery of ambivalent mutants. Irwin Tessman at Purdue had been isolating *r*II mutants in connection with his studies on mutagenesis, and I asked him to give me samples of any *r*II mutants that indicated previously unobserved map sites. Some of those he gave me did not behave like *r*II mutants at all—they multiplied happily on my strain of K12(λ). It turned out that Tessman's K12(λ) strain was different from mine, and that some of my "good" *r*II mutants would lyse *his* strain. This immediately suggested to me that there might be differences in the genetic code in the two bacterial strains. Sewell Champe and I then demonstrated, by genetic experiments, that one such suppressor mutation in a bacterium could modify the genetic code of the cell in such a way that a nonsense codon in the phage genome would be changed to sense.

The place to look for the physical basis of these changes in the code was in the sRNA and the amino acid-activating enzymes. So I became, at last, really embroiled in biochemistry. Soon I was grinding up cells and extracting sRNA and getting involved with various combinations of collaborators, such as Bernard Weisblum, Gunter von Ehrenstein, Robert Holley, Fritz Lipmann, and François Chapeville.

It turned out that sRNA and activating enzymes were, indeed, subject to genetic modification, and that an organism may contain several sRNA varieties that accept the same amino acid. The validity of Crick's idea that sRNA acts as an adaptor in protein synthesis was demonstrated in two ways. The first was done by attaching an amino acid A to its normal sRNA, then changing it, while still attached, to another amino acid B. When transferred into protein, amino acid B went into the position where A normally belonged. The second proof was that a given amino acid, when attached to two different sRNA molecules, would go into different positions in protein. The latter experiment also established a physical basis for degeneracy in the code.

This all got more and more exciting, until it dawned on me how many people were doing the same things. I had almost gone down the biochemical drain. Delbrück saved me, when he wrote to my wife to tell me to stop writing so many papers. And I did stop.

R. S. EDGAR
Division of Biology, California Institute of Technology, Pasadena, California

Conditional Lethals

Conditional lethal mutants are now quite popular among microbial geneticists. They are simply mutations which result in the loss of some vital function, but only under conditions over which the experimenter has some control. Temperature sensitive mutants, for example, survive and multiply at one temperature (25°C) but not at another (42°C). Most such mutations probably affect the heat lability of one or another protein essential for growth. Amber mutations, another type of conditional lethal, are lethal mutations whose effects may be reversed by certain suppressor genes. It has been shown that amber mutations create a nonsense triplet in the mutant gene, permitting it to produce only a fragment of the protein product. The suppressor genes act by permitting the nonsense triplet to be read as sense, resulting in the production of a functional gene-product protein. In the case of amber mutations in phages, the amber suppressor genes reside in the host bacterium rather than the virus. Thus the amber mutants can propagate in strains carrying a suppressor gene but not in strains lacking one. These are but two examples of conditional lethal mutations. They permit the study of essential genes in haploid organisms.

What follow are my rather dim recollections concerning the exploitation of conditional lethal mutants in phage. It is claimed that when Max Delbrück was first told that the newly discovered *amber* mutations occur in many new and different genes, he replied, "That's too bad." Clearly, he paid little attention to the development of conditional lethal systems which to a large extent occurred under his nose at Caltech. Nevertheless, as will become clear later, in a sense he started it all, and in his usual way, by asking N. H. Horowitz a trenchant question.

The discovery of the *amber* mutants of T4 (after six years, the original paper is still to appear in print), which led to the notion of conditional lethal mutations, came about in a circuitous way. In 1960, R. H. Epstein, then at Caltech, was studying the phenomenon of multiplicity reactivation (MR) in phage T4. The "vulnerable center" was a name given to one of the parameters in mathematical formulations to account for the shapes of survival curves of complexes of bacteria multiply-infected with ultraviolet-irradiated phage. At that time, it was thought that these hypothetical targets, the vulnerable centers, could have biological meaning in terms of specific genes whose functions were essential before the restoration of the UV-damaged phage could occur through recombination and replication.

Experiments by D. R. Krieg (1959) had suggested that the *r*II genes of T4 should behave as vulnerable centers in the bacterial strain K(λ) where they perform early and essential functions, but should not behave as vulnerable centers in strain B (or K) where the *r*II genes are not essential for normal phage growth. This line of reasoning suggested to Epstein that MR experiments in B and K(λ) should reveal differences in the vulnerable center component of the survival curves.

He found, however, that the kinetics of MR in K(λ) and B were the same. Nevertheless, he reasoned that the *r*II genes could still be vulnerable centers in K(λ), but not in B, if there existed in the phage genome a corresponding pair of genes of similar UV-sensitivity which, however, perform essential functions in B but not in K(λ). If this were the case, he argued, mutants defective in such genes could be isolated that would grow on K(λ) but not on B.

In the initial search for these hypothetical "anti-*r*II" mutants, he was joined by C. M. Steinberg, and together they persuaded Harris Bernstein, then a graduate student in Neurospora genetics, to help in the "mutant hunt." They promised Harris Bernstein that if any mutants were found they would be named after his mother. In the first experiment about twenty mutants were found, and the promise to Bernstein was kept. (Amber is the English equivalent of the German "Bernstein.")

Subsequent work showed, however, that the *amber* mutants were not really of the expected type. They grow not only in K(λ) strains but also in many strains of K not lysogenic for lambda, and thus are not truly "anti-*r*II" mutants. In fact, if the "mutant hunt" had been more carefully designed, by using isogenic strains of K and K(λ), instead of B and K(λ), *amber* mutants would not have been found in the first place.

Complementation and mapping analysis soon showed that the *amber* mutations are located in a large number of different and heretofore undiscovered genes, scattered throughout the T4 genetic map. Preliminary experiments on the physiological properties of the mutants clearly indicated that different mutants are blocked in growth in the restrictive host at a variety of different stages.

Epstein was struck by the similarity of the *amber* mutants and the so-called *hd*, or *host-defective* mutants, previously described for lambda by Allan Campbell (Campbell and Balbinder, 1958; Campbell, 1959). Epstein found that the *amber* mutants of T4 and the *hd* mutants of lambda shared common growth properties on a wide variety of strains and thus were likely to be the same type of mutant. At this point, sufficient information was available to indicate that the *amber* and *hd* mutations could not readily be explained, by analogy with the *r*II mutations, as occurring only in a special class of genes. It became obvious that they are, instead, a special class of mutations that can occur in a great variety of different genes.

Who first had the idea that *amber* mutations are a general class of "suppressor-sensitive" mutations, I don't recall. As is true of many ideas, it probably arose during a blackboard discussion, with no single person responsible. One notion that was entertained during this period, as a possible "for instance," was that the *amber* mutations resulted in the production of mutant proteins highly sensitive to heavy metals and that the permissive and restrictive strains differed with regard to their permeability to heavy metals.

I first became excited by the notion of conditional lethals as a general and useful class of mutations as a consequence of a discussion with Allan Campbell at Cold Spring Harbor in the summer of 1960. It was while describing to him the views of Epstein and Steinberg regarding the nature of *amber* and *hd* mutants that the example of temperature-sensitive and pH-sensitive mutations as further examples of a general class of conditional lethals came up. (The term conditional lethal was later suggested to me by my wife, to replace the less euphonious term, facultative lethal, which I had been using.) As a consequence of that discussion, on my return to Caltech, I started to exploit the temperature-sensitive mutant system of T4 (Edgar and Lielausis, 1964). That year, Campbell (1961) did further experiments to show that the *hd* mutants (then renamed *sus*, for suppressor-sensitive) were in fact responding to a bacterial suppressor gene, and that temperature-sensitive and pH-sensitive mutants of lambda could be isolated.

(It is amusing to note that Campbell found it necessary to rename his mutants after learning more about them, whereas the name *amber* is just as meaningless, and thus just as useful, now, as when the mutants were first discovered and named for Mrs. Bernstein.)

From an historical point of view, perhaps the most curious fact concerning the exploitation of conditional lethals is an embarrassing parallel (in miniature) to the rediscovery of the Mendelian laws. Our notions concerning the nature of conditional lethals and their applications arose in an historical vacuum, almost exclusively from direct considerations of the *amber* mutants. Yet in genetic studies of other organisms, notably Drosophila, lethals had been studied intensively (Hadorn, 1961) for many years for precisely the same reasons that conditional lethals are now being used in haploid organisms. Conditional lethals had also been recognized as a special and useful class of lethals and had been named such (Hadorn, 1961). Temperature-sensitive lethals have been known for at least 30 years (Whiting, 1934). More than fifteen years ago Horowitz and Leupold (1951) had clearly enunciated and used the principle of conditional lethals as specifically applied to temperature-sensitive mutations. They used temperature-sensitive lethals in Neurospora and *Escherichia coli* as a way of meeting an objection Max Delbrück had raised to previous tests of the one gene—one enzyme hypothesis. Delbrück had pointed out that since the mutants of biochemical genetics had been generally selected for their ability to respond to simple growth supplements, those requiring complex growth supplements or affecting indispensable functions would have gone undetected. The studies of Horowitz and Leupold showed that this objection, though logically valid, did not apply since the mutants they isolated, which had been selected simply for inability to grow at the restrictive temperature, generally showed also only a single functional defect. Horowitz's work was well known to all self-respecting microbial geneticists in the 1950's. (I recall being asked about this work during my Ph.D. final examination and at that time not being able to describe it.) Whether the relevance of Hadorn's or of Horowitz's work occurred to Epstein or Steinberg I don't know, but it was sometime after I had been working on temperature-sensitive mutants of T4 that it occurred to me.

Why was there a thirteen-year lag between the papers of Horowitz and Leupold and of Edgar and Lielausis? Unlike the rediscovery of Mendel's work, we cannot blame ignorance. One might argue that during the 1950's the "time was not ripe" for physiological genetic studies of systems other than those concerned with intermediary metabolism. Yet Jacob, Fuerst, and Wollman had by 1957 exploited a rather intractable conditional lethal system in lambda—the prophage defectives—for physiological studies of some value. Sydney Brenner, at about this time, started but abandoned a study of minute mutants of T4. His notion was that these mutants, which form small plaques, represent "leaky" mutations in diverse essential genes. (He was probably correct. Many temperature-sensitive mutants make small plaques at intermediate temperatures.) In both these rather awkward and limited systems the general notion of conditional lethality was implicit.

But independently of physiological studies, conditional lethals are of great value purely as markers in formal genetic studies. They are selective in nature and widely distributed over the genome. Fifteen years ago, phage workers and other geneticists, such as those working with non-prototrophic bacteria like Pneumococcus and Hemophilus, were desperately in need of good markers for formal genetic analysis. Why did the relevance of Horowitz's work not occur to them?

Scientific publications are, in a sense, "fabrications," pieced together to create pleasing stories which, although they are sometimes reflections of nature, are rarely mirrors of the scientist at work. In any case, they are often ignored, except in retrospective construction of bibliographies.

REFERENCES

CAMPBELL, A. 1959. Ordering of genetic sites in bacteriophage λ by the use of galactose-transducing defective phages. Virology, *9*: 293–305.

———1961. Sensitive mutants of bacteriophage λ. Virology, *14*: 22–23.

CAMPBELL, A., and E. BALBINDER. 1958. Properties of transducing phages. Carnegie Institution of Washington Year Book, *57*: 386–389.

EDGAR, R. S., and I. LIELAUSIS. 1964. Temperature sensitive mutants of bacteriophage T4D: Their isolation and genetic characterization. Genetics, *49*: 649–662.

HADORN, E. 1961. Developmental genetic and lethal factors, John Wiley and Son, New York.

HOROWITZ, N. H., and U. LEUPOLD. 1951. Some recent studies bearing on the one gene-one enzyme hypothesis. Cold Spring Harbor Symp. Quant. Biol., *16*: 65–72.

JACOB, F., C. FUERST, and E. WOLLMAN. 1957. Recherches sur les bactéries lysogènes défectives. Ann. Inst. Pasteur, *93*: 724–753.

KRIEG, D. R. 1959. A study of gene action in ultraviolet-irradiated bacteriophage T4. Virology, *8*: 80–98.

WHITING, P. W. 1934. Mutants in Habrobracon, II. Genetics, *19*: 268–269.

IV. *Bacterial Genetics*

Laboratories at Cold Spring Harbor, Long Island

S. E. LURIA
Department of Biology, Massachusetts Institute of Technology, Cambridge, Massachusetts

Mutations of Bacteria and of Bacteriophage

The two studies whose history I am going to retrace—the analysis of bacterial mutation to phage resistance and the early study of phage mutations—have several features in common. Both these studies stemmed from my apprenticeship and collaboration with Max Delbrück. Both had an almost accidental inception, a quality that explains why at the onset they were carried out in blissful and protective ignorance of the past history of the problems involved. And both studies were destined by the peculiar rules of the history of science to play significant roles in the rise of molecular biology, of which they came to constitute one of the main roots, together with the study of nutritional genetics in Neurospora, of type transformation in Pneumococcus and, of course, of the nature and reproduction of viruses, especially phage. More specifically, the studies of bacterial and phage mutations in the years 1942–1945 provided a stimulus and experimental material that attracted Demerec and Hershey, among others, to the feast table of microbial genetics.

PHAGE RESISTANT MUTANTS

Max Delbrück and I met for the first time on December 28, 1940, at a Physical Society meeting in Philadelphia. He was teaching then at Vanderbilt University. After a few hours of conversation (and a dinner with W. Pauli and G. Placzek, during which the talk was mostly in German, mostly about theoretical physics, mostly above my head) Delbrück and I adjourned to New York for a 48-hour bout of experimentation in my laboratory at the College of Physicians and Surgeons. I had received from Dr. Bronfenbrenner of St. Louis two *coli* phages active on the same host: P28, later called α, later T1, and PC, later called γ, later T2. The reason for choosing the names α and γ was that my typewriter had the signs α, β, and γ; β was left out for reasons of symmetry, the common host being called B for bacterium.

Delbrück, who had published with Ellis the first of his papers on phage growth (Ellis and Delbrück, 1939), had recently been struggling with a staphylococcus phage which supposedly could not make plaques, and he was interested in the phage plating methods that I had learned from

Eugène Wollman in Paris. In the two days in New York we made lovely plaques of T1 and T2 and planned to do mixed infection experiments with these two phages, using "secondary growth," that is, phage resistant variants, as indicators for each phage in a mixture.

We also debated where we could carry out our planned collaboration, in Nashville or New York. On January 20, 1941, Delbrück wrote me that he had been invited to attend the Symposium at Cold Spring Harbor and to spend the summer there. Could we work together? "If that could be arranged satisfactorily at C. S. H., I might overcome my antipathy to the place." (This was written, of course, before he spent his honeymoon there that very summer.)

Before summer, I had to isolate the necessary phage resistant strains as indicators for the mixed infection experiments. The origin of these phage resistant strains, which did not adsorb the phage in whose presence they had grown, was puzzling me at that time. In a note from Delbrück dated April 2, 1941, replying to a nonextant letter of mine, I find the remark: "I agree with you that the rise of the secondary culture would be an interesting and attackable problem for future collaboration."

It was not until the fall of 1942, in Nashville, that we returned to this problem, after almost two years busily spent on mutual exclusion, electron microscopy, and even effects of sulfonamides on phage. Due to a number of false leads, Delbrück and I vacillated between the idea that the secondary growth consisted of spontaneous mutants, naturally resistant to phage attack, and the idea that they were survivors of a massive phage attack giving progeny cells that lacked phage receptors. (In retrospect, it is interesting to note that the latter mechanism has some resemblance to the mutation-like loss of cell wall in penicillin-induced L forms.) An experimental decision, based on the properties or the kinetics of formation of resistant bacteria in a growing phage-infected culture, proved hard to reach. The literature (Bail, 1923; Burnet, 1929) provided support for the mutation hypothesis mainly by the correlation of S-R variation with phage sensitivity changes; but this evidence was confounded by the unclear distinction between resistant variants and lysogenic derivatives (which, however, do adsorb the phage to which they are immune).

It was in January 1943, after I had moved to Indiana University in Bloomington, that I thought of a critical choice between spontaneous and phage-induced resistance. The first written statement of it I find in a letter to Delbrück dated January 20: "I thought that a clean cut experiment would be to find out how the fluctuations in the number of α-resistants depend on the culture from which they come. That is: If I plate with α ten samples of the *same* culture of B, I find numbers of resistants which fluctuate according to Poisson's law. If I plate 10 samples of 10 *different* cultures of B, all containing the same amount of B, I find much larger fluctuations. If the resistants were produced on the plate, after contact with α, they should show the same fluctuations in both cases." The idea of this experiment came to me, in fact, while watching the fluctuating returns obtained by various colleagues of mine gambling on a slot-machine at the Bloomington Country Club, where faculty dances were then held one Saturday a month. The first experiment was done on the following Sunday morning. (In a letter dated January 21, Delbrück exhorted me to go to church.)

The U.S. Post Office works swiftly. On a postcard dated January 24, Delbrück replied: "You are right about the difference in fluctuations of resistants, when plating samples from one or from several cultures. In the latter case, the number of *clones* has a Poisson distribution. I think what this problem needs is a worked out and written down theory, and I have begun doing so." The MS of the theory arrived on February 3, accompanied by a letter which also described the first impressions that Delbrück had of Hershey, who had just visited Nashville: "Drinks whiskey but not tea. Simple and to the point. Likes living in a sailboat for three months, likes independence."

So much for historical details. The fluctuation analysis was carried out experimentally on the mutants of *Escherichia coli* B resistant to phage T1 (Luria and Delbrück, 1943). In my judgment, its significance resided in three contributions: (1) It provided adequate evidence that phage-resistant mutants originated by spontaneous mutations, not as a result of phage attack. (This evidence, however indirect, was of the type to convince geneticists accustomed to statistical arguments, although it was long considered inadequate by some physical chemists.) (2) It provided methods for measuring mutation rates through analysis of the distribution of mutants in bacterial cultures. (3) Most important, it made available a quantitative approach to measurements of mutation rates lower by several orders of magnitude than those which had hitherto been open to investigation except possibly in the study of reversions from nutritional auxotrophy in Neurospora. This last contribution was probably the most important, because by opening new approaches to the study of mutations and mutagenesis it attracted attention to the remarkable possibilities of bacterial genetics.

It may be of interest to restate here those essential features of the fluctuation analysis that distinguished it, for example, from the deterministic theory of the accumulation of bacterial mutants in a growing culture under mutation pressure, balanced or not by reverse mutation and selection pressures (Bunting, 1940; Shapiro, 1946). The accumulation of mutants within a clone under a constant mutation pressure only becomes a nearly deterministic function of the cell number N and the mutation rate m when $mN \gg 1$, so that the coefficient of variation $\sqrt{mN}/mN$ is small. For $m = 10^{-7}$,

N must be around 10^9 to satisfy the requirement. Two difficulties arise, however. Cultures should be used such that each sample tested contains at least N bacteria and, more important, such cultures should be started with inocula in which the number of pre-existent mutants is not greater than the "fair share" number, in order not to mask the orderly increase in mutant numbers. But, if the inoculum is small and samples are taken only at late times, then the numbers of mutants are subject to the chance of early fluctuations, which either magnifies or reduces the contribution of the early generations. If the inoculum is large, then its composition will reflect the fluctuations that have occurred in the culture from which it comes. The value of the fluctuation method is precisely that *it makes the fluctuations themselves the basis for the measurement of mutation rates*: because, on the one hand, it measures the number of mutational events from the presence or absence of mutant clones and, on the other hand, it utilizes the average number of mutants by comparing it to the number expected if the fluctuations were not "too bad" (the likely average method). Some limitations of these methods, such as phenomic lag and delay in growth of mutants, have been well recognized later. The chemostat technique (Novick and Szilard, 1951) coming several years after the fluctuation test provided the improvement required for precise quantitative studies on mutation rates.

A little anecdote provides an appropriate conclusion to this section. In July 1947, as we met at the Copenhagen Airport, Delbrück handed me a letter from a Brazilian scientist requesting an airmail answer and pointing out that in our 1943 paper there were some errors in variance calculation: specifically, one value should have been 6883 instead of 6620. This is a good place for me to make amends: For the 1943 paper I calculated all variances using the nearest integer to the mean rather than the mean itself, and then forgot to make the necessary correction. To our Brazilian correspondent and other conscientious readers, my warmest apologies.

PHAGE MUTATIONS

While using phage resistant bacteria as phage indicators I first noticed the occurrence of a few phage plaques when large amounts of phage T2, then called phage γ, were plated on B/2. These plaques yielded phage mutants, which attacked the resistant bacteria. (The same phenomenon was at first not noticed with phage α (= T1) and B/1 for the reason that in most of the early work with mixed phages the indicator for α was a strain that later proved to be B/1,5, not attacked by any mutant of T1.)

The analysis of this phenomenon, which I also initiated in Nashville in the fall of 1942, was delayed by the work on bacterial mutations. It was completed and published some years later (Luria, 1945a). This analysis established the main features of the host range mutants of phage. It had the good fortune of impressing Hershey with the remarkable possibilities of phage genetics in the course of a seminar I gave in St. Louis the day after Thanksgiving, 1943. I remember his comments and that visit to St. Louis most vividly, because I returned to Bloomington with a pneumonia whose sequels kept me in bed for several months. This illness gave me the necessary leisure to read Topley and Wilson's *Principles of Bacteriology and Immunology*, just as service in the Italian army in 1936–1937 had given me a chance to study calculus, and a month's wait for the American visa in Marseilles in 1940 had permitted me to read G. N. Lewis' *Physical Chemistry*.

In my first study of the host-range mutants of phages T1 and T2 (Luria, 1945a) I noticed that some phage cultures contained mutant clones smaller than the total phage burst size, an indication that phage mutants did arise in the course of phage multiplication. Because of technical reasons (low efficiency of plating), I was unable at that time to analyze the precise distribution of host-range mutants in individual cultures or bursts. In the course of a visit to Nashville in 1944, Delbrück pointed out that the distribution of mutants should be like the one of bacterial mutants only if the phage multiplied logarithmically, like bacteria in a culture. If the phage genome were reproduced by successive acts of "stamping" from a unique template, the distribution of mutants in single bursts should be a Poisson distribution. Other distributions could also be considered (Luria, 1945b).

It was only several years later that I succeeded in carrying out a detailed analysis of phage mutant distribution, using the *r* mutants of phage T2 (Luria, 1951). This distribution proved to be precisely the one predicted by the logarithmic replication hypothesis; even more precisely, in fact, than I was able to judge at first, because a complete theory of the expected distribution of clone sizes had to await studies on the kinetics of the phage "genetic pool," generated by replication and limited by phage maturation. The complete mathematical theory was finally worked out by Steinberg and Stahl (1961).

The time sequence of these studies is of some interest. In 1945, when the clones of phage mutants were first noted, and even in 1951, when the distribution of mutant clones was determined, the mechanism of phage reproduction was uncertain. Not only had the Watson and Crick model of DNA structure and replication not yet been proposed, but even the identification of the phage genome with the phage DNA was still questionable, and was actually questioned by myself as well as by others. Only in 1952 did Hershey and Chase perform their classical experiment on injection of phage DNA and its separation from the protein capsid (Hershey and Chase, 1952). Thus, we knew that the phage genome was replicated by a logarithmic process before we knew what it was made of. Despite the clearly recognized fact that phage multiplication occurred by replication of subviral components followed by assembly and maturation, all biological evidence insisted, to the very last, in directing our attention to a "fission-like" rather than a "template-like" process of replication. Only with the Watson-Crick (1953) model was the fission vs. template antinomy dialectically resolved.

TWENTY-FIVE YEARS LATER

Twenty-five years ago, when Delbrück and I first met, we were probably the only two people interested in phage from the point of view of "molecular biology." Our correspondence between 1940 and 1943 dealt, more often than not, with the problem of attracting the interest of geneticists, biochemists, and cell physiologists to the dimly glimpsed green pastures of this promised land. The first phage meeting, in Nashville, Tennessee, March 1947, attracted eight people (M. H. Adams, T. F. Anderson, S. S. Cohen, Max Delbrück, A. H. Doermann, A. D. Hershey, M. Zelle, and myself). Today, annual phage meetings at Cold Spring Harbor attract literally hundreds of participants. Bacterial and phage genetics occupies an increasing share of courses of general genetics as well as of general microbiology. More grimly, these and allied fields have flooded the scientific literature to an extent that constitutes a threat comparable only to that of the population explosion. The growth rate of molecular biologists is much higher than that of humanity.

Is this success? Certainly so, by most standards of measurement. Is it good? By which standard should this or any other human activity be judged? Let it here be judged only from the test of the emotional and intellectual rewards. Of the joy of solid friendships and mutual respect, of the excitement of discovering and teaching ever-growing numbers of brilliant young minds, and even of the sweet torment of friendly competition, we have had our full share. Seldom has a group of men been so richly rewarded as have we, the molecular biologists whom the physicist Max Delbrück, more than any one else, guided to the exploration of the deep mysteries of life.

REFERENCES

BAIL, O. 1923. Versuche über die Vielheit von Bakteriophagen. Zeitschr. f. Immunoforsch. I (Orig.), *38*: 57–164.

BUNTING, M. I. 1940. The production of stable populations of color variants of *Serratia marcescens* No. 274 in rapidly growing cultures. J. Bacteriol., *40*: 69–81.

BURNET, F. M. 1929. "Smooth-rough" variation in bacteria in its relation to bacteriophage. J. Path. Bact., *32*: 15–42.

ELLIS, E. L., and M. DELBRÜCK. 1939. The growth of bacteriophage. J. Gen. Physiol., *22*: 365–384.

HERSHEY, A. D., and M. CHASE. 1952. Independent functions of viral protein and nucleic acid in growth of bacteriophage. J. Gen. Physiol., *36*: 39–56.

Luria, S. E. 1945a. Mutations of bacterial viruses affecting their host range. Genetics, *30*: 84–99.

———1945b. Genetics of bacterium-bacterial virus relationship. Ann. Missouri Bot. Garden, *32*: 235–242.

———1951. The frequency distribution of spontaneous bacteriophage mutants as evidence for the exponential rate of phage reproduction. Cold Spring Harbor Symp. Quant. Biol., *16*: 463–470.

Luria, S. E., and M. Delbrück. 1943. Mutations of bacteria from virus sensitivity to virus resistance. Genetics, *28*: 491–511.

Novick, A., and L. Szilard. 1951. Experiments on spontaneous and chemically induced mutations of bacteria growing in the chemostat. Cold Spring Harbor Symp. Quant. Biol., *16*: 337–343.

Shapiro, A. 1946. The kinetics of growth and mutation in bacteria. Cold Spring Harbor Symp. Quant. Biol., *11*: 228–235.

Steinberg, C., and F. Stahl. 1961. The clone-size distribution of mutants arising from a steady state pool of vegetative phage. J. Theor. Biol., *1*: 488–497.

Watson, J. D., and F. H. C. Crick. 1953. A structure for deoxyribose nucleic acid. Nature, *171*: 737–738.

ROLLIN D. HOTCHKISS
The Rockefeller University, New York, New York

Gene, Transforming Principle, and DNA

Our present knowledge of gene action derives from two revolutions: that in which gene and gene product were recognized as material, and that in which linear template control of catalysis was made manifest. Like all revolutions, they required the surrender of some previous orthodoxy, and have in turn substituted their own, at an accelerating pace. The individualist mistrusts orthodoxy, and the conformist, finding his security in fellowship, embraces it. However much the comforts of scientific togetherness, future advances will likely come from uncovering flaws in the present orthodoxies. The changes can come progressively in quiet logical advances, as did the first part of the present revolution. But when, as now, optimistic philosophies and fashions are presented to a wide and interested public, the togetherness and joys of ready communication constitute a kind of intellectual inertia. Public, scientist, and teacher alike—and even the usually-to-be-counted-on young scholar—find it difficult to pause in the dance, when the rhythm is so compelling. The more impressively the doctrines have been inculcated —whether in a personal or a rational way—and the more widely, the more the next changes seem to have the quality of "discovery," or revolution.

It may be of value to us in our present prospect to reconsider how some of the past steps were achieved, so that we can the better sense by what path, or in what guise, the next discoveries may come. Perhaps those who had only to listen if they would understand gene chemistry and action, and only to find an untried combination of parameters if they would investigate it, can gain insight by trying to reconstruct how the improbable, or unpredictable, early experiments came to be done. This may have somewhat the flavor of restaging old battles, the strategic aspects of which have already been told and retold by the new historians. We shall here mainly indulge in campaign stories, for this veteran is not temperamentally inclined to produce either an epic history or an exhortatory recruiting speech.

To recapture the state of biochemical genetics around 1940, one must think separately of the biochemists' genetic knowledge, which was negligible, and of the geneticists' biochemical knowledge, which, if sometimes marked with great insight, was mainly not functional. To be sure, interest in genetic modification of flower, and animal skin, eye, or hair pigments as enzyme products had developed, and Sewall Wright in particular, as early as 1917, had laid brilliant groundwork for later attention to enzymes. But even for a decade after the work of Beadle and Tatum, the biochemistry studied was mainly the altered chemistry of mutants, not the chemistry of genetic material.

Most important of all, classical genetics was the outstanding example of formal, or theoretical, development in biology. Gene, linkage, recombination, mutation, and even "position" were formal concepts, sharply enough defined that language was self-consistent, so that pure, exact statements could be made. This fact is not sufficiently emphasized, I think. The recognition that a material entity such as *chromosome* manifests the properties assigned to a formalism such as *linkage group*, brought, however, new temptations and opportunities for inexact statement. (Do chromosomes "recombine," or do they exchange segments? Do genes "make" enzymes?) I shall return later to this point, which became more acute when DNA entered the picture.

Biochemistry was outgrowing its primarily descriptive early years, and increasingly learning to "explain," though seldom to predict, biological events. And I say *events* rather than *processes*, for even in the insight of exceptional individuals, there were relatively few levels at which chemistry could be linked to basic, general, biological principles. At a time when the finest details were being added to the understanding of sugar oxidation and various fermentations, for biochemists *biosynthesis* quite universally meant only gain in weight or nitrogen retention, and was ignored by most (with the special exception of Clifton, who concerned himself with both energy release and polysaccharide synthesis). Respiration and nutrition were coming in for refined and broad analysis, and in immunology and enzymology interactions of numerous proteins could be described, though not predicted. Thereby, a concept of information content, "specificity," became well established—a recognition that "globular" proteins could display most precisely adjusted affinities for specific small and large molecules. Periodicities and order, on the other hand, were the domain of the "structural" proteins, along with the polysaccharides. But at that time, proteins that were structural were more or less explicitly not functional, and as for the functional proteins, no enzyme chemist who spoke of denaturation could ever again hold up his head, except by way of cursing. Structure at the subcellular level (30s, 70s, and the whole inventory) was a taboo subject. I still wince when I remember how, about 1944, a truly magnificent biochemist, a constant user of cell-free extracts—as turbid as pea soup and for much the same reason—privately advised me that I was performing a "dirty" kind of experiment when I told him that Albert Claude and I were separating just such extracts of liver by centrifugation into crude mitochondrial and microsomal fractions which concentrated the phosphatase, transaminase, cytochrome oxidase, etc. (If only we had thought of measuring biosynthesis and managed it somehow—within two years I was beginning to measure amino acid uptake into protein; but that was in respiring washed staphylococci instead of cell particles!)

I don't know who, in 1940, could possibly have foreseen that microbiology would furnish the links connecting biochemistry ever after with genetics. Certainly, some excellent biochemists were microbiologically inclined, and the schools of Kluyver and Neuberg had shown that bacteria are vigorous, pure-hearted and diverse in breaking down carbohydrates. And the productive group that started in Wisconsin with Petersen (Tatum was a student there), together with the workers in bacterial nutrition, showed the world that coenzymes are similar in mycetes and man, and were likely to show up as vitamins for any organism that needed help in making them. Certainly, too, bacteria could change their habits with their habitat—but the microbiologists no less than the geneticists considered that they almost always adapted, cell and cell alike, if sometimes inefficiently or slowly. Thus, cultures were "trained" to grow in new media, or in presence of drugs, by successive multiple passages. In spite of their frequent experience that morphological "variation" showed up in single clones or colonies, microbiologists did not find it easy to think that growth properties or fermentations might be clonal mutations. The fantastic potentials of proliferation when there can be a doubling every half hour take these things out of the realm of the purely intuitive understanding that a biologist was likely to rest upon.

The possibility of making an operational distinction between adaptation and mutation with selection only began to seem real to microbiologists around the mid-forties. And then it might only be half learned. As late as 1950, when I was trying to locate genetic markers other than capsule synthesis for possible transfer by DNA, I was still obsessed with a need to keep the new selective traits in full flower. Looking for penicillin resistance, I passed two different stocks, a rough and a smooth one, through penicillin for over seventy passages, until the much insulted pneumococci had a degree of resistance I have never seen again. This helped indirectly because I was able to test a DNA that carried the marks of several successive, and separable, mutations. But the pair of markers I had planned to test for separateness were capsule and penicillin resistance; it was merely overanxiety that led me to keep the cells trained for resistance. Likewise—and here I paid a price—my first mannitol-using pneumococcal strain was at this time nervously maintained in mannitol instead of glucose media. It grew delicately for several weeks, but one long weekend it surrendered its fragile life in the icebox. We didn't get around again to developing and testing this marker until three years later when Julius Marmur had come to the laboratory. That one was destined to show the first linkage, to streptomycin resistance. Of course, I soon reversed my policy of selection, and now it is a hardfast rule of the laboratory that the stock marker strains shall only be exposed to specific selection once in their entire history, for each marker they contain. If purification is necessary, it is done by simple colony selection on indifferent media.

So, back in 1940, most microbiologists were not mathematician enough, and also not biologist enough, to feel called upon to think in genetic terms. The systematic facts and lore of microbiology had been built upon the preponderant principle that when cells divide, like begets like. For any small number of not very selective traits this is true enough.

Geneticists, too, could not get past the objection that one whole bacterial cell took part in making two daughter cells; so they found no sign of the channeling of genetic determinants through such a concentrated stage as a chromosome. It may have seemed to some unfair for them to ask for a mating test to demonstrate the genes in bacteria—but without it, where was the evidence that specific determinants exist which sometimes do, and sometimes do not, manage to gain access to a particular cell? This was of course to be supplied later, first with transformation in 1944, then with bacterial mating, in two more years, and finally, with transduction, in 1952. Mating, in particular, eventually would show that bacterial chromosomes are linear arrays carrying a considerable number of genes. But preceding all these was another giant step, the demonstration by Luria and Delbrück in 1943 that mutations of bacteria are discrete, spontaneous events.

To many of the general attitudes I have mentioned there were, of course, individual exceptions; nevertheless, I have tried and will try to describe attitudes and expectations that were expressed in private and speculative discussion, where people were presumably most free. But it is time to concentrate more closely on the development of DNA transformations, and I shall be ambitious enough to try to portray Dr. O. T. Avery's views on the subject he did so much to open up, as I saw them, intermingled with views and colorations that will probably be more peculiarly my own. (For a more detailed essay on Avery's contributions see Hotchkiss, 1965.)

The record of Griffith's transformation of pneumococcal types within the infected mouse is adequately available (Griffith, 1928; Neufeld and Levinthal, 1928). It seemed at first to Avery a contradiction of the type stability he had painstakingly demonstrated, but Neufeld's quick confirmation is said to have convinced him. In a series of investigations in his laboratory (Dawson, 1930; Dawson and Sia, 1931; Alloway, 1932, 1933) Martin

Dawson, who had previously been working on the selective effect of antisera on pneumococcal variation, collaborated on the conversion of the in vivo to an in vitro transformation. Alloway, who first produced a cell-free active transforming extract, was by all accounts more of a quiet, lone worker, but working in the same room with Avery, it is beyond question that he was, like all of us, spurred, stimulated and advised at many major turns in the road by his "Professor."

It is noteworthy that each of the successive investigators evaluated his transforming agent in terms of the concepts then significant in Avery's evolving picture of the pneumococcus. The antiserum selection was an outgrowth of attempts to see the forces acting on the bacteria in an animal host. Griffith, working in England, used it to seek "remnants" of the capsule forming system in killed bacteria (and brought to light the genetic determinants instead). Dawson felt that the active agent might be the capsular substance itself; Alloway a little later felt that it might be the protein-containing full capsular antigen. By 1935, when I arrived in the laboratory to help Walther Goebel with the immunochemistry of polysaccharides, Avery was telling his intimates of the separation of this protein antigen too, from the transforming activity. His engaging discourses on this topic fired me with a strong wish to work on transformation, and in 1938, returning from a year abroad, I asked permission to do so. But Avery asked me to wait, and soon World War II was occupying our attention and efforts.

For a few years little was done on the topic but it was always prominent among the subjects of the brilliant hour-long, highly organized monologues he would bestow on his colleagues, perhaps in place of public lectures, which he assiduously avoided. Others were on such topics as the nature of the intact type-specific antigen of pneumococcus, virulence, serum changes during infection, and serological specificity. In all of these years, the discourse on transformation was flavored like good cooking with opposite tastes, optimism and frustration, the latter expressing the delicate irreproducibility of the early transformations ("many are the times we were ready to throw the whole thing out of the window!").

AVERY ON TRANSFORMATION

By 1941–1942, Colin MacLeod had further refined the dilution end point assay of the transforming agent so that purification could be followed. Then Maclyn McCarty, a pediatrician with excellent biochemical skill arrived to help with the chemistry. The attitude which prevailed when the nature of the transforming agent began to clear up was one of extraordinary caution and responsibility. The manuscript (Avery, MacLeod, and McCarty, 1944) which finally was submitted in November 1943, was handed around to a few associates—for months, it seems to me now—with a genuine eagerness for adverse criticism. I hope that a rather extended quotation from Avery himself will convey the spirit of that period.

In May 1943, Avery wrote his brother Roy Avery, a medical bacteriologist working in Vanderbilt University Medical School, about the state of his work on transformation. (I am indebted to Max Delbrück for a copy of this letter.) For a letter between brothers, it shows the caution of his reasoning and even his informal speech, and perhaps will suggest the graceful touches of fantasy that lay close below his elegant expression.

Indicating that he was going to continue his research and not yet retire to join his brother, Avery continues:

I have not published anything about it—indeed have discussed it only with a few because I am not yet convinced that we have (as yet) sufficient evidence.

It is the problem of the transformation of pneumococcal types. You will recall that Griffith in London, some fifteen years ago, described a technique whereby he could change one specific type into another specific type through the intermediate R form. For example: Type $II_R \rightarrow$Type III. This he accomplished by injecting mice with a large amount of heat killed Type III cells together with a small inoculum of a living R culture derived from Type II. He noted that not infrequently the mice so treated died and from their heart blood he recovered living encapsulated Type III pneumococci. This he could accomplish only by the use of mice. He failed to obtain transformation when the same bacterial mixture was incubated in broth. Griffith's original observations were repeated and confirmed both in our laboratory and abroad by Neufeld and others. Then you remember Dawson with us reproduced the phenomenon in vitro by adding a dash of anti-R serum to the broth culture. Later Alloway used filtered extracts prepared from Type III cells in the absence of formed elements and cellular debris and induced the R culture derived from Type II to become a typical encapsulated III pneumococcus. This you may remember involved several and repeated transfers in serum broth, often as many as five or six before the change occurred. But it did occur and once the reaction was induced, thereafter without further addition of the inducing extract, the organisms continued to produce the Type III capsule that is to say the change was hereditary and transmissable in series in plain broth thereafter.

For the past two years, first with MacLeod and now with Dr. McCarty, I have been trying to find out what is the chemical nature of the substance in the bacterial extract which induces this specific change. The crude extract of Type III is full of capsular polysaccharide, C (somatic) carbohydrate, nucleoproteins, free nucleic acids of both the yeast and thymus type, lipids, and other cell constituents. Try to find in the complex mixtures the active principle! Try to isolate and chemically identify the particular substance that will by itself, when brought into contact with the R cell derived from Type II, cause it to elaborate Type III capsular polysaccharide and to acquire all the aristocratic distinctions of the same specific type of cells as that from which the extract was prepared! Some job, full of headaches and heartbreaks. But at last *perhaps* we have it. The active substance is not digested by crystalline trypsin or chymotrypsin, it does not lose activity when treated with crystalline ribonuclease which specifically breaks down yeast nucleic acid. The Type III polysaccharide can be removed by digestion with the specific Type III enzyme without loss of transforming activity of a potent extract. Lipids can be extracted . . . and . . . the extract can be deproteinized . . . until protein free and biuret negative. When extracts, treated and purified to this extent, but still containing traces of protein, lots of C carbohydrate, and nucleic acids of both the yeast and thymus types are further fractionated by the dropwise addition of absolute ethyl alcohol, an interesting thing occurs. When alcohol reaches a concentration of about 9/10 volume there separates out a fibrous substance which on stirring the mixture wraps itself about the glass rod-like thread on a spool and the other impurities stay behind as granular precipitate. The fibrous material is redissolved and the process repeated several times. In short, this substance is highly reactive and on elementary analysis conforms *very* closely to the theoretical values of pure desoxyribose nucleic acid (thymus) type (who could have guessed it). This type of nucleic acid has not to my knowledge been recognized in pneumococcus before, though it has been found in other bacteria.

Of a number of crude enzyme preparations from rabbit bone, swine kidney, dog intestinal mucosa and pneumococci and fresh blood serum of human, dog, and rabbit, only those containing active depolymerase capable of breaking down known authentic samples of desoxyribose nucleic acid have been found to destroy the activity of our substance—indirect evidence but suggestive that the transforming principle as isolated may belong to this class of chemical substance. We have isolated a highly purified substance of which as little as 0.02 of a microgram is active in inducing transformation. In the reaction mixture (culture medium) this represents a dilution of one part in a hundred million—potent stuff that—and highly specific. This does not leave much room for impurities but the evidence is not good enough yet.

In dilutions of one to a thousand the substance is highly viscous as are authentic preparations of desoxyribose nucleic acid derived from fish sperm. Preliminary studies with the ultra centrifuge indicate a molecular weight of approximately 500,000—a highly polymerized substance.

We are now planning to prepare new batches and get further evidence of purity and homogeneity by use of the ultracentrifuge and electrophoresis. This will keep me here for a while longer. . . .

If we are right, and of course that is not yet proven, then it means that nucleic acids are not merely structurally important but functionally active substances in determining the biochemical activities and specific characteristics of cells and that by means of a known chemical substance it is possible to induce predictable and hereditary changes in cells. This is something that has long been the dream of geneticists. The mutations they induced by X ray and ultra violet are always unpredictable, random, and chance changes; if we prove to be right—and of course that is a big if—then it means that both the chemical nature of the inducing stimulus is known and the chemical structure of the substance produced is also known, the former being thymus nucleic acid, the latter Type III polysaccharide, and both are thereafter reduplicated in the daughter cells and after innumerable transfers without further addition of the inducing agent and the same active and specific transforming substance can be recovered far in excess of the amount originally used to induce the reaction. Sounds like a virus—may be a gene. But with mechanisms I am not now concerned. One step at a time and the first step is what is the chemical nature of the transforming principle? Some one else can work out the rest. Of course the problem bristles with implications. It touches the biochemistry of the thymus type of nucleic acids which are known to constitute the major part of chromosomes but have been thought to be alike regardless of origin and species. It touches genetics, enzyme chemistry, cell metabolism and carbohydrate synthesis. But today it takes a lot of well documented evidence to convince anyone that the sodium salt of desoxyribose nucleic acid, protein free, could possibly be endowed with such biologically active and specific properties and that is the evidence we are now trying to get. It is lots of fun to blow bubbles but it is wiser to prick them yourself before someone else tries to.

Most of what follows will center around the two principal questions we were concerned about in 1945, the true chemical nature of transforming agent, and the genetic significance of bacterial transformation. Something ought to be said, however, about the conditions and efficiency of the transformation process, which affect one's idea of mechanism. I am sure that Harriett Ephrussi-Taylor, who has been as interested in the theoretical meanings of transformation as I have been and occupied with the genetic-chemical side of the work even a little longer, agrees that this has been a constant concern throughout the years, in spite of the questions one wants to ask of the system. Many others starting out on transformation (as well as Tomasz and I within the last year) have found efficiency and competence to demand their attention. I well recall my early scepticism about this. Once Maclyn McCarty and I were going home from our laboratories resplendent in Navy uniform; it must have been in the winter of 1943–44 while their classic paper was still in press. I implied with some feeling that I thought it was time to test the ability of DNA to transfer markers other than capsular antigen type. Mac, with his unvarying good nature, said probably so; he had been thinking of trying a virulence trait, also a glutamine independence in some strains, but it was really necessary to work out more things about the protein requirements of the transforming environment. Two or three years later I was working on transformation (after McCarty turned toward rheumatic fever investigation) and did not feel so condescending having come around to his point of view. For two years I was greatly occupied with showing that serum albumins are the principal cofactors for pneumococcal transformation.

Another tale hangs here. Since many serum samples, human and animal, were not effective in furthering transformation, while pathological serous exudates and some tissue extracts usually were, McCarty had been looking for an abnormal protein. Crude globulin fractions had been more active than crude albumin fractions. By the time proof was read on their article (McCarty, Taylor, and Avery, 1946), however, I had shown with Harriett Ephrussi-Taylor that normal serum albumins, crystallized or adequately separated from some (possibly lipid) antagonist, are active, and hence active material could be prepared even from inactive whole sera.

Their paper frequently mentioned the earlier active globulin preparations. I was able to convince Avery that these statements, while correct in fact, were wrong in principle. He refused to borrow our facts to decorate their paper, but neatly protected the readers by changing every "globulin fraction" in the manuscript to "protein fraction." We never published more than an abstract (Hotchkiss and Ephrussi-Taylor, 1951) on our albumin factor itself, and for some years people spoke either in frightened or scornful tones, according to temperament, about the supposed complexity of our environments.

TRANSFORMING AGENT AND DNA

The association of transforming activity with DNA brought two very different kinds of confusion or disagreement, which had larger semantic components than it first seemed. One derived from the difference between inclusive and exclusive connotations of the designation DNA, and the other from the difficulty mentioned before of equating a material, DNA, with a concept, the gene. Both troubles come from the little verb, *to be*. I am sure we conscientiously tried even in speech to speak only of "activity associated with DNA" or "an agent of DNA nature" (Avery et al wrote, "largely, if not exclusively, composed of DNA"). But if we said, or were misquoted as saying, the "transforming agent *is* DNA," who (except Charity herself) would feel that maybe only the *inclusive* meaning (of DNA nature or DNA-containing) was intended, when it was so much fun to criticize the *exclusive* meaning (made of DNA and nothing else)? In law courts, of course, it is acceptable to argue about another man's mere words, and to try to show that they mean something inconsistent with both nature and manliness, but in science it wastes time and space.

The statement "The gene is DNA" is even worse because, I think, an abstract formalism can never be made *identical with* a piece of matter—except as a figure of speech ("life is just a bowl of cherries," they used to say). I think I have been fairly consistent in trying merely to *relate* abstractions to concrete nouns (gene-like activity; linkage manifested by a DNA preparation; even mutation resulting in an altered protein—certainly not "producing" it). At a later period, I maintained that "messenger" was the last great genetic abstraction and ought not to be equated to both collective hypothetical and individual actual pieces of RNA. As for DNA and gene, we don't even know how big they are and where they terminate. Such a thing as "a DNA" is either another formalism, or possibly a very individual molecular fragment.

Some of the objections to transforming agent as DNA were concrete enough. The one that it might be a protein dependent for its activity or stability on DNA had operational meaning but required a gratuitous extension beyond anything already known about the subtlety of proteins. We in the Avery laboratory were concerned throughout with the possibility that traces of very active protein might account for transformation. My own respect for proteins owed very much to long hours of fascinating learning from Alfred Mirsky during the thirties. Quite on my own, then, I felt the same doubts he did: that the nitrogen-phosphorus atom ratios of nucleic acid and protein could vary only as much as the phosphorus—that DNase, purified, in fact all but discovered by McCarty out of a proteinase-rich pancreas fraction, might still have mild proteinase action. Mirsky spoke about these objections, but not very much to Avery's group or he would have learned as I did how eager they were to see the search for traces of protein continued.

In 1947, I began the group analysis of transforming agent, looking for purines, pyrimidines, amino acids, and deoxysugar, instead of elementary analysis. Amino acid microanalysis was not yet very specific, so I determined total ninhydrin-reactive compounds. There was very little, and I could show that that little was liberated on acid hydrolysis slower than amino acids from peptides, in fact at the same rate as glycine is formed from the decomposition of adenine. The fast liberated amino acid could amount to no more than 0.2% of the nitrogen, if indeed there was any. But when I looked for specific amino acids by the brand-new qualitative techniques of paper chromatography, I couldn't put enough hydrolysate on the paper to show even the glycine spot! But all of that white paper bothered me, and I reasoned that the butanol must have carried something down the paper strip. I cut it up into pieces and sure enough there were UV-absorbing substances; in fact four pretty good spots. Our germicidal UV lamp did not give me any useful photographs or other visualization, and anyway, cutting out strips blindly and soaking out the purines and pyrimidines gave me a quantitative spectrophotometric assay. Just then the first note from Chargaff's laboratory appeared, but I judged that they were not interested in quantitation, for they used UV-absorbing solvents, and visualized purines only as mercury or lead salts. So I went on for a while to develop a quantitative paper chromatographic method for purines and pyrimidines —and so did Chargaff.

Crude purine and pyrimidine base patterns for transforming agent of pneumococcus, quite different from those of calf thymus DNA (which showed the first "minor base" constituent, 5-methyl cytosine), were presented in some detail at a symposium in Paris (Hotchkiss, 1949). The adenine analyses were later seen to be low, but the characteristic slow acid decomposition of adenine to make glycine was made a virtue, too, for it gave a kinetic model for the slight, slow release of amino acid from the transforming preparations. I was also fortunate enough to have by now

Kunitz' crystalline DNase to inactivate transforming agent. That conference, arranged by Lwoff and Ephrussi, included early reports of Monod's studies of enzyme induction, of Ephrussi's cytoplasmic petite variants of yeast, of Harriett Ephrussi-Taylor's two-step transformations, of Brachet and Jeener's fractionation of cytoplasmic particles, and was punctuated by Delbrück's theoretical model of alternative metabolic pathways which by cross-inhibition would seem "adaptive"—and other good things. I was naive enough to suppose that a story translated into flowing French and sponsored by the Pasteur Institut and the C.N.R.S. amounted to a veritable publication. A year later, before I had realized that I almost never would find anyone who had read the symposium article, I was distressed to find that a tired abstractor for Chemical Abstracts had covered my own and also Ephrussi's conference papers in two short words, "a review"! I suppose the considerable, mostly new, data in both papers were too concrete to ignore and too concentrated to abstract, so there you are: a review. Ephrussi soon published elsewhere (in French, so the question of language is controlled), but I did not until another symposium (Hotchkiss, 1952) forced me to.

Sometimes the so-called "simplest" hypothesis is the offspring of a wedding of enthusiasm with ignorance. If genetically active DNA was heresy, it was accompanied by evidence, and was opposed by ideas that had become ingrained with little evidence (heresy *versus* hearsay, almost). Avery's own earlier career had centered considerably around the demonstration, against prevailing doctrine, that immunological specificity of antibodies could sometimes be directed toward polysaccharide as well as toward protein antigens. But the antibody is of protein nature, and all textbooks of the 30's and 40's told one that biological catalysts are proteins, and that nucleic acids are obscure but repetitious assemblages of sugar, nitrogen bases, and phosphate. How could something as specific as a transforming agent for antigen type be other than a protein? Many knew of the pitfalls—of the surprise and embarrassment good chemists have experienced at various times when an enzyme they have been pursuing is suddenly purified so much that its solution now seems to be "protein-free" to ordinary test, although its activity is higher than ever. A non-protein form of pepsin had once been reported by Willstätter, for example.

It would take such a small amount of "active protein" if a few molecules could function as transforming agent! I had been stating in all my lectures since 1949 that the protein content of one of our purified active preparations was below 0.02% (at most). But I was to face many a biologist or chemist who, with the authority of textbooks on his side, would demonstrate once again how you can get a large number by multiplying almost any tiny fraction by Avogadro's number. The flavor of one such confrontation is partly preserved in a condensed version of the interchange between the cytologist Kenneth Cooper and myself, published with a symposium talk on transformation at Oak Ridge (Hotchkiss, 1955). Others have told me that this was just about the last appearance of the protein hypothesis. Cooper was persistent; but to me, he seemed professional and gentlemanly throughout. Perhaps I was all but on his side after he summarized his discussion with the question that has recurred to me ever since as the haunting, baffled outcry of the conservative rationalist. Speaking of the association of genetic activity with DNA, he asked whether this was really based upon evidence or was merely "a voting agreement." How often I have wondered which of our ideas take root merely because it becomes impractical and then impolitic to take up the effort of questioning them!

At the Phosphorus Symposium of 1952 I presented the 1949 data showing that the only amino acid released from transforming agent is quantitatively accountable as glycine, from decomposition of the adenine. Zamenhof at the same symposium also described evidences that *Hemophilus* transforming agent is inactivated approximately in keeping with the loss of viscosity brought on by graduated doses of heat, acid, alkali, or DNase.

It is interesting that T-coliphages (by 1952, a support for genetic roles of DNA) were earlier used against transforming DNA. If one phage particle could, as shown by Delbrück and Luria, establish an infection, the gambit ran, why could not one or a few (unknown) protein molecules? Luckily, Alfred Hershey began to ask an almost opposite question; if phage contained DNA—and that was known!—why couldn't that be its active component? I remember that Hershey borrowed a couple of my lantern slides showing how little amino acid there was in transforming DNA, in about 1950, for a lecture somewhere on DNA as genetically active substance. He was getting ready for his classic demonstration with Martha Chase, that DNA is the active component of coli T-phage. Of course, those results became available in 1952, and brought the stimulating value of an independent demonstration. I have always felt that their case was more supported than ours by the textbook simplifications: one could read that proteins contained sulfur and no phosphorus, while DNA contained phosphorus and no sulfur. The first statement is of course too broad, and the exotic DNA's some had invented to explain transformation might well have contained sulfur! Hershey simply stuck to what he knew, and that was good enough: not more than 20% of the sulfur, and at least 85% of the phosphorus, constituted the inside story of phage. Both components were later separated in considerably purer form, something like 97% DNA in the naturally injected form. We dealt with DNA of much higher purity—even this last figure allows room for 150 times more protein than our maximum —but isolated DNA acts by random mass action laws and we could not approach the high efficiencies of infection obtained with phage.

Most people at that time had learned—too well—the textbook doctrine that DNA was a simple repeating polymer. But I must testify that I do not recall ever seriously believing that DNA's were composed of Levene's assumed "tetranucleotide unit." It always seemed obvious that in group analysis for purines or pyrimidines, the organic chemist might round off a 0.6 or a 1.2 frequency ratio to 1.0, but to do so for large molecules circularly involved a part of this assumption of simple constitution. The differences in base composition which my 1948 chromatography indicated for pneumococcal and calf thymus DNA's did not enchant me for long, for two reasons: Chargaff and collaborators were very actively accumulating much more data on DNA of different species; also I doubted whether I would be likely to measure significant differences in base composition within the DNA of differently marked strains of the same species.

My personal conception of the differences in DNA's were that they were conformational like those of proteins, made possible by compositional differences and brought about by precise control (perhaps by 3-dimensional templates) of secondary structure. From 1946–48 I spent considerable effort looking for subtle shifts in UV absorption spectra of differently marked DNA's (within and between species) occasioned by addition of various divalent and trivalent metal cations and organic bases. The most pronounced shift I observed was with dilute mercuric salts—a slowly developing effect, not entirely reproducible. I eventually could attribute it to slow penetration of sulfur-containing New York City laboratory air, forming some type of mercuric sulfide, colloidal when DNA was present. The principal interaction I did discover in about 1947 was DNA denaturation by alkali and acid—the hyperchromic shift, concomitant with biological inactivation, and also being discovered by Kunitz for DNase hydrolysis, used in his eventual crystallization of McCarty's purified pancreatic enzyme. I found that the DNase and alkali hyperchromicities were the same, mutually exclusive rather than additive, and discussed eagerly for a while the idea that the secondary structure of DNA was "hiding" the UV absorbing groupings behind each other. Not finding anybody very interested in my slender evidences as I talked hesitantly about them, I characteristically drew the conclusion that it would not be much use to publish it. I recall how Alexander Todd on one occasion easily suppressed what confidence I had by assuring me that DNA certainly would be found to contain covalent bonds labile to pH. The lesser effect (after cooling) of heat denaturation also disturbed me. By 1951–52, Thomas, Shack, Frick and others had begun to report spectrometric monitoring of DNA denaturation and later the several structural denaturations could be identified with, or related to, each other.

The Watson-Crick structural theory for DNA seemed attractive and acceptable when announced, and impressively convincing after one could learn something of the reasoning and the numerous unsatisfactory models discarded. But their corollary genetic theory of semi-conservative replication seemed for me prematurely accepted. I had had a personal experience with complementary models just before that. When returning on the *Liberté* from the Biochemistry Congress in Paris in 1952, I was invited (by Steve Zamenhof) to the captain's dinner up in cabin class. The ship's busy master of organized fun tried to get the party going by a pretty device. One man and woman were asked to dance and at a musical signal the pair(s) would repeatedly separate and each time choose a new partner from the energy-rich audience. The room rapidly filled with exponential replicas of the original pair. For me it was one of those proverbial moments of revelation—the perfect mechanism for genetic replication! The complementary halves were, of course, DNA and protein; each was gene-specific and able to induce formation of its opposite. Protein as transforming agent from outside would have a harder time than DNA and of course was not yet demonstrated; it was still miraculous enough that DNA would sometimes work. Almost none of us knew until later that complementary replication models, even one involving DNA and protein, had already been speculatively discussed, by Pauling and Delbrück among others. But I was sufficiently aware of the charm of my rumination to be rather amazed when less than a year later the simplicity of the same principle seemed to be taken as virtual demonstration that the Watson-Crick DNA strands actually replicate this way. Perhaps I was stubborn because I was pleased with the structural model and thought my experience with DNA denaturation indicated how relatively difficult strand separation would be. Max Delbrück himself did not feel this block: he notified the expected participants that the just published genetic theory of Watson and Crick would be required reading before the Cold Spring Harbor Symposium of June 1953 and passed out reprints. The genetic theory got off to a good start and discussion was stimulating. It did take a few years before Meselson and Stahl were able to do the experiment attempted by Mazia and to contribute elegant support to the genetic model. The relief generated by their demonstration was noticeable and gratifying, even though the intervening period was hardly one of anxiety.

TRANSFORMING AGENT AS GENE

It has always seemed to me that the first stage in making a discovery is the optimistic one in which faint evidence and speculation are encouraged; a stage shared with one's research associates and those in a position to judge (but maybe not the public!), and playing the role of guiding one into profitable channels, developing new ideas, etc. The second is consolidation by the investigator himself, the one best able to see some of the objections and alternatives, a stage during which he should be over-critical, in essence preparing the communication which will inform, but responsibly not misinform or overinform, all people able to use or enjoy the conclusions but not fully able to evaluate them. Of course, if one is patient or thorough, the startling quality of "discovery" will tend to disappear.

It was one thing to recognize the heritable and gene-like aspect of the capsule transforming agent, and quite another to know what biological generalization could responsibly be drawn. Were traits transferable in certain bacteria only, all bacteria, or potentially all organisms; most urgently, were other than antigenic traits transmissible, and if so, would DNA be the active agent for all? Avery, MacLeod, and McCarty (1944) mentioned the gene and virus analogies to transforming agents by brief reference to other's comments, but would not before the public dwell on the possibilities. Avery collected with great interest literature notes on the reactions evoked by Griffith's and his own later paper, and in 1954 gave me these notes. Briefly, they record some early speculations on the specificity of nucleic acids and several references to Griffith's experiments (mostly those given in their paper), and show that within months Sewall Wright (1945), Marshak and Walker (1945), and G. E. Hutchinson (1945) had pointed to the DNA as a chromosomal fragment *acting a genetic role*. Sonneborn about then (1943) held that the agent might activate different antigenic states produced by genes already present. Muller (Muller et al, 1947) thought that a gene might reflect both a specific DNA and a protein, but was worried about the protein contamination still reputed not to be ruled out of the DNA. Later on, Beadle (1948) spoke of transformation as an effect upon genes, a "specific transmutation." Much later, Lederberg (1956) retrospectively listed seven possible interpretations of transformation, but some of them—plasmagene, action through a distance, virus conversion, and unique idiosyncrasy—had not actually had sufficient weight to be appreciably considered or to persist.

The autocatalysis of capsule synthesis by polysaccharide was easiest to rule out. Not only did polysaccharide fail to transform, or to be demonstrable by sensitive test in the DNA, but it should not have worked anyway! Known primers furnish suitable end groups for starting a biosynthetic chain. The structure of the product is determined by the enzyme, and in some cases by a template (the whole of its structure—not as in a primer by its ends). Recognition of template function—the second great revolution referred to at the beginning—was appropriate enough to DNA, but it was to have another meaning. And even there confusion could be noticed: for "priming" of DNA polymerase, the best workers used the best high molecular DNA. It was really a good template, but for a long time the purest polymerase had to carry an ingenious DNase, needed to convert part of it into a primer but also making inroads upon the product. The semantic distinction leading to the step of adding both a small primer and a large template was resisted until 1964. I pressed the lore of polysaccharide priming and purity of our DNA upon M. Stacey in 1947 as he visited Avery's laboratory, but the next day in a seminar he still "explained" transformation as autocatalysis; I was amused, but speechless! The counterarguments had to be given again at a symposium in England next year by Ephrussi-Taylor, and that helped to end the matter.

All of the alternative ideas about transforming agents, including the abstruse ones, were in large measure ultimately eliminated by obtaining other specific transformations—but the path to that was by no means as easy as nowadays. MacLeod and McCarty tried to transfer pneumococcal virulence for rabbits and found it somatic (meaning that it was not co-transferred with the capsule trait). Occasional transfer of M-protein antigen specificity was observed by Austrian and MacLeod (1949) but at first only in vivo, where the capsule trait was also involved; salicin fermentation (Austrian and Colowick, 1953) seemed to be transferred but could only be detected after serial passage. These traits, and those first used in *Hemophilus* by Alexander and Leidy (1951), or in *E. coli* by Boivin (1947), as well as the pneumococcal intermediate capsule factors of Ephrussi-Taylor (1951) allowed at best only indirect selection of the (mostly antigenic) traits. Certain of the cell types could be counted on plates after a vaguely selective growth, but untransformed cells could not be efficiently separated or eliminated.

We found that the classical rough to smooth transformation required the delicate agglutination of rough cells by anti-R antibodies—not too soon, or they would carry down potential transformants, and not too late or transformants would be suppressed. The inoculum and conditions specified by Avery gave spontaneous agglutination at a favorable time, but by experimental control of the agglutination (Table 1) the yield of transformants could either be increased considerably or apparently blotted out. The principles thus learned were applied (Figure 1) in 1950 in what I believe is the first attempt to measure the "expression" process of an introduced gene. It suggested that between 1 and 2 hours were required to complete a capsule coat rendering a cell insusceptible to anti-rough cell antiserum.

William Atchley, a fundamentally oriented M.D., the first postdoctoral associate who chose to work in my laboratory, simultaneously chose *E. coli* as his target. We agreed to allow only three months to try transforming it, but during 1950–52, Atchley persevered for some year and a half, long after all easy experiments had been done. This singleness of purpose was displayed just at the time pneumococci, in the same laboratory room, became temptingly available to quantitative transformations with our new resistance markers.

Table 1

SELECTIVE ACTION OF ANTI-ROUGH CELL ANTIBODY IN CAPSULAR TRANSFORMATION

Treatment of culture	*Amount of capsular transformation found after 24 hours*			
	Undisturbed culture	*Culture mixed at time indicated*	*Lightly centrifuged supernate*	*Sedimented cells resuspended in 7 hour supernate*
Culture manipulated at				
3½ hours	+	+	–	
5 hours	+	–	–	–
6 hours	+	–	±	–
7 to 11 hours	+	–	+ +	–
12 hours	+	(–)	+ + + +	±
14 hours	+	+ +	+ + + +	+ ±
7 hour supernate mixed			+ +	
At 7 hours, mixed and			Notes:	
centrifuged at 7 hours		+ + +	+ + etc. = C growth	
centrifuged at 8 hours		+ +	from $10^{5.5}$ to 10^{8} cells/ml	
centrifuged at 9 hours		–	(–) = 10^{3} to 10^{5} C cells/ml	
not centrifuged		–	– = less than 10^{3} cells/ml	
Antibody added late at			Reaction with DNA:	
0 to 2 hours	+ +		aliquots of rough culture	
3 hours	±		all treated with C (capsule)	
4 to 7 hours	–		DNA at 4 hours	

Conclusions: Anti-R is needed throughout the standard Avery et al. (1944) transformation for adequate selection of C (capsule) transformants (column 2). If after transformation has begun (col. 3) at any time until after 12 hours, the clumped R cells are redistributed, they interfere with development of C cells, unless promptly recentrifuged (bottom col. 3), or unless growth is so heavy that spontaneous resedimentation is prompt (col. 5). If centrifuged after mixing (col. 4) potential C cells are carried down at early times; after about 7 hours (3 hours after DNA exposure) C cells will be left in the supernate in increasing amounts and yield considerably surpasses that of standard transformation. (Hotchkiss, 1950, unpublished.)

Our brave and heroic early attempts to get DNA into *E. coli* were armed with good DNA prepared from cells lysed with warm deoxycholate. We used three lines—B, K, and W—as both donors and recipients, in all sorts of environments, looking for about six biochemical and phage resistant markers from the growing catalogue. To offset suspicions of mutation we were prepared to support the transformations that never happened by making two-factor transfers. (The other half of the lab was just about to show that most factors transformed singly!) The most heroic of all our plans was to carry DNA in simultaneously with phage—but we were not destined to be the discoverers of transduction or virus-helped transformation, for the only viruses we thought of trying were lytic virulent phages. Our hopes that DNA might save the infected cells, or catch hold of those in process of mutation to virus resistance, went to roost in dreamland with other discarded, naive hopes.

Effect of Centrifugation and Resuspension of Transformed Culture

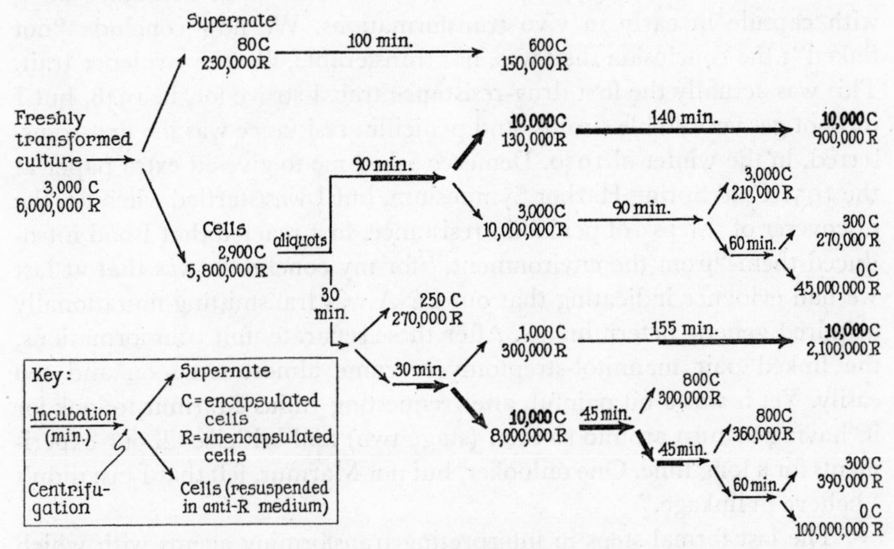

FIGURE 1. An early experiment on the rate of expression of transformation for the capsule marker, Type III, in Avery transformation medium containing rough cell agglutinins. The heavy lines and numerals indicate where the bulk of the Type III transformants (C cells) are to be found, in this and supplementary experiments. The nascent transformants were remixed at various times with the rough (R) recipient cells and then the R- anti-R clumps were lightly centrifuged out at various intervals. In general it appears that between 60 and 90 minutes after DNA, the C transformants begin to escape antiserum agglutination with the R cells and pass into the supernate, but some entrapment continues until nearly 180 minutes. When held in the presence of large numbers of R cells in either precipitate or supernate, the C cells fail to develop or may virtually disappear (from Hotchkiss, 1950, unpublished).

Since we did find out that markers were for the most part transformed singly by DNA, it is now difficult to recapture how the DNA was viewed in relation to the genome back in the late 1940's. For those who thought other traits would be transferable, I think it is fair to say that most supposed each transformant would acquire all the donor's traits. I ran into that assumption often, and I do not recall ever hearing doubts about the question, until unit genetic transfers were described in 1951. I can remember describing to Harriett Ephrussi-Taylor just before she left Avery's laboratory in 1947, my intentions to get a doubly or triply marked donor to test this. I recall well her responding in pleasure saying "That's interesting—

so you think the factors may separate in the DNA?" Memory does not now inform me whether I answered perhaps or probably—or whether her surprise seemed to be because I asked this question or perhaps because she had been thinking of it too. She did have two step transformations of rough or filamentous cells to encapsulated within a couple of years—recombinations we now know, but qualitative and in markers sufficiently obscure in origin and nature that the mechanism seemed questionable at the time.

Langvad-Nielsen (1944) tried to introduce sulfonamide resistance along with capsule in early in vivo transformations. We now conclude "not linked"; the conclusion then was, not transferable, like the virulence trait. This was actually the first drug-resistance trait I strove for, in 1948, but I did not get very stable strains, and penicillin resistance was the first transferred, in the winter of 1950. Demerec asked me to give an extra paper at the 1951 Cold Spring Harbor Symposium, but I was startled when he, the discoverer of "steps" of penicillin resistance, first reacted that I had introduced them "from the environment," for my conclusion was that at last we had evidence indicating that our DNA was transmitting mutationally acquired genetic determinants. After these separate unit transformations, the linked pair mannitol-streptomycin came almost too soon and too easily. Yet it was a bit painful, after requesting Julius Marmur to look for it, having to turn around so soon (stage two) and criticize all our experiments for a long time. One onlooker, but not Marmur, felt that I just didn't "believe in linkage."

The last formal steps in interpreting transforming agents with which I was connected were probably those which involved the "inevitability" of linkage. Dorothy Lane and I showed that both alleles of mannitol (+ and −) and streptomycin (resistance and sensitiveness) could be linked, in the three assessable combinations. Finally, we built up a series of some 15 strains bearing this pair of markers, with various other transformations interposed between. I wanted to see whether "linkage" would associate any last two introduced markers no matter what they were—as it might if external DNA went to an adventitious site (the end of the chromosome, or perhaps some cytoplasmic or membrane site). Probably this possibility sounded wilder in the 1950's than it does now, after a multiple resistance determining particle has been found in coliform bacteria. Anyhow, all of our mannitol-streptomycin pairs showed equivalent linkage (Hotchkiss, 1956), no matter to how many or to what sequences of DNA's the parent cells had been exposed. So, bacteria have self-respecting chromosomes and the new heterodoxy was assimilable into the old orthodoxy. Perhaps sometimes one feels disappointed to have something one is at pains to prove come very quickly to be taken as "obvious"—but as Ephraim Katchalski once remarked, that is really one of the best things one can ever hope to do.

REFERENCES

ALEXANDER, H. E., and G. LEIDY. 1951. Determination of inherited traits of *H. influenzae* by desoxyribonucleic acid fractions isolated from type specific cells. J. Exptl. Med., *93*: 345–359.

ALLOWAY, J. L. 1932. The transformation *in vitro* of R pneumococcus into S forms of different specific types by the use of filtered pneumococcus extracts. J. Exptl. Med., *55*: 91–99.

———1933. Further observations on the use of pneumococcus extracts in effecting transformation of type in vitro. J. Exptl. Med., *51*: 265–278.

AUSTRIAN, R., and M. S. COLOWICK. 1953. Modification of the fermentative activities of pneumococcus through transformation reactions. Bull. John Hopkins Hospital, *92*: 375–384.

AUSTRIAN, R., and C. M. MACLEOD. 1949. Acquisition of M protein by pneumococci through transformation reaction. J. Exptl. Med., *98*: 451–460.

AVERY, O. T., C. M. MACLEOD, and M. MCCARTY. 1944. Studies on the chemical nature of the substance inducing transformation of pneumococcal types. J. Exptl. Med., *79*: 137–158.

BEADLE, G. W. 1948. Genes and biological enigmas. Amer. Scientist, *36*: 71.

BOIVIN, A. 1947. Directed mutation in colon bacilli, by an inducing principle of desoxyribonucleic nature: its meaning for the general biochemistry of heredity. Cold Spring Harbor Symp. Quant. Biol., *12*: 7–17.

DAWSON, M. H. 1930. The transformation of pneumococcal types. II. The interconvertibility of type-specific S pneumococci. J. Exptl. Med., *51*: 123–147.

DAWSON, M. H., and R. H. P. SIA. 1931. In vitro transformation of pneumococcal types. I. A technique for inducing transformation of pneumococcal types in vitro. J. Exptl. Med., *54*: 681–699.

EPHRUSSI-TAYLOR, H. 1951. Genetic mechanisms in bacteria and bacterial viruses. III. Genetic aspects of transformations of pneumococci. Cold Spring Harbor Symp. Quant. Biol., *16*: 445–456.

GRIFFITH, F. 1928. The significance of pneumococcal types. J. Hyg. (Cambridge, England), *27*: 113–159.

HOTCHKISS, R. D. 1949. Etudes chimiques sur le facteur transformant du pneumocoque. Colloq. intern. centre natl. recherche sci. (Paris), *8*: Unités biol. douées contin. génét. 57–65.

———1952. The role of desoxyribonucleates in bacterial transformation, p. 426–436. *In* W. D. McElroy and B. Glass [Ed.] Phosphorus Metabolism, vol. II. Johns Hopkins Press, Baltimore.

———1955. Bacterial transformation (Oak Ridge Symposia). J. Cell. Comp. Physiol., *45*: 1–22.

———1956. The genetic organization of the deoxyribonucleate units functioning in bacterial transformations, p. 119–130. *In* O. H. Gaebler [Ed.] Enzymes: Units of biological structure and function. Academic Press, New York.

———1965. Oswald T. Avery. Genetics, *51*: 1–10.

HOTCHKISS, R. D., and H. EPHRUSSI-TAYLOR. 1951. Use of serum albumin as source of serum factor in pneumococcal transformation. Fed. Proc., *10*: 200.

HUTCHINSON, G. E. 1945. The biochemical genetics of pneumococcus. Amer. Scientist, *33*: 56–57.

LANGVAD-NIELSEN, A. 1944. Change of capsule in the pneumococci. Acta Pathol. Microbiol. Scand., *21*: 362–369.

LEDERBERG, J. 1956. Genetic transduction. Amer. Scientist, *44*: 264–280.

MARSHAK, A., and A. C. WALKER. 1945. Mitosis in regenerating liver. Science, *101*: 94–95.

McCarty, M., H. E. Taylor, and O. T. Avery. 1946. Biochemical studies of environmental factors essential in transformation of pneumococcal types. Cold Spring Harbor Symp. Quant. Biol., *11*: 117–183.

Muller, H. J., C. C. Little, and L. H. Snyder. 1947. Genetics, Medicine and Man, p. 6. Cornell University Press, Ithaca, N.Y.

Neufeld, F., and W. Levinthal. 1928. Beiträge zur Variabilität der Pneumokokken. Z. Immunitätsforsch., *55*: 324–340.

Sonneborn, T. M. 1943. Gene and cytoplasm. Proc. Natl. Acad. Sci., *29*: 338.

Wright, S. 1945. Physiological aspects of genetics. Ann. Rev. Physiol. *1*: p. 79, 83.

WILLIAM HAYES
Medical Research Council, Microbial Genetics Research Unit, Hammersmith Hospital, Ducane Road, London, England

Sexual Differentiation in Bacteria

In 1950 I was primarily engaged in the teaching and practice of medical bacteriology, but my research interests were of a much more academic nature. My main preoccupation at that time was the mechanism of antigenic phase variation in *Salmonella*, and especially of the type involving the somatic complex 12 (Kauffman, 1941; Hayes, 1947) which alternates so rapidly from one phase to the other that the hypothesis of forward- and back-mutation, then deemed adequate to explain flagellar diphasic variation (see Stocker, 1949), seemed scarcely tenable. Stimulated by the successful demonstration of genetic recombination in *Escherichia coli* by Lederberg and Tatum (1946; Tatum and Lederberg, 1947; Lederberg, 1947, 1949), as well as by Luria's (1947) review of the progress of bacterial genetics, I recorded in my research notes of March, 1950, the design of an experiment involving growth in mixed culture of a pair of genetically marked *Salmonella* strains, stabilized in each phase, and the selection and examination of recombinants for restoration of the variation.

This experiment was never attempted. Before the stabilized strains could be made, I attended a summer school in bacterial chemistry, organized by E. F. Gale at Cambridge University. There I had the good fortune to meet L. L. Cavalli-Sforza, who had recently discovered his famous Hfr strain (Cavalli, 1950) and was then actively engaged in studying recombination in *E. coli* K12. This system was then regarded as homothallic—i.e., it was thought that the two parental types of bacteria are equal partners whose fusion leads to formation of fully diploid zygotes. But the frequency of recombinants, being only about 10^{-6}, was much too low to permit any direct observation of conjugation, and the formation of complete bacterial zygotes remained no more than a reasonable working assumption.

Confronted with the opportunity of acquiring the basic *E. coli* strains with the minimum of trouble, I thought that a working acquaintance with their recombination might provide clues which would help to solve some of my own research problems. So, thanks to Cavalli-Sforza's tuition and kindness, I embarked on the study of conjugation in *E. coli* K12.

In retrospect and from the vantage of the sophistication of present-day molecular biology, my instinctive reaction to the naiveté of some of my early experiments would be that they had better be hidden forever from the critical eyes of scientific colleagues, were it not for the fact that one of them offered an unforeseen and gratuitous clue to the true nature of bacterial conjugation. This experiment was designed to reveal the kinetics of conjugation as a prelude to subsequent microscopic examination of mating mixtures. In particular, it was my plan to determine the time after mixing and spreading the washed parental cultures on minimal agar at which recombinants begin to appear.

ONE-WAY GENETIC TRANSFER

My method was to spread a mixture of a streptomycin-resistant (58-161: *met*$^-$, *str*^{r}) and a streptomycin-sensitive (W677: *thr*$^-$, *leu*$^-$, *thi*$^-$, *str*^{s}) parental strain on a series of minimal agar (+thiamine) plates and then, after various times of incubation, to respread the plates with a solution of streptomycin. Assuming a homothallic system, I thought that the addition of streptomycin prior to genetic transfer would prevent the emergence of prototrophic recombinants by killing one of the parents; but once genetic transfer has occurred, a proportion of those *str*^{r} parental bacteria which had participated in formation of zygotes would yield *str*^{r} recombinants in the presence of streptomycin. It turned out that addition of streptomycin at any time during the first two hours of mating completely suppresses the appearance of prototrophic recombinant colonies; after that time recombinant colonies begin to appear and then continue to increase in number.

If this system were really homothallic, involving formation of complete zygotes, this experiment should yield the same results, irrespective of which parent is streptomycin-resistant. However, when similar kinetic experiments were performed with crosses reversed with respect to the *str*^{r} marker (58-161:*str*^{s} × W677:*str*^{r}) the outcome was quite different. Now, roughly the same number of recombinant colonies arose on every plate, without regard to the time of addition of streptomycin, even when the parental strains were initially mixed on minimal agar containing the drug. This result implied that the potentiality of the two parental strains for genetic transfer was unequally affected by streptomycin, although in their *str*^{s} state, the viability of both proved to be equally sensitive to the lethal action of the drug. The effects on recombinant production of virtual sterilization of one or the other parental culture by treatment with streptomycin, as judged by a million-fold or greater reduction in colony-forming capacity, was therefore investigated. Crosses between sterilized strain 58-161 and untreated strain W677 (whether *str*^{r} or *str*^{s}) invariably yielded recombinants, usually at between one and ten per cent of the normal frequency, whereas crosses involving sterilized strain W677 were, without exception, completely sterile.

These findings seemed explicable only on the hypothesis that the viability of strain W677 is essential for recombinant production, implying that this strain acts exclusively as a gene *recipient* in which the entire process of recombination and segregation takes place. In contrast, strain 58-161 behaves as a gene *donor*, which becomes dispensable once its function has been fulfilled (Hayes, 1952a). On the grounds that donor bacteria sterilized by treatment with 1000 μg/ml streptomycin in aerated broth for as long as 18 hours at 37° could hardly be expected to perform any strictly *cellular* function and yet are fertile in crosses, the additional hypothesis was made that genetic transfer is mediated by a streptomycin-resistant "gamete," which is extruded on to the surface of donor bacteria and thence transferred by contact to recipients (Hayes, 1952a). This subsidiary hypothesis, though almost certainly erroneous in its details, did suggest that the donor gamete might be a temperate virus with novel properties.

THE EFFECT OF ULTRAVIOLET LIGHT ON FERTILITY

At this time the phenomenon of transduction had not yet been discovered, but it was known that certain temperate viruses carried by lysogenic bacteria can be induced to enter the lytic cycle by irradiation with ultraviolet (UV) light (Lwoff, Siminovitch and Kjelgaard, 1950; Lwoff, 1951), and that *E. coli* K12 is lysogenized by the inducible phage λ (Weigle and Delbrück, 1951). Moreover Haas, Wyss and Stone (1948) had shown that UV irradiation of mixtures of fertile *E. coli* strains produces an increased yield of recombinants. An analysis of this effect was therefore made in terms of the postulated donor-recipient relationship. Washed suspensions of the donor and recipient strain were each irradiated to about 50% survival, incubated in broth at 37° for one hour, mixed with an unirradiated culture of its mating partner and finally, after washing, plated for recombinants. Those crosses in which the donor strain had been irradiated showed a 5- to 20-fold, or even greater, increase in the yield of recombinants over the yield from unirradiated control crosses. In contrast, irradiation of the recipient parent produced a decrease in the number of recombinants, which was roughly proportional to the fall in viable count (see Table 1).

As in the case of prophage induction, incubation in broth following irradiation of the donor proved essential for the development of the stimulatory effect. If the donor bacteria were mixed with the recipient suspension and plated immediately after irradiation, or were incubated in fully supplemented minimal medium instead of in broth, the recombinant frequency did not increase and indeed was sometimes observed to fall (Hayes, 1952b). Finally, the possibility that the stimulatory effect of UV might be secondary to the induction and liberation of phage λ was excluded by the finding that the fertility of each of two non-lysogenic donor strains was similarly enhanced by UV irradiation (Hayes, 1953a). The absence of any correlation between lysogenicity and fertility was also demonstrated by Lederberg, Cavalli and Lederberg (1952).

THE SEX FACTOR, F

The concept of one-way genetic transfer from donor to recipient bacteria, together with the marked stimulation of donor fertility by UV light under conditions closely comparable to those required for prophage induction, suggested that transfer might be mediated by occasional association between the bacterial genetic material and some non-pathogenic, virus-like vector which could be extruded by donor cells and thence transferred to recipient cells by contact. Accordingly, a search was made among a

Table 1

DIFFERENTIAL EFFECT OF ULTRAVIOLET LIGHT ON THE FERTILITY OF F^+ AND F^- STRAINS

F^+ parent = *met*$^-$ auxotroph, 58-161
F^- parent = *thr*$^-$*leu*$^-$*thi*$^-$ auxotroph, W677.
Crosses were made in duplicate on minimal agar + thiamine.
/UV = young broth cultures centrifuged, resuspended to volume in buffered saline and 5 ml aliquots irradiated for the same time. Irradiated suspensions were suspended in broth at 37 degrees for 65 minutes before mixing with unirradiated heterologous suspension, washing and plating.

Cross	Average no. recombinant prototrophs per plate
$F^+ \times F^-$	57.5
$F^+ \times F^-$/UV	30.5
F^+/UV $\times F^-$	260.
*F^+/UV $\times F^-$	67.5

*The F^+ strain was here irradiated through the glass lid of the Petri dish.

From Hayes, 1953b

large number of isolated colonies of the donor strain 58-161 for an infertile derivative clone. If the infertility of such a clone were due to loss of a virus vector, it was hoped that fertility might prove restorable by infection from the wild type donor. No infertile isolates were found in these experiments, but while they were being performed I learned by chance that a pair of originally fertile strains (58-161 and W677) held by Clive Spicer had failed to yield any recombinants in crosses following a year's storage at 4°. Analysis of these strains, by crossing each against its partner of known fertility, revealed that their infertility was due to a defect in the original 58-161 donor strain. However, as the hypothesis predicted, this defective donor behaved as an efficient recipient in crosses with other donors, such as wild type *E. coli* K12 and various prototrophic recombinants.

To test for infective restoration of the donor state, a streptomycin- and sodium azide-resistant mutant of the defective donor was selected and grown overnight in mixed broth culture with the sensitive, fertile donor. Colonies of the doubly-resistant strain were then re-isolated and tested in crosses against the standard recipient. Thirty-six per cent were found to have recovered normal fertility (Hayes, 1953a). After this experiment had been set up on January, 26 1952, but before the result was known, I wrote to Cavalli-Sforza telling him what I had done and the results I hoped to obtain, pointing out that if the experiment was successful the hypothesis of an infective agent as the determinant of genetic transfer and the donor state would be established. He replied that he already knew the outcome, since the Lederbergs and he had performed the same experiment about three weeks previously! (See Lederberg et al, 1952.)

The high efficiency with which fertility can be restored to the infertile strain 58-161 by infection suggested that the standard recipient strain, W677, might similarly be converted to the donor state. In fact this conversion turned out to be even more efficient since, after overnight growth in mixed culture with the fertile donor 58-161, 75% of reisolated W677 colonies proved fertile in crosses with the defective 58-161 (recipient) strain. However, the number of prototrophic recombinants generated in the mixture was only of the order 10^{-5} of the W677 population, and none of the converted W677 donor clones had incorporated any one of ten recognizable genetic characters carried by strain 58-161.

All this led to the hypothesis that the donor state, and its potentiality for genetic transfer, are determined by a freely transmissible, extrachromosomal factor (Hayes, 1953a), which was termed F (for fertility). Strains which harbor F were called F^+, and those lacking the factor F^- (Lederberg et al., 1952). Extensive analysis of the properties of the F factor, especially by Cavalli, Lederberg and Lederberg (1953; see also Hayes, 1953a, b), showed that it is very stable in F^+ strains and freely transmissible through an indefinite series of F^- strains. Its transmission depends absolutely on cellular contact, however; filtrates, or cell-free lysates, or extracts of donor cultures cannot convert recipient into donor cultures.

THE GENETIC EFFECTS OF SEXUAL DIFFERENTIATION: INCOMPLETE GENETIC TRANSFER

Although the Lederbergs and Cavalli, on the one hand, and I, on the other, were in virtually complete accord concerning the experimental facts which we had independently discovered, our interpretation of these facts became a matter for keen controversy. So far as my own rather limited data went, not only was there a complete correlation between the F^+ and donor states, but whereas all recombinants from the standard 58-161.$F^+ \times$ W677.F^- cross were F^+, only about four per cent of non-recombinant W677 bacteria were converted to the F^+ state under the same conditions. I therefore regarded the F factor as the agent directly responsible for mediating genetic transfer by those rare cells in which it had become "effectively associated" with the genome: on this hypothesis the effect of UV light on the fertility of donor populations was ascribed to stimulation of the formation of such effective associations. The Lederbergs and Cavalli-Sforza, however, were not convinced by the evidence I had to offer, partly on account of its limited scope, and partly because recombinants arising from highly fertile crosses involving Cavalli-Sforza's Hfr strain were invariably F^-, suggesting that the F factor in this strain is *not* acting as a genetic vector (Lederberg et al., 1952; Cavalli et al., 1953). They favored the view that the F factor determines mating compatibility within the *E. coli* K12 system, such that mating only occurs when one of the parents is F^+. The variable, but usually much lower, fertility of $F^+ \times F^+$, as compared with equivalent $F^+ \times F^-$ crosses, and the very high fertility of Hfr $\times F^-$ crosses, suggested to them the possible operation of a system of relative sexuality, under which the combined action of the F factor and the residual bacterial genotype determines the mating potential: the greater the potential difference between two strains, the greater the fertility of their cross (Lederberg et al., 1952; Cavalli et al., 1953).

Since donor (F^+) and recipient (F^-) strains of both parental stocks were now available, it was clearly important to see whether reversal of the donor-recipient relationship affects the genotype of recombinants. The recombinant patterns of the two crosses involving such reversal turned out to be dramatically different (Cavalli et al., 1953; Hayes, 1953a, b). Although the data of Cavalli et al. were considerably more extensive than my own, my results were so clearcut that the analysis of only a small number of prototrophic recombinants for inheritance of various unselected markers was required to reveal the essential features. Such an analysis with respect to four unselected markers is given in Table 2. Two striking features of reversing the "F polarity" are apparent.

1. The genotype of recombinants is predominantly that of the F^- parent. Thus in the A.$F^+ \times$ B.F^- cross, 70.6% of recombinants inherit all the tested unselected markers of the F^- parent. Similarly in the reversed cross, if the F^+ marker *thi*⁻ is excluded from consideration (see below), 85.2% of recombinants possess all the other markers of the F^- strain.

2. If we look at the unselected markers of the F^+ parent which do appear

among recombinants, we find that in the A.F$^+$ × B.F$^-$ cross only *lac*$^+$ (24.7%), and in the reverse cross only *thi*$^-$ (57.6%), are inherited with significant frequency.

Table 2

EFFECT OF REVERSAL OF F POLARITY ON THE GENETIC CONSTITUTION OF RECOMBINANTS FROM OTHERWISE SIMILAR CROSSES

Strain A = 58-161 = *met*$^-$ *lac*$^+$ *thi*$^+$ *mal*$^+$ *str-r*
Strain B = W677 = *thr*$^-$ *leu*$^-$ *lac*$^-$ *thi*$^-$ *mal*$^-$ *str-s*
Crosses were made on minimal agar + thiamine. Three hundred prototrophic recombinants from each cross were analyzed.

Cross		Percentage frequency among recombinants of genotypes:		
		mal$^+$.*str-r*	*mal*$^-$.*str-s*	*mal*$^+$.*str-s* or *mal*$^-$.*str-r*
A.F$^+$ × B.F$^-$	*lac*$^+$.*thi*$^+$	0	1.7	1.0
	lac$^+$.*thi*$^-$	0	23.0	
	lac$^-$.*thi*$^+$	0	3.7	
	lac$^-$.*thi*$^-$	0	70.6	
B.F$^+$ × A.F$^-$	*lac*$^+$.*thi*$^+$	35.2	2.0	4.9
	lac$^+$.*thi*$^-$	50.0	6.0	
	lac$^-$.*thi*$^+$	0.3	0	
	lac$^-$.*thi*$^-$	1.3	0.3	

From Hayes, 1953b.

These findings seemed to me precisely those expected on the hypothesis that F$^+$ bacteria are exclusively donors, and F$^-$ bacteria exclusively recipients, of genetic material, *provided* the further assumption was made that the donor transfers only part of its genome to the recipient. Thus strain B(W677) requires threonine and leucine for growth, so that when this strain is the recipient (Table 2: cross A.F$^+$ × B.F$^-$), the development of prototrophic recombinants requires transfer of genes *thr*$^+$ and *leu*$^+$ from the donor strain A.F$^+$(58-161); but previous analysis had shown that *lac*$^+$ is located on the same linkage group as *thr.leu*, whereas *thi* is much less closely linked (see Lederberg et al, 1951), so that inheritance of *lac*$^+$, but not of *thi*$^+$, among *thr*$^+$ *leu*$^+$ recombinants would be expected. Alternatively strain A requires methionine for growth so that, in the reversed cross where this strain is the recipient, selection is made for prototrophic recombinants arising from those A.F$^-$ bacteria which have received from the B.F$^+$ parent a fragment of chromosome carrying the *met*$^+$ gene. But this gene had already been shown to be closely linked to *thi*⁻, so that inheritance of *thi*⁻ as an unselected marker could be predicted in this cross (Hayes, 1953b).

One further peculiarity of the effect of the F factor on recombinant pattern was the outcome of $F^+ \times F^+$ crosses. Not only were these crosses usually much less fertile than $F^+ \times F^-$ crosses, but the genotypes of recombinants issuing from them mimicked those expected from a mixture of two crosses of reversed F polarity (see Table 3). Following the discovery that F^+ cells might form temporary "F^- phenocopies" under certain physiological conditions of growth (Lederberg et al, 1952), this result became explicable if young cultures of F^+ bacteria, grown under normal conditions, contained a small proportion of phenotypically F^- cells (Lederberg et al, 1952; Hayes, 1953b). Since the ability of F^- phenocopy preparations of F^+ bacteria to act as recipients is destroyed by streptomycin treatment, I thought that this hypothesis could be directly tested by treating each

Table 3

EFFECT OF STREPTOMYCIN TREATMENT OF EACH PARENT IN AN $F^+ \times F^+$ CROSS ON THE DISTRIBUTION OF UNSELECTED MARKERS AMONG RECOMBINANTS

Strain A = 58-161 = *met*⁻ *lac*⁺ *mal*⁺
Strain B = W677 = *thr*⁻ *leu*⁻ *thi*⁻ *lac*⁻ *mal*⁻
Crosses were made on minimal agar + thiamine.
/ST = young broth cultures were treated with 1000 µg/ml streptomycin for 2 hours at 37 degrees (static) and then washed three times. The viable count of each strain was reduced to about 5×10^{-5} of its initial value by this treatment.

Cross	No. of recombinants having genotype:				No. of recombinant colonies analyzed
	lac⁺.*mal*⁺	*lac*⁻.*mal*⁻	*lac*⁺.*mal*⁻	*lac*⁻.*mal*⁺	
A.F⁺ × B.F⁻	0	14	1	0	15
A.F⁻ × B.F⁺	14	0	1	0	15
A.F⁺ × B.F⁺	24	24	9	3	60
A.F⁺/ST × B.F⁺	0	22	8	0	30
A.F⁺ × B.F⁺/ST	18	0	1	1	20
A.F⁺/ST × B.F⁺/ST	No recombinant colonies arose				

From Hayes, 1953b

parent of an $F^+ \times F^+$ cross, in turn, with streptomycin, and then analyzing the recombinants for a shift of genotypic pattern toward that of a pure $F^+ \times F^-$ cross. As Table 3 shows, "these theoretical considerations were unambiguously vindicated by experiment." The crosses behaved as if the treated F^+ parent was homogeneously F^+ and the untreated parent was pure F^- (Hayes, 1953b). In fact, and in retrospect, this experiment shows what had been assumed when it was devised, namely, that phenotypically expressed F^+ cells do not mate with one another. Curiously enough, the physiological basis of F^- phenocopy formation still remains a mystery (see Clark and Adelberg, 1962), though there is now evidence that the probable cause of the absence of mating between F^+ cells is their same surface charge, which renders them unable to make intimate contact (see Maccacaro and Comolli, 1956).

Perhaps I have presented here the hypothesis of one-way, partial transfer of genetic material from F^+ donors to F^- recipients as a highly controversial issue, hotly contested on both sides. As I reread my early correspondence with Joshua Lederberg and Luca Cavalli-Sforza, in which we freely exchanged our views, and once again peruse their papers, I am impressed by their mature judgment and their refusal to be rushed into acceptance of new theories on insufficient evidence, whereas my own role seems that of an enthusiastic amateur—which I was! In fact they were quite prepared to accept the possibility of one-way, partial transfer. But mainly because of the complicated data on persistent heterozygous diploids, which suggested that a specific chromosomal segment from the F^+ parent is eliminated from the zygotes—a phenomenon for which analogies exist in higher organisms—they preferred to explain the predominance of the F^- genotype among recombinants on the basis of post-zygotic exclusion (Lederberg et al, 1951; Lederberg et al, 1952; Cavalli et al, 1953). Proof of the validity of my hypothesis awaited the studies of Wollman and Jacob several years later.

ISOLATION AND ANALYSIS OF AN Hfr STRAIN

I first reported the hypothesis of one-way, partial transfer, together with all the supporting experimental evidence related above, to the 2nd European Symposium on Microbial Genetics, held at Pallanza, Italy in September, 1952. Cavalli-Sforza also presented the results of his and the Lederbergs' studies at this meeting. It was there that I first met James Watson, then working with Francis Crick on the structure of DNA. Due to Watson's interest in this novel interpretation of the conjugation phenomenon, and his close liaison with Max Delbrück, I was invited to participate in the 1953 Cold Spring Harbor Symposium on "Viruses" organized by Delbrück. And in the autumn of that year I went to work for six months in Delbrück's laboratory at the California Institute of Technology.

The interval between the Pallanza and the Cold Spring Harbor meetings, which I had planned to use for consolidating and extending my existing experiments, was marked by two diversionary episodes. The more important was the accidental isolation of an Hfr clone from a culture of the F^+ donor strain 58-161. Preliminary investigation showed that, in all essentials, this strain was similar to the prototype isolated from the same F^+ strain by Cavalli-Sforza (1950; Cavalli et al, 1953). In addition, some new and significant features of Hfr behavior were discovered (Hayes, 1953b).

1. The fertility of the Hfr strain proved highly resistant to treatment with streptomycin. Thus treatment which reduced the viable count by a factor 10^7 resulted in only a five-fold reduction in the number of prototrophs obtained in the subsequent cross. The strain was therefore a "donor," as defined in the case of F^+ bacteria.

2. In the Hfr $\times$ F^- cross, selection for the donor markers *thr*$^+$*leu*$^+$ or *lac*$^+$ yielded about 1000 times more recombinants than in an analogous $F^+ \times F^-$ cross. But when Hfr markers belonging to what then appeared to be other linkage groups (*thi—met* or *str—mal*) were selected in the crosses, the frequency of recombinants fell to the level characteristic of F^+ crosses. This was taken to indicate a special association of the F factor with the *thr.leu—lac* linkage group and to establish the independent transfer of this fraction of the genome, at least in the Hfr $\times$ F^- cross.

3. Although selection for Hfr markers on the *thr.leu—lac* linkage group always produced F^- recombinants, recombinants for other Hfr markers, inherited at low frequency, were often Hfr donors. Thus about 75% of recombinants inheriting the Hfr marker *thi*$^+$, and 25% of those inheriting *met*$^+$, were themselves Hfr strains which behaved similarly to the parent strain. This finding permitted the recombinational synthesis of a variety of Hfr strains of different genotypes, but its significance in pointing to a chromosomal location of the F factor was not fully realized at the time.

4. Unlike F^+ strains, treatment of the Hfr strain with UV light failed to increase its fertility, so far as *thr*$^+$ *leu*$^+$ or *lac*$^+$ recombinants were concerned. This led to the view that, in Hfr strains, effective association between the F factor (or a "mutant" factor, HF) and the *thr.leu—lac* linkage group was already maximal and, therefore, could not be increased further.

This Hfr strain, Hfr*H*, later became famous through the definitive work of Wollman and Jacob on the mechanism of conjugation, and I still bask a little in its reflected glory. For instance, at a recent dinner I was introduced to an American biochemist who, on hearing my name, asked "Are you, by any chance, *Hfr* Hayes?"

The second diversion during this period was the development by James Watson of the idea, which seemed a good one at the time, that the three linkage groups, as then defined in *E. coli*, might represent separate chromosomes any one of which, but not more than one, was usually transferred by any F^+ cell during a particular mating event. More extensive data on the effects of reversed F polarity were obtained and, with the collaboration of Cavalli-Sforza, a plausible case for the hypothesis was made (Watson and Hayes, 1953). If it was supposed that the F factor occasionally promoted concomitant transfer of more than one linkage group, the hypothesis seemed to explain neatly the apparent linkage but absence of linearity of markers on different linkage groups observed by Lederberg et al (1951). The hypothesis therefore formed the basis of a formal model of genetic transfer that I put forward at the Cold Spring Harbor Symposium (Hayes, 1953b). At that meeting, support for the model came from an unexpected quarter, when E. D. DeLamater (1953), as a result of his controversial conclusions (since withdrawn) from cytological studies of alleged mitosis in bacteria, claimed that at least three chromosomes could be visualized in *E. coli* K12.

For the progress of molecular biology, this symposium was particularly notable for two events: the presentation by Watson (Watson and Crick, 1953) of the Watson-Crick structure for DNA, which was to revolutionize biology to a degree unknown since the discoveries of Mendel; and the report by G. R. Wyatt (1953) that hydroxymethyl cytosine replaces cytosine in the DNA of the T-even phages, a finding which was to pave the way for subsequent studies of the kinetics of phage DNA synthesis in infected bacteria.

THE KINETICS OF MATING

The final stage of these early studies, which preceded Wollman's and Jacob's brilliant series of experiments on the mechanism of conjugation, took place at Caltech, where I had anticipated further collaboration with James Watson. This collaboration did not materialize, however, because Watson, fresh from the conquest of DNA, had become engrossed in the problem of RNA structure. So I worked by myself again, but now with the great advantage of many discussions with Jean Weigle, Giuseppe Bertani, George Streisinger, Dale Kaiser, Robert DeMars and, of course, Max Delbrück, whose general ideas undoubtedly influenced the trend of my experiments. Having survived the ordeal of a five-hour seminar under Delbrück's direction that failed to generate any new ideas, I decided to return to the study of the kinetics of conjugation, which now seemed feasible with the advent of Hfr donor strains.

The first project was to follow the kinetics of zygote formation. For this purpose, samples of a mixture of Hfr and F^- bacteria were infected at various times with a high multiplicity of a virulent phage, to which only the Hfr parent was susceptible, and then plated for recombinants. I had anticipated that at some specific time after the commencement of mating, the Hfr donor chromosome fragments would be transferred rather quickly as "packages" to the F^- recipients; hence, killing of the donor by phage infection before this time should prevent formation of zygotes, but have no effect thereafter. The first phage strain to be used, T1, gave variable and ambiguous results, probably for at least two reasons: the failure of T1 to arrest rapidly the metabolism of the Hfr bacteria and the presence of T1 host-range mutants which often lysed the T1-resistant F^- recipient population. However, use of phage T6 produced a clearcut and reproducible picture: when untreated samples were plated, recombinants began to appear immediately after mixing the parental cultures; when phage T6-treated samples were plated, no recombinants appeared during the first 8 to 10 minutes after mixing, but thereafter the number of recombinants rose linearly with time, until a plateau was reached at about 40 minutes after mixing.

In these experiments selection was always made for recombinants inheriting the *thr*$^{+}$ *leu*$^{+}$ genes from the Hfr parent. On several occasions, however, recombinants arising from samples treated with phage T6 10 and 15 minutes after the commencement of mating were scored for inheritance of the unselected linked markers *lac*$^{+}$ and *T6-s*. Such recombinants were almost invariably found to be missing these donor markers. I failed to follow up, or to realize the significance of this finding until Wollman and Jacob (1956) published the results of their famous interrupted mating experiments. Instead, I ascribed this apparently paradoxical result to contraselection by phage T6 of the *T6-s* recombinant segregants on the plates.

The ability to prepare "zygote suspensions," from which the parental Hfr bacteria has been eliminated by treatment with phage T6, suggested further experiments on the kinetics of segregation, and of the expression of genes determining resistance to sodium azide and phage T1, which are closely linked to the selective markers *thr*$^{+}$ *leu*$^{+}$ on the Hfr chromosome. For example, if a suspension of newly formed zygotes is diluted and incubated in fresh broth and samples then plated at intervals for recombinants, the time at which the number of recombinants begins to increase indicates the commencement of division among the recombinant segregants, so that the time of segregation can be assessed. Furthermore, if the samples are plated on selective media containing inhibitory or lethal concentrations of sodium azide or phage T1 respectively, only those zygotes or recombinants can yield colonies which inherit the appropriate resistance gene from the Hfr parent and in which this gene has already expressed itself. These new methods were initiated at Caltech and the definitive experiments completed during 1954, but they were not published until two years later (Wollman, Jacob and Hayes, 1956; Hayes, 1957).

I will conclude this reminiscence with the story of an observation which was forced on me in a rather dramatic way and which caused me considerable anxiety. When I arrived at Caltech I inherited a number of culture media constituents, including an impressive bottle, containing what was affirmed to be highly purified agar powder, from Marguerite Vogt, who had been studying some quantitative aspects of the cellular contacts which mediated bacterial conjugation. When crosses involving either Hfr or F^{+} donors were plated on minimal agar prepared from these reagents, no recombinants at all arose. Was I not only incompetent but a fraud as well?

Further investigation showed that the medium supported normal growth of previously isolated prototrophic recombinants, as well as the normal development of recombinant colonies, when a suspension of preformed zygotes was plated on it; some step in the process of conjugation itself was clearly being prevented on the Caltech medium. It then turned out that crosses became normally fertile when asparagine (see Gray and Tatum, 1944), but especially aspartate and, to a lesser extent, pyruvate, fumarate or succinate were added to the medium; addition of glutamic and other amino acids had no effect. Apart from prompting a seminar on the value of aspartate as a bacterial aphrodisiac, the direct outcome of these observations was the first analysis of the energetic requirements for conjugation by K. W. Fisher, who became my first Ph.D. student in bacterial genetics in 1954 (Fisher, 1957a, b).

RETROSPECTIVE ASSESSMENT

Assessment of the importance or significance of one's own work is a difficult and dangerous business, which is always best left to an admirer more prejudiced and less modest than one's self. All the more so in my own case, since most of my early key experiments were also done independently and at the same time by others—a circumstance of value to all of us, since no one thought to disbelieve the experiments themselves. I think that the main novelty of my own approach lay in the directness and plausibility of my interpretation of these experiments, for even the hypothesis of one-way genetic transfer remained unproven for several years, while much of the speculative detail turned out to be wrong. Indeed, I am sure that the important features of this interpretation could not have remained dormant for long anyway, for they were obvious possibilities deducible from the genetic data alone and, in fact, were explicitly stated as possibilities in the early papers of the Lederbergs and Cavalli-Sforza.

It is, I think, more profitable to look at the work done at this time as a whole and, as in the evolution of all scientific progress, to regard the subsequent clarifications and new horizons as a natural product of the dialectic which this work generated. In the years 1952–1954, the time was indeed ripe for new observations and ideas, for research into the mechanism of conjugation in *E. coli*, conducted almost exclusively through recombinant analysis, was rapidly becoming bogged down in a mass of contradictory and paradoxical genetic data of increasing complexity from which there seemed no escape. Discovery of the sex factor and the effects of F polarity, together with the idea of partial genetic transfer, permitted most of the paradoxes to be explained in a simple and satisfying manner, and so cleared the way for future progress. Of course, many complexities still remained. The great advances and simplifications were to come later; perhaps the real evaluation of the work reported here depends upon the extent to which it served as a necessary intermediate in the evolution of the fundamental and highly original research of François Jacob and Elie Wollman.

There is, however, one aspect of the work of the Lederbergs, Cavalli, and myself which warrants independent assessment. This is the discovery of the sex factor, which has turned out to be the prototype of an entire new class of extra-chromosomal elements determining important, if sometimes non-essential, bacterial functions. Within this class is a group of conjugation factors, including the sex factor itself, resistance transfer factors, and certain colicin factors, which are able to promote their own conjugal transfer between bacteria of not only the same but also different species and genera; some of these factors also mediate chromosome transfer, though with low efficiency. In many respects these factors behave like viruses, with a novel mode of infectivity. Such factors—and so far only a few of them have been recognized—clearly contribute greatly to the evolutionary flexibility of bacteria. Whatever their origin, the theoretical and practical importance of these factors is growing rapidly, and for bacteria, at least, they have already an established place in the concept of the cell.

REFERENCES

CAVALLI-SFORZA, L. L. 1950. La sessulita nei batteri. Boll. Ist. Sieroter. Milano., *29*: 281.

CAVALLI, L. L., J. LEDERBERG, and E. M. LEDERBERG. 1953. An infective factor controlling sex compatibility in *Bacterium coli*. J. Gen. Microbiol., *8*: 89.

CLARK, A. J., and E. A. ADELBERG. 1962. Bacterial conjugation. Ann. Rev. Microbiol. *16*: 289.

DELAMATER, E. D. 1953. The mitotic mechanism in bacteria. Cold Spring Harbor Symp. Quant. Biol., *18*: 99.

FISHER, K. W. 1957a. The role of the Krebs cycle in conjugation in *E. coli* K-12. J. Gen. Microbiol., *16*: 120.

———1957b. The nature of the endergonic processes in conjugation in *Escherichia coli* K-12. J. Gen. Microbiol., *16*: 136.

GRAY, C. H., and E. L. TATUM. 1944. X-ray induced growth factor requirements in bacteria. Proc. Natl. Acad. Sci., *30*: 404.

HAAS, F., O. WYSS, and W. S. STONE. 1948. The effect of irradiation on recombination in *Escherichia coli*. Proc. Natl. Acad. Sci., *34*: 229.

HAYES, W. 1947. The nature of somatic phase variation and its importance in the serological standardisation of *O* suspensions of *Salmonella*. J. Hyg., Camb., *45*: 111.

———1952a. Recombination in *Bact. coli* K-12: unidirectional transfer of genetic material. Nature, *169*: 118.

———1952b. Genetic recombination in *Bact. coli* K-12: analysis of the stimulating effect of ultraviolet light. Nature, *169*: 1017.

———1953a. Observations on a transmissible agent determining sexual differentiation in *Bact. coli*. J. Gen. Microbiol., *8*: 72.

———1953b. The mechanism of genetic recombination in *Escherichia coli*. Cold Spring Harbor Symp. Quant. Biol., *18*: 75.

———1957. The kinetics of the mating process in *Escherichia coli*. J. Gen. Microbiol., *16*: 97.

KAUFFMAN, F. 1941. A typhoid variant and a new serological variation in the *Salmonella* group. J. Bacteriol., *41*: 127.

LEDERBERG, J. 1947. Gene recombination and linked segregations in *Escherichia coli.* Genetics, *32*: 505.

———1949. Aberrant heterozygotes in *Escherichia coli.* Proc. Natl. Acad. Sci., *35*: 178.

LEDERBERG, J., L. L. CAVALLI, and E. M. LEDERBERG. 1952. Sex compatibility in *Escherichia coli.* Genetics, *37*: 720.

LEDERBERG, J., E. M. LEDERBERG, N. D. ZINDER, and E. R. LIVELY. 1951. Recombination analysis of bacterial heredity. Cold Spring Harbor Symp. Quant. Biol., *16*: 413.

LEDERBERG, J., and E. L. TATUM. 1946. Novel genotypes in mixed cultures of biochemical mutants of bacteria. Cold Spring Harbor Symp. Quant. Biol., *11*: 113.

LURIA, S. E. 1947. Recent advances in bacterial genetics. Bact. Rev., *11*: 1.

LWOFF, A. 1951. Conditions de l'efficacité inductrice du rayonnement ultraviolet chez une bactérie lysogène. Ann. Inst. Pasteur, *81*: 370.

LWOFF, A., L. SIMINOVITCH, and N. KJELGAARD. 1950. Induction de la production de bactériophages chez une bactérie lysogène. Ann. Inst. Pasteur, *79*: 815.

MACCACARO, G. A., and R. COMOLLI. 1956. Surface properties correlated with sex compatibility in *Escherichia coli.* J. Gen. Microbiol., *15*: 121.

STOCKER, B. A. D. 1949. Measurements of rate of mutation of flagellar antigenic phase in *Salmonella typhimurium.* J. Hyg., Camb., *47*: 398.

TATUM, E. L., and J. LEDERBERG. 1947. Gene recombination in the bacterium *Escherichia coli.* J. Bacteriol., *53*: 673.

WATSON, J. D., and F. H. C. CRICK. 1953. The structure of DNA. Cold Spring Harbor Symp. Quant. Biol., *18*: 123.

WATSON, J. D., and W. HAYES. 1953. Genetic exchange in *Escherichia coli* K-12: evidence for three linkage groups. Proc. Natl. Acad. Sci., *39*: 416.

WEIGLE, J. J., and M. DELBRÜCK. 1951. Mutual exclusion between an infecting phage and a carried phage. J. Bacteriol., *62*: 301.

WOLLMAN, E. L., F. JACOB, and W. HAYES. 1956. Conjugation and genetic recombination in *Escherichia coli.* Cold Spring Harbor Symp. Quant. Biol., *21*: 141.

WYATT, G. R. 1953. The quantitative composition of deoxypentose nucleic acids as related to the newly proposed structure. Cold Spring Harbor Symp. Quant. Biol., *18*: 133.

ELIE L. WOLLMAN
Service de Physiologie Microbienne, Institut Pasteur, Paris, France

Bacterial Conjugation

When, in the fall of 1948, I arrived in Pasadena as a Rockefeller Fellow, I felt very intimidated indeed. This was not only because I was coming to as famous an institution as the California Institute of Technology, but even more so because it was with Max Delbrück that I was coming to work. Although I had already met him on two occasions, these meetings, instead of lessening my apprehensions, had only increased them. During the preceding years, thanks mainly to Max Delbrück and Salvador Luria, bacterial genetics had come into existence and study of the bacteriophage had undergone a profound revolution. Taking a strictly cartesian attitude, Delbrück and Luria had swept away the facts and interpretations accumulated by their predecessors over twenty years and started anew. Within a few years they, and the small group of other workers they had attracted by the simplicity, the precision, and the elegance of their new departure, had made tremendous advances. I was particularly sensitive, then, to these happenings. Born in the year of the discovery of bacteriophage and raised with its early development, I had known in my childhood and youth many of the people who had first worked on it. When I joined André Lwoff's laboratory, at the end of the War, the advances that had taken place in the United States during the War were only just becoming known in France. Upon first reading the papers of the new American Phage Group, I found myself admiring their revolutionary progress and, at the same time, surprised by the carelessness with which they treated historical matters. Ready to add new portraits to the Gallery of Great Bacteriophage Workers, I felt somewhat distressed at seeing how the older portraits had been simply removed. I recall that once, looking into a bibliographical index, at Caltech, I came across a reference to one of my parents' papers (Wollman and Wollman, 1937, 1938). This paper reported that no infective phage can be recovered when *Bacillus megatherium* is infected with bacteriophage and later lysed with lysozyme. My parents' conclusion was that phage entered a noninfectious intracellular phase after infection. Mature phage particles which are later released from the infected cell are therefore not the direct progeny of the infecting particles. Instead, they are synthesized by the bacterium from the noninfectious form. The comment on the Caltech index card, referring to this paper, was "Nonsense."

One of the most striking discoveries of just that time was Doermann's (1948) demonstration that shortly after infection of *E. coli* by phage T4, no infective phage particles can be recovered by artificial, induced lysis of the infected bacteria. The definition of the eclipse period based on this discovery gave a new impetus to the analysis of the reproductive cycle of bacteriophages. In the summer of 1949, nearly the whole Phage Group had gathered at Caltech for several days in Delbrück's laboratory in order to try to fit the data available at the time into a general scheme. This reunion was the source of the "Syllabus on procedures, facts and interpretations in phage" included in the proceedings of the small, but pregnant, Viruses 1950 conference held in Pasadena the next March (Delbrück, 1950).

One of the categories of the Syllabus assigned to me was lysogeny, a chapter of bacteriophage research which had been neglected (and even been pronounced non-existent) by the Phage Group, but one on which my parents had worked for many years. As it happened, André Lwoff, in one of his several avatars, had taken over the problem of lysogeny in the meantime. On my return to Paris in the fall of 1950, I found our laboratory greatly animated by Lwoff's new-found insights. After completing some work undertaken at Caltech with Gunther Stent on adsorption and adsorption co-factors of phage T4, I finally began, as I had intended all the while, the study of lysogenic bacteria.

THE GENETIC DETERMINATION OF LYSOGENY

By 1950, the concept of prophage was well-established and the induction of prophage development had just been demonstrated. Lwoff was now studying the physiological conditions required for induction, and Siminovitch was comparing the metabolism of infected bacteria with that of induced lysogenic bacteria. Both were working with *Bacillus megatherium*, whereas a newcomer to our laboratory, Francois Jacob, was extending these studies to *Pseudomonas pyocyanea*. One of the most important problems of that time was the relationship of the prophage to its lysogenic host bacterium. This was clearly a genetic problem, which became tractable in 1951 when Esther Lederberg discovered that the K12 strain of *Escherichia coli*, the very strain in which genetic recombination of bacteria had been found (Lederberg and Tatum, 1946a, b), is lysogenic, and that nonlysogenic variants of this strain can be isolated (Lederberg, 1951). Soon thereafter, it was found that the temperate phage, λ, whose prophage is carried by *E. coli* K12, is inducible (Weigle and Delbrück, 1951). It was now apparent that *E. coli* K12 and its λ prophage are eminently suitable materials for the study of lysogeny, and this is the organism with which I decided to work.

Within a few years of its discovery, Joshua Lederberg and his colleagues had accumulated an impressive body of knowledge on genetic recombination in *E. coli*; but, as the amount of information increased, the situation became more and more confusing (Lederberg et al., 1951). That was because at that time only "low frequency" crosses could be performed and sexual differentiation in bacteria had not yet been discovered. Nevertheless, the first genetic crosses of lysogenic and non-lysogenic bacteria showed that λ lysogeny is linked to the galactose fermentation markers on the *E. coli* "chromosome" (Lederberg and Lederberg, 1953; Wollman, 1953).

A new era in the study of bacterial recombination dawned in 1952 when the first indications of sexual polarity and sexual differentiation were uncovered (Cavalli, Lederberg, and Lederberg, 1953; Hayes, 1952). In the spring of that year I visited William Hayes for the first time in his laboratory at the Postgraduate Medical School in London. His working conditions were then so modest that they made our musty attic in the Pasteur Institute look almost luxurious by comparison. I was particularly impressed by his tiny petri plates, 3 to 4 centimeters in diameter and cut out from the bottom of vials, and by the watchmaker's eye-lens with which he counted the minute colonies of recombinants appearing on these plates. Shortly after this visit, I gave an account of the new developments in recombination in bacteria, and of the genetic basis of lysogeny, at the first international conference on "Bacteriophage" held at Royaumont. Max Delbrück, who was present at Royaumont and who had been all along somewhat suspicious of genetic recombination in bacteria, listened with interest to my description of Hayes' work. Though still far from convinced that there was anything to this sexual polarity business at all, he decided to invite Hayes to give a paper at the following Cold Spring Harbor Symposium on Viruses.

From the results of genetic crosses performed hitherto between lysogenic and non-lysogenic bacteria, it appeared that, whereas lysogeny segregates among the recombinants produced by mating a non-lysogenic "donor" to a lysogenic "recipient" bacterium, lysogeny is not transmitted to recombinants produced by mating a lysogenic donor to a non-lysogenic recipient bacterium. In order to obtain better data on this asymmetry of inheritance of lysogeny, I proceeded to prepare a multiply marked F^- recipient strain which was to be non-lysogenic and galactose-negative. In 1953, such a project still took time and effort. The result was a strain derived from Lederberg's strain W677 and called P678. Further crosses involving this strain then confirmed the asymmetric distribution of lysogeny.

In the meantime, François Jacob and I had joined efforts in the study of lysogeny of *E. coli* K12. When Hayes (1953) described the isolation of his Hfr strain that produces 10^3 to 10^4 times more recombinants than the previously known F^+ donors, it seemed promising to reinvestigate the genetic determination of lysogeny with this system. Thanks to Hayes, who sent us his Hfr strain, we could then unravel the cause of the asymmetry observed in crosses between lysogenic and non-lysogenic bacteria. Whereas the crosses between non-lysogenic Hfr and lysogenic F^- or between lysogenic Hfr and lysogenic F^- lead to segregation of the prophage character among the recombinants, and hence demonstrate the chromosomal location of prophage λ (Wollman and Jacob, 1954, 1957), crosses between lysogenic Hfr and non-lysogenic F^- result in the induction of phage development upon transfer of prophage to the zygote (Jacob and Wollman, 1954, 1956a). We called this phenomenon "erotic induction" at first, but later renamed it more decorously "zygotic induction." It showed that immunity devolves from a cytoplasmic substance, later called immunity repressor, which also is responsible for maintenance of the phage genome in the prophage state.

The discovery of zygotic induction, furthermore, allowed an operational distinction and definition of the different steps of the process of bacterial conjugation.

SEXUAL CONJUGATION AND ITS DIFFERENT STEPS

Just as bacteriophage research had long been obliged to focus on the last part of the reproductive cycle, i.e. lysis of the bacteria and release of mature progeny particles, study of genetic recombination in bacteria had been necessarily restricted to the appearance of genetic recombinants. In phage research the definition of the stages of intracellular growth, such as eclipse period and appearance of the first intracellular progeny particles, was essential for sorting out the different steps of phage multiplication. Similarly, zygotic induction now offered a means for defining the zygote and therefore for distinguishing the processes which take place *before* transfer, from those taking place *after* transfer, in the zygote or in its progeny. These steps could be operationally defined and the main features of the conjugation process could be analyzed by comparing crosses in which no zygotic induction takes place with crosses in which it does take place (Wollman, Jacob and Hayes 1956; Wollman and Jacob 1957).

The main conclusions of these experiments were the following:

(1) Different genetic markers of the donor are transmitted to recombinants with different frequencies, and can therefore be ranked according to the frequencies of their transmission.

(2) The frequency of formation of recombinants is neither equal to the frequency of transfer of the corresponding genes to the zygotes, nor, *a fortiori*, equal to the frequency of conjugation, which in such crosses could be estimated to be close to 100%.

(3) The unequal frequency of transmission to recombinants of different

donor markers is probably the consequence of an unequal frequency of transfer of the corresponding genes to the zygote.

Three main steps of the recombination process were therefore defined:

(a) conjugation, involving a specific pairing between male and female bacteria;

(b) transfer of genetic material, from male to female bacteria leading to the formation of partial zygotes, or merozygotes;

(c) genetic recombination in the merozygotes, leading to the formation of genetic recombinants (Jacob and Wollman, 1955).

When bacterial conjugation was thus proved to be a high frequency phenomenon, it became clear that it ought to be observable in both light and electron microscope. In 1955–1956 we were lucky to have T. F. Anderson spend his sabbatical leave with us. During that time, he succeeded in obtaining electron-micrographs of conjugating bacteria (Anderson, Wollman, and Jacob, 1957).

THE MECHANISM OF GENETIC TRANSFER

Having defined and distinguished the various steps involved in the conjugation process, it became possible to investigate separately the mechanisms of conjugation and genetic transfer. The first such experiments were made by Hayes (1955, 1957), who studied the kinetics of zygote formation by destroying the male donor bacteria with a virulent phage. In these experiments he looked at the transfer of those markers of the Hfr donor strain which are transmitted at the highest frequency. He found in this way that the kinetics of conjugation do not seem to differ significantly from the kinetics of zygote formation.

In the two years that we had been working together at Caltech, Gunther Stent had succeeded in giving me a nodding acquaintance with the principles of kinetic analysis. It seemed appropriate to make use of this knowledge in studying conjugation and genetic transfer by looking at the time course of different manifestations of the mating process. For this purpose, we decided to separate male Hfr donors from female F^- recipients at various times after the onset of conjugation, by taking samples from a mating mixture and submitting them to strong shearing forces generated in a Waring blendor. This shear treatment had already been used in bacteriophage studies by T. F. Anderson (1949) and by Hershey and Chase (1952), and had been found not to affect the viability of either normal bacteria or infectious centers.

The results of our experiments on the kinetics of the conjugation process itself showed that, whatever may be the timing of the later stages of the process, conjugation starts at the moment of mixing donor and recipient bacteria. The maximum number of recombinants reached depends on the particular genetic marker examined, and different genetic markers of the donor are transferred to zygotes at different times, the time of entry of the marker into the zygote being related to its position on the genetic map.

Mating between Hfr donor and F^- recipient bacteria was, therefore, seen to involve a peculiar mechanism of genetic transfer, different from any known to obtain in other organisms. Here, the chromosome of the male enters the female from a particular point, or "origin" (we gave it that name because Hayes [1953] had spoken of a "vector" of genetic transfer in conjugation), and transfer of the other genetic markers of the donor chromosome proceeds in a linear order, some hundred minutes being required for transfer of the whole donor chromosome. The spontaneous separation of the conjugating bacteria, and the resulting interruption of the transfer process, occurs with a fixed probability per unit time and accounts for the observed gradient of transmission of the donor markers. Most of the zygotes formed by this conjugation mechanism are thus partial zygotes, or merozygotes, that harbor only segments of the donor chromosome nearest to the origin. The shear-induced rupture of the donor chromosome during its transfer provided means to formulate a measure of genetic distance in a temporal dimension. Further, it was found that the genetic determinant of the Hfr character itself is transferred last of all donor characters, i.e., that it is located at the utmost distal end of the Hfr chromosome (Wollman and Jacob, 1955, 1958a; Wollman, Jacob, and Hayes, 1956).

It appeared, therefore, that the peculiarities hitherto observed in bacterial crosses were not, as had been thought, the consequence of some aberrations of genetic structure, but reflect the very nature of the process of chromosome transfer. Thus, as genetic systems, bacteria fell back into line with other, higher organisms, their peculiarity having been shown to reside merely in their method of intercellular genetic exchange.

SEXUAL TYPES, CHROMOSOMES, AND EPISOMES

As soon as the main features of the conjugation process in high frequency Hfr × F^- crosses had been clarified, it became of interest to reinvestigate the "classical" low frequency crosses between F^+ and F^- bacteria. It was quite apparent that in such low frequency crosses conjugation is as frequent as in Hfr × F^- crosses, since here the transfer of the sex factor itself proceeds at high frequency. However, the low frequency of recombinant formation is a consequence of the low frequency of transfer of the chromosomal markers (Jacob and Wollman, 1955). The situation, therefore, is very different in Hfr × F^- crosses and in $F^+ \times F^-$ crosses, with respect to the transfer of both the chromosomal markers and the genetic determinant of sex. Hence two questions were raised regarding this situation: What is the origin of genetic recombinants in $F^+ \times F^-$ crosses? And, what is the nature of the $F^+ \rightleftarrows$Hfr variation?

The first question could be epitomized as follows: is a population of F^+ bacteria homogeneous with respect to those cells which are responsible for the formation of recombinants? This problem then became essentially the one that Luria and Delbrück had faced thirteen years earlier when considering the origin of bacterial mutants resistant to the action of bacteriophage. By means of the methods available for demonstrating the spontaneous origin of bacterial mutants we could find the answer: recombinants are formed by spontaneous Hfr variants appearing in the F^+ population which, furthermore, can be isolated at will (Jacob and Wollman, 1956b). In retrospect, it seems surprising that this obvious question, posed by the low frequency of genetic recombination in bacteria, had not been considered for ten years.

The clue to the answer to the second question was that, concomitant with the formation of an Hfr variant of a given polarity, transferability of the extrachromosomal sex factor is lost. It was clear that this phenomenon could be most easily interpreted by assuming that the difference between F^+ and Hfr males is determined by the state of the sex factor: extrachromosomal in F^+ bacteria and chromosomal in the case of Hfr bacteria.

In the early spring of 1956, Sol Spiegelmann happened to be in Paris and on learning of our results wrote immediately to M. Demerec, asking him to include a paper on recombination in bacteria in the program of the 1956 Cold Spring Harbor Symposium on Genetic Mechanisms. By then, our results had been published only partially in preliminary form, and moreover only in French. Similarly Hayes, with whom we had been in close contact, and who had played a major role in putting forward the concept of unidirectional transfer, also had not yet published his experiments on the kinetics and the physiology of zygote formation and on the expression of transferred genetic characters. Thus the Cold Spring Harbor Symposium seemed an appropriate forum for giving a complete account of the actual status of the process of sexual conjugation in bacteria. This was the origin of the paper entitled "Conjugation and genetic recombination in *E. coli* K12" (Wollman, Jacob, and Hayes, 1956). The general picture of the mating system of *E. coli* which was drawn in that paper was however not readily accepted.

Another conclusion which was also first received with some scepticism was our picture of the structure of the bacterial chromosome. The genetic analysis of a number of Hfr strains we had isolated showed that each Hfr donor could be described as having a linear chromosome with its own particular origin of transfer and its own terminus where the integrated sex factor resides. All of these different genetic linkage groups appeared however to be circular permutations of each other. Hence we drew the conclusion that the F^+ bacteria from which all these Hfr variants originated, as well as F^- bacteria, possess a closed, or circular, chromosome. Female, F^- bacteria, are devoid of sex factor; male F^+ bacteria carry an autonomous, extrachromosomal sex factor; and Hfr bacteria derive from F^+ bacteria by insertion of the sex factor into the chromosome, which endows this chromosome with a defined polarity (Jacob and Wollman, 1957; Wollman and Jacob, 1958b).

These alternate states of the bacterial sex factor reminded us strongly of the behavior of the temperate bacteriophages that we had studied previously. Both of these kinds of genetic elements are dispensable; when they are present, they can be either autonomous or integrated into the chromosome, and when they are absent they can be acquired only from an external source. We called such genetic elements *episomes* (Jacob and Wollman, 1958).

Upon transition of an episome from the integrated, chromosomal state, to the extrachromosomal, autonomous state, neighboring genes may become incorporated into the episome. This occurs in the specific transduction by λ bacteriophage of the galactose (Morse, 1954) or biotin (Wollman, 1963) markers. It might have been anticipated that this could also be the case for an integrated sex-factor.

While spending the year 1958–59 in Berkeley with Gunther Stent, I was told by Edward Adelberg of the peculiar behavior of a substrain of an Hfr he had brought from Paris. This strain had properties intermediate between those of an F^+ and those of an Hfr donor. It transferred its chromosome in a specific oriented way, and also transferred at high frequency the intermediate donor character. The most likely interpretation seemed to me that upon leaving its chromosomal location the sex factor had incorporated a chromosomal fragment, thus acquiring a specificity for integration into the homologous chromosomal region of the recipient (Adelberg and Burns, 1959, 1960).

This interpretation proved to be correct and, shortly after, sex factors carrying known bacterial genes were isolated from appropriate Hfr strains (Jacob and Adelberg, 1959; Hirota, 1959). Such recombined sex factors offered the possibility, hitherto lacking for bacteria, of forming persistent diploid heterogenotes for any segment of the bacterial chromosome, thus greatly facilitating the analysis of the physiological genetics of bacteria (Jacob and Wollman, 1961).

Sexual conjugation in bacteria was thus seen to involve peculiar features of determination of sexual types and mechanism of genetic transfer. Once these peculiarities were appreciated, the importance of the phenomenon discovered by Lederberg and Tatum became even more conspicuous. Being the most highly evolved of the mechanisms of genetic transfer in bacteria bacterial conjugation has permitted a unitary view of the other known processes of partial genetic transfer, such as transformation, transduction, and sexduction; the term *meromixis* has been proposed for such processes of partial genetic transfer (Wollman and Jacob, 1959; Jacob and Wollman, 1961). Genetic elements other than the sex factor, such as certain colicinogenic factors or antibiotic resistance transfer factors, have since been shown to be able to determine conjugation and sometimes genetic transfer, although less efficiently than sexual conjugation.

Despite the versatility of the mechanisms of genetic transfer in bacteria, the studies on bacterial conjugation have proved that, from the genetic point of view, bacteria do not differ all that much from other organisms, and even offer a number of advantages for the analysis of the function of the genetic material (Jacob and Wollman, 1961). If, therefore, bacterial conjugation has by now long ceased to be an exciting mystery, it has become a rather useful, workaday tool for the study of more basic and important problems.

REFERENCES

ADELBERG, E. A., and S. N. BURNS. 1959. A variant sex factor in *E. coli*. Genetics, *44*: 497.

———, ———1960. Genetic variation in the sex factor of *E. coli*. J. Bacteriol., *79*: 321–330.

ANDERSON, T. F. 1949. The reactions of bacteria viruses with their host cells. Botan. Rev., *15*: 477.

ANDERSON, T. F., E. L. WOLLMAN, and F. JACOB. 1957. Sur les processus de conjugaison et de recombinaison génétique chez *E. coli*. III. Aspects morphologiques en microscopie électronique. Ann. Inst. Pasteur, *93*: 450–455.

CAVALLI-SFORZA, L. L., J. LEDERBERG, and E. M. LEDERBERG. 1953. An infective factor controlling sex compatibility in Bacterium coli. J. Gen. Microbiol., *8*: 89–103.

DELBRÜCK, M. 1950. Ed., Viruses, California Institute of Technology, Pasadena, 1950.

DOERMANN, A. H. 1948. Intracellular growth of bacteriophage. Carnegie Inst. Wash. Yearbook, *47*: 176–186.

HAYES, W. 1952. Recombination in *Bact. Coli* K12: unidirectional transfer of genetic material. Nature, *169*: 118–119.

———1953. The mechanism of genetic recombination in *E. coli*. Cold Spring Harbor Symp. Quant. Biol., *18*: 75–93.

———1955. A new approach to the study of kinetics of recombination in *E. coli* K12. Proc. Soc. Gen. Microbiol. J. Gen. Microbiol. 13, ii.

———1957. The kinetics of the mating process in *E. coli*. J. Gen. Microbiol., *16*: 97–119.

HERSHEY, A. D., and M. CHASE. 1952. Independent functions of viral protein and nucleic acid in growth of bacteriophage. J. Gen. Physiol., *36*: 39–56.

HIROTA, Y. 1959. Mutants of the sex factor in *E. coli* K12. Genetics, *44*: 515.

JACOB, F., and E. A. ADELBERG. 1959. Transfert de caractères génétiques par incorporation au facteur sexuel d'*E. coli*. Compt. Rend. Acad. Sci., *249*: 189–191.

JACOB, F., and E. L. WOLLMAN. 1954. Induction spontanée du développement du bactériophage λ au cours de la recombinaison génétique chez *E. coli* K12. Compt. Rend. Acad. Sci., *239*: 455–456.

JACOB, F., and E. L. WOLLMAN. 1955. Etapes de la recombinaison génétique chez *E. coli* K12. Compt. Rend. Acad. Sci., *240*: 2566–2568.
———, ———1956a. L'induction par conjugaison ou induction zygotique. Ann. Inst. Pasteur, *91*: 486–510.
———, ———1956b. Recombinaison génétique et mutants de fertilité chez *E. coli* K12. Compt. Rend. Acad. Sci., *242*: 303–306.
———, ———1957. Analyse des groupes de liaison génétique de différentes souches donatrices d'*E. coli* K12. Compt. Rend. Acad. Sci., *245*: 1840–1843.
———, ———1958. Les épisomes, éléments génétiques ajoutés. Compt. Rend. Acad. Sci., *247*: 154–156.
———, ———1961. Sexuality and the genetics of bacteria. Academic Press, New York.
LEDERBERG, E. M. 1951. Lysogenicity in *E. coli* K12. Genetics, *36*: 560.
LEDERBERG, E. M., and J. LEDERBERG. 1953. Genetics study of lysogenicity in *E. coli.* Genetics, *38*: 51–64.
LEDERBERG, J., E. M. LEDERBERG, N. D. ZINDER, and E. R. LIVELY. 1951. Recombination analysis of bacterial heredity. Cold Spring Harbor Symp. Quant. Biol., *16*: 413–441.
LEDERBERG, J., and E. L. TATUM. 1946a. Novel genotypes in mixed cultures of biochemical mutants of bacteria. Cold Spring Harbor Symp. Quant. Biol., *11*: 113–114.
———, ———1946b. Gene recombination in *E. coli.* Nature, *158*: 558.
MORSE, M. L. 1954. Transduction of certain loci in *E. coli* K12. Genetics, *39*: 984–985.
WEIGLE, J. J., and M. DELBRÜCK. 1951. Mutual exclusion between an infecting phage and a carried phage. J. Bacteriol., *62*: 301–318.
WOLLMAN, E., and Mme E. WOLLMAN. 1937. Les phases des bactériophages (facteurs lyogènes). Compt. Rend. Soc. Biol., *124*: 931–934.
———, ———1938. Recherches sur le phénomène de Twort d'Hérelle (bactériophagie ou autolyse hérédo-contagieuse). Ann. Inst. Pasteur, *60*: 13–58.
WOLLMAN, E. L. 1953. Sur le déterminisme génétique de la lysogénie. Ann. Inst. Pasteur, *84*: 281–294.
———1963. Transduction spécifique du marqueur biotine par le bactériophage λ. Compt. Rend. Acad. Sci., *257*: 4225–4226.
WOLLMAN, E. L., and F. JACOB. 1954. Lysogénie et recombinaison génétique chez *E. coli* K12. Compt. Rend. Acad. Sci., *239*: 455–456.
———, ———1955. Sur le mécanisme du transfert de matériel génétique au cours de la recombinaison chez *E. coli* K12. Compt. Rend. Acad. Sci., *240*: 2449–2451.
———, ———1957. La localisation chromosomique du prophage λ et les conséquences génétiques de l'indiction zygotique. Ann. Inst. Pasteur, *93*: 323–339.
———, ———1958a. Le mécanisme du transfert de matériel génétique. Ann. Inst. Pasteur, *95*: 641–666.
———, ———1958b. Sur le déterminisme génétique des types sexuels chez *E. coli* K12. Compt. Rend. Acad. Sci., *247*: 536–539.
———, ———1959. La sexualité des bactéries. Masson ed. Paris.
WOLLMAN, E. L., F. JACOB, and W. HAYES. 1956. Conjugation and genetic recombination in *E. coli.* Cold Spring Harbor Symp. Quant. Biol., *21*: 141–162.

J. WEIGLE
Division of Biology, California Institute of Technology, Pasadena, California

Story and Structure of the λ Transducing Phage

"A serious ape whom none take seriously obliged in this fool's world to earn his nuts by hard buffoonery."—GEORGE ELIOT

Zinder and Lederberg had discovered (1952) that a phage of *Salmonella* can transfer single bacterial genes from one bacterium to another. This transfer of genetic material by a phage vector became known as transduction (although at first Lederberg, who invented the name, intended it to cover a wider range of phenomena). The bacterial gene transduced by the phage is probably enclosed in the phage coat; almost any gene of the donor bacterium in which the transducing phage had grown can be transferred to the recipient bacterium. How the gene gets inside the phage coat, what its relations are to the phage genome, how it becomes integrated into the recipient bacterial genome, these were questions which were (and mostly still are) unanswered.

Two years later, Morse (1954), working in Lederberg's laboratory, discovered that phage λ, for which the K12 strain of *Escherichia coli* is normally lysogenic, can transduce the *gal* loci, a series of genes adjacent to the prophage site, controlling the utilization of galactose. In the prophage state, this phage is inserted into the chromosome of the lysogenic bacterium between the *gal* loci and the gene for biotin synthesis. It can transduce only these two markers, but never both together. Transducing λ phages are found only in lysates obtained after induction of lysogenic cultures and never in lysates from sensitive bacteria infected with λ. This *specialized* transduction has other remarkable properties which were discovered by Morse and the Lederbergs. They published their observations in two beautiful papers in 1956 (Morse, Lederberg and Lederberg, 1956 a, b) in which they showed that the recipient bacteria which receive a *gal* gene by transduction are heterogenotes (or, more generally, syngenotes); that is, they carry both their own *gal* gene and, in addition, the donor *gal* gene brought in by λ. Moreover, these heterogenotes are lysogenic, and when induced with UV light they produce nearly as many transducing as plaque forming phages. Specifically, when cultures containing 10^9 bacteria per ml were induced, about 4×10^8 per ml plaque formers and about 4×10^7 per ml transducers were found. It seemed that upon UV induction each heterogenotic bacterium liberates, on the average, less than one infective and one transducing phage particle.

Before 1956, transducing lysates obtained from either *Salmonella* or *E. coli* contained no more than about one transducing particle (for a given gene) in 10^6 plaque formers. But the lysates of induced heterogenotes, called HFT (high frequency of transduction) in contrast to LFT (low frequency of transduction), contained so many transducing particles that it became possible to study the nature of the transducing phage. For that reason, I shall restrict myself here mostly to a description of transducing λ phages.

I remember being in a hospital room when I first saw the paper of Morse and the Lederbergs. Someone (it may have been George Streisinger) was keeping me posted on the activities of the laboratory at Caltech and brought me this paper after a seminar had been given on it. For some reason I can now no longer remember even after re-reading that paper, I was immediately convinced that the transducing λ particles are a sort of mutant of λ and that the UV-induced heterogenotes should yield a normal burst of 100 particles. I knew that, to use Max Delbrück's disdainful indictment of phage work which did not come up to his standards, the Lederbergs, the specialists *par excellence* of bacterial crosses, "had never taken the phage course." Thus, I did not believe that only about one transducing particle is liberated per induced bacterium (although Morse and the Lederbergs were careful not to attribute any importance to that one-to-one relationship). Anyway, a simple experiment would have proven that the transducing particles multiply as phages: infect a sensitive culture of bacteria with an HFT lysate and show that after lysis of the infected bacteria there are many more transducing phages than before infection.

Thinking back upon those days, I wonder if my strong conviction did not stem from an idea which had been imprinted on my mind during an evening conversation with James Watson in Geneva in 1952. We had just returned from a meeting in Bellagio, where Zinder and Lederberg's discovery of phage-mediated transduction had been discussed, and that evening we had said "let us play with the wildest ideas we can imagine." It must have been Watson who proposed that during phage multiplication the phage chromosome necessarily pairs with the bacterial chromosome and occasionally crosses over. This idea led us to envisage various consequences that seemed to contradict what we then knew, and soon we were forced to discard it. However, during the following winter I happened to plate UV-inactivated λ phages on UV'd bacteria and observed that such bacteria reactivate a very large fraction of the "dead" phages. And among these reactivated phages a large proportion are mutants. So I thought that perhaps the UV-damaged DNA of the phage is replaced by undamaged pieces of the bacterial genome, irradiation of the bacterium having increased recombination between phage and host and replacement of a phage gene by a bacterial gene causing a mutant genotype. Just for laughs, as Hershey would say, I looked to see whether the reactivated phage lysates contained transducing particles, unfortunately choosing streptomycin resistance as the test marker for transduction. Not surprisingly, I failed to find any transduction of streptomycin resistance by λ, but the idea of phage genomes patched with pieces of bacterial genome lingered in my mind.

When I left the hospital and came back to the laboratory in 1956 I told Guiseppe Bertani that I was certain that a piece of bacterial DNA, namely the *gal* region, could be incorporated into the λ phage chromosome and that in this state it would multiply at the rate of phage multiplication. Since the prophage incorporated into the bacterial chromosome multiplies by courtesy of whatever controls the rate of multiplication of the bacterial DNA, it seemed obvious to me that a piece of bacterial DNA incorporated into the phage chromosome would multiply by courtesy of the mechanism of phage DNA duplication. This idea, which today seems trivial, filled me then with such enthusiasm that it must have been contagious, for Bertani immediately made me do the experiments that would show that transducing particles multiply. Had it not been for Bertani I might never have done the experiments, so strong was my conviction that I was right. And so I made my first heterogenote by transducing the K12-112 gal^- strain of Elie Wollman to gal^+/gal^- with an LFT lysate from induced C600 (λ). Upon induction of the heterogenote, I obtained an HFT lysate in which the number of λ transducing particles was at least 10 times (and probably 50 times) the number of induced bacteria. Infection of a λ-sensitive culture with the HFT lysate at a multiplicity of one plaque-former per bacterium produced a lysate which contained about 50 times more transducing particles than had been used for infection. Thus, quite in accord with my expectation of phage-like multiplication of the bacterial genes, each bacterium infected with one transducing particle gave a burst of 50 of these particles. Happy with these modest results, I wrote a small paper and sent it to *Virology*, and felt free to go further into the study of the structure of the transducing phage λ. But the paper came back with criticisms from Luria, one of the editors. I made some revisions, but the paper was again returned. I made further revisions, but the paper was returned for a third time. At this point I complained to Delbrück saying that I did not want to spend all my time revising one little paper, since doing experiments was more amusing. Delbrück put on the angelic smile he sometimes uses and asked me whether it was not true that each new version of my paper was better written than the previous one. And I had to admit that this was true. I like to think, though, that Luria's very careful reading of my successive efforts finally had a double effect: first, it led to a better paper (Weigle, 1957), and second, it may have led Luria and his collaborators to repeat with the transduction of the lactose genes by phage P1 what had been done with the transduction of the galactose genes by phage λ.

Morse and the Lederbergs had noted that the bacteria transformed by λ transduction were almost always immune to λ and often carried a defective prophage, in that after UV-induction such bacteria would lyse but produce no phages. At that time, Werner Arber was studying defective prophages for his doctoral dissertation at the University of Geneva, and I suggested to him that he should also investigate this new source of defectives. Grete Kellenberger, Arber, and I worked on λ transduction in Geneva the whole summer of 1956, and when I returned to California they continued this work, while I was mostly busy constructing homogenotes and other necessary strains. Allan Campbell had come to Pasadena to work with Bertani that summer; I had found "something funny" in transductions when the λ phages carried the host range, or *h*, marker, and Campbell began to investigate this anomaly. Ever since those days, both Arber and Campbell have been working with phage λ and have made meanwhile some most fundamental discoveries. The "suppressor-sensitive" mutants of Campbell (1961) and the clarification of the host specificity phenomenon by Arber (1958) were not only important technically but opened new vistas of understanding of the function of the genetic material. Indirectly, these discoveries were consequences of the work on transducing λ.

Let us call the transducing λ by the name which Arber gave it: *λdg* (*d* for defective, *g* for galactose). With hindsight, it now seems obvious that the transducing particles must be defective and that incorporation of the *gal* loci into the λ chromosome engenders deletion of at least the *h* marker of the phage. We did not immediately see that *λdg* is defective for two reasons: 1) *λdg* was assayed by transduction involving a lysogenization, whereas λ was assayed as plaque former, and 2) *λdg* alone, without the help of λ, lysogenizes sensitive bacteria with the very low frequency of about 1%, compared to 20% for λ or for *λdg* assisted by λ. I had been lucky in my first experiments on the multiplication of *λdg*, for I had used a high enough multiplicity of infection to insure that many sensitive bacteria were infected with both λ and *λdg*. Had I been more thorough and measured the yield of *λdg* as a function of the multiplicity of infection, I would have found immediately that *λdg* is defective and requires the presence of a normal phage for its multiplication. This very type of measurement was done by Arber and Grete Kellenberger during the winter of 1956–57, while I measured the killing titer of the HFT lysate. So, by the next summer, we, as well as Campbell, had arrived at an almost complete understanding of the structure of *λdg*. We wrote then: "One must think that the heterogenotes liberating phages are double lysogens, they carry an active and a defective prophage, the defect being due to the insertion of Gal into the genome of the prophage" (and we show further that a region involving h is deleted). "The heterogenotes, after induction, thus produce the two sorts of phages in approximately equal quantities. The defective heterogenotes are lysogenic for the defective Gal prophage only. After induction they are not able to produce phages because of the defect, although lysis occurs (lysozyme synthesis is not blocked by the defect). The active phage brings in the functions lost by the deletion in the defective phage and allows a normal yield by complementation" (Arber, Kellenberger and Weigle, 1957). Arber gave ample experimental proofs of these assertions in his dissertation.

Just at that time, Meselson, Stahl, and Vinograd had developed the CsCl density gradient equilibrium centrifugation at Caltech, and by its use Meselson and Stahl had shown the correctness of the Watson and Crick model for duplication of the DNA. But Delbrück was not really convinced by this experiment and had invented explanations other than semi-conservative replication to account for the origin of hybrid DNA molecules. When Stahl left Pasadena, Meselson worked feverishly, running the analytical ultracentrifuge 24 hours a day, to show that Delbrück was wrong. It was then that I realized that *λdg*, harboring both a deletion and an insertion in its genome, might contain a different amount of DNA, and hence have a different density from normal *λ*. So I asked Meselson to put an HFT lysate in the centrifuge. At first Meselson was not very responsive, not only because he was so busy showing that Delbrück was wrong, but also because he and Stahl had already put T4 phage in CsCl gradients and found such a broad distribution that he thought the density heterogeneity of normal phages would obscure whatever differences we could hope to observe with *λ* and *λdg*.

Finally, Meselson did put *λ* into the centrifuge, and to our surprise (which was not to be the last) we found a very narrow distribution, almost as narrow as the molecular weight of *λ* would have led us to expect. Then we centrifuged the HFT lysate obtained by induction of my first heterogenote. Lo and behold, we found four bands of phages! The weekend had come by then and on the way to a camp site in the desert Delbrück asked me what the result of our experiment had been. Upon learning that there were four phage bands, his comment was brief: "Life is complicated, Jean, isn't it?" Next, we made a preparative CsCl gradient equilibrium run of the HFT lysate and collected successive layers of the gradient dropwise by piercing the bottom of the centrifuge tube. Analysis of the fractions revealed that the two most dense bands were infective *λ* phages, the two others transducing *λdg*'s. When we later studied another HFT lysate, we found only two bands, one of *λ* and one of *λdg*. Finally, following an inspiration by Kenneth Paigen, we made many independent heterogenotes and upon centrifugation of their HFT lysates found only two bands in each. The four bands of the very first heterogenote never occurred again, although I must have studied about 50 independent heterogenotes (Weigle, Meselson and Paigen, 1959). The λ's always had the same density and the λdg's were either more or less dense than λ; some λdg's contained only about ¾ of the DNA of λ. But in the original four band-lysate one of the λ bands was denser than wild type λ; later studies showed that this dense character of λ is genetically stable, the gene responsible for the high density being located in the C region of the λ genome. The two λdg's of this lysate differed in density by the same amount as the two λ's; in other words, they carried the same deletion and the same substitution. Although it still remains unexplained, this observation opened our eyes to the fact that λ's of different densities, and presumably of different DNA contents, can be found. This realization became important in our later studies of the mechanism of recombination and lysogenization. I wonder in retrospect if we would have bothered to study the density distribution of a second HFT lysate, had it not been for this mystery of the four bands. And not knowing that there exist λ phages of different density would we have studied recombination and found that it can occur by breaks and reunions?

This is about all that is known of the structure of λdg. We may now consider what is known about its formation and its integration after transduction. The main ideas we have concerning these problems are due to Allan Campbell, and although they are still partly speculative, they do provide a nice picture of what *could* happen. On infecting a bacterium the chromosome of the phage is thought to become a circle; phage chromosome and bacterial chromosome may then synapse in a region of assumed homology within which a single crossover would insert the phage genome into the bacterial chromosome and render the bacterium lysogenic. The crossing over occurs between the λ genes *C* and *h*, so that after insertion, the prophage has its *C* marker toward the *gal* genes of the bacterium and the *h* marker away from them. This is the picture of lysogenization. Induction would be the inverse process: a loop including the prophage forms, and a single crossing over within the region of homology releases the prophage and allows it to multiply independently. Very rarely the releasing crossing over occurs at the wrong place, leaving out the *h* gene from the λ genome and introducing the *gal* genes instead, thus giving rise to a λdg genome. The insertion of λdg upon lysogenization of a recipient bacterium thus produces a duplication of the gal loci in the bacterial chromosome, and the very extensive homology of λdg with another nearby part of the bacterial chromosome and the attendant high probability of looping explains the instability of the λdg lysogens. There are still some difficulties with this model, however, when one tries to make it too precise.

So far I have spoken of the nature of the λ transducing phage. I now turn to some later research, which was affected, directly or indirectly, by our work on λdg.

The separation of λdg from λ by centrifugation in CsCl gradients was being made when Aaron Novick came to visit us in Pasadena. He was quick to see that the effect of gene dosage on enzyme synthesis could be studied with λdg, since by infection with these phages one could introduce into a bacterium as many *gal* operons as one wanted. The same idea probably occurred to many other workers, and especially to Peter Starlinger and Kenneth Paigen, who were with us at the time and who, having a then still unusual bent for biochemistry, carried out such experiments. Starlinger's results showed that the rate of synthesis of the enzyme galactokinase does not depend on the number of *gal* operons in the cell but decreases when the cell is simultaneously infected with λdg and λ (Starlinger, 1963). The interpretation of these results is still difficult, because one does not know the extent of λdg DNA duplication after infection, and because nothing is known at the present time about the mechanism by which the phage chromosome gets an advantage over the chromosome of the bacterium in determining protein synthesis. Buttin later studied the function of the *gal* operon by means of λdg, and Adler and Kaiser used λdg for the genetic mapping of the *gal* operon, finding that the gene order is galactokinase, transferase, operator, epimerase and then, further away, λ prophage (Adler and Kaiser, 1963).

Let us turn to another field of research which λdg opened, although its main interest today has nothing to do with *d* or *g*. Kaiser and Hogness decided to prepare the DNA of λdg. I imagine (although I have not inquired into the truth of this surmise) that the ultimate goal of their experiments was to duplicate in vitro the *gal* DNA by means of Kornberg's DNA polymerase and to show that the in vitro DNA is biologically active; that is, once introduced into a bacterium it would be capable of synthesizing the *gal* enzymes. A prerequisite, however, to these experiments was to be able to introduce the naked DNA into bacteria without the help of the phage coat. Knowing that λdg is defective and lysogenizes very poorly without the help of a complementing nondefective phage, Kaiser and Hogness must have decided to increase their chances of observing λdg DNA incorporation into a bacterium by infecting the bacterium with what they called a helper-phage having most of the genetic markers, except the immunity of λ. Their source of λdg was a heterogenote which after induction gave an HFT lysate containing four times as many λdg's as λ's, making their separation in CsCl very easy (the reason why different heterogenotes give different proportions of $\lambda dg/\lambda$ is still unknown). They then infected gal^- recipient bacteria with λdg DNA and the helper phage and plated the infected cells on galactose indicator agar. In this way they observed transformed cells that were gal^+/gal^- heterogenotes; that is, cells which carried their original gal^- marker and the gal^+ marker brought in by the donor DNA. What happened to the presumed attempts to synthesize active DNA I don't know, but I do know that what was initially a side issue of this "transformation" experiment, the need for a *helper* phage to make *E. coli* K12 cells competent to receive DNA, turned out to be most important. By use of a helper, infection with λ DNA is relatively effective, about one molecule of DNA in 1,000 succeeding in entering the bacteria, whereas infection without helper is hardly measurable.

I believe that Allan Campbell discovered the *Sus* mutants of λ because of *λdg*, although I never asked him how it came about. He probably thought that some of the UV'd phages, reactivated by plating on UV'd bacteria, may have been repaired by insertion of *gal* genes into their genome. To test this point, what would be simpler than to pick the plaques of phages reactivated on the gal^+ strain C600 and spot test them on a gal^- indicator strain for gal^+ transduction? If some of the reactivated phages were *λdg*, they should transform the gal^- bacteria to gal^+ bacteria. I am sure that Campbell did not find any *λdg* in this way, but he did observe that the phages in many plaques of UV'd λ reactivated on C600 do not grow on the gal^- indicator strain that he used. After having shown that they are genetically stable, he called these mutants "suppressor sensitives," or *Sus*, because he showed that C600 carries a suppressor gene which restores function to the gene defective in the phage mutant, a bacterial gene which is absent from the gal^- strain.

So now, we are back to Watson's notion of recombination between phage and bacteria. I will take advantage of the forum provided to me here and now to publish for the first time some results pertaining to this idea that Grete Kellenberger and I obtained in 1958. Perhaps we never sent them to *Virology* for fear of another rejection. Anyway, presenting those data here seems quite appropriate, since not long ago Delbrück told me that he supposed that in honor of his 60th birthday "they" will produce a Festschrift in which everyone will publish papers that have been repeatedly rejected by many journals!

The problem is this: Does a phage which is known to have part of its genome homologous to the bacterial genome recombine with it during lytic development? In the *gal* operon different genes are known: $gal\ 1^-$ for instance is deficient in transferase, while $gal\ 2^-$ is deficient for galactokinase. So we can put the problem in this way: When a *λdg* which is $gal\ 1^-$ $gal\ 2^+$ multiplies in a $gal\ 1^+$ $gal\ 2^-$ bacterium, does it produce progeny which are $gal\ 1^+$ $gal\ 2^+$ (or $gal\ 1^+$ $gal\ 2^-$)?

For brevity, in what follows we shall omit the *gal* and call a bacterium 1^+2^- when it carries the $gal\ 1^+$ and the $gal\ 2^-$ genes, etc., and also call the

λdg simply $1^- 2^+$. The experiment is easily carried out since λdg $1^+ 2^+$ can be detected by plating the lysate (and helper phage) with $1^- 2^-$ bacteria on the appropriate indicator plates and counting the *gal*$^+$ transductants.

A number of precautions were taken and controls done: for instance, the stock of helper phage λ was made by induction of $1^- 2^-$ (λ), to make sure that it did not contain any λdg *gal*$^+$. The λdg's came from HFT lysates of induced homogenotes, the bacterium containing the same *gal* markers as the defective prophage, either $1^+ 2^-$ or $1^- 2^+$. Growing λdg $1^- 2^+$ in a $1^- 2^+$ or a $1^- 2^-$ bacterium gave no phage able to transform a $1^- 2^-$ bacterium into a *gal*$^+$. The same test was performed on a λdg $1^+ 2^-$ on the corresponding bacteria. For the experiment itself, the bacteria were infected with a multiplicity of 3λ and of 0.15 λdg per bacterium.

To make sure that any new λdg's found were the result of recombination in the *gal* regions of phage and bacterium, either the bacteria or the phages were given different doses of UV radiation, a treatment that is known to increase recombination. A typical result is given in Table 1, where it can be seen that recombinant λdg are indeed produced.

Table 1

Dose of UV (in seconds) on bacteria	*Concentration in the lysate of phage transducing different bacteria to Gal $1^+ 2^+$*			*Ratio λdg $1^+ 2^+ / 1^- 2^+$*
	Bacteria $1^+ 2^-$	$1^- 2^+$	$1^- 2^-$	
0	2.4×10^6	8.5×10^2	8.3×10^2	3.5×10^{-4}
45	2.6×10^6	4.4×10^3	5.3×10^3	2.0×10^{-3}
90	1.2×10^6	5.6×10^3	5.9×10^3	4.9×10^{-3}
120	9.7×10^5	6.7×10^3	5.7×10^3	5.9×10^{-3}
150	9.5×10^5	6.7×10^3	6.4×10^3	4.7×10^{-3}
210	4.5×10^5	2.8×10^3	2.8×10^3	4.2×10^{-3}

Cross of phage λdg *gal* 1^- *gal* 2^+ with bacteria *gal* 1^+ *gal* 2^- having received different doses of UV. The burst size of λdg is about 20 phages per bacterium.

A similar experiment, but irradiating λdg instead of the bacteria, gives an even larger increase in the number of recombinant λdg $1^+ 2^+$. And the experiments with λdg $1^+ 2^-$ crossed with bacteria $1^- 2^+$ again give similar results. All these experiments show clearly that during the vegetative growth of λdg the phage recombines with the bacterium in the *gal* region of homology. Since the efficiency of transduction on $1^- 2^+$ and $1^- 2^-$ are not necessarily equal, it is not possible to decide, from the values given in the table, whether the lysate contains, in addition to the $1^+ 2^-$ parents, only $1^+ 2^+$ recombinants, or if a mixture of $1^+ 2^+$ and $1^- 2^+$ recombinants appears in the progeny. It is not possible either to calculate a frequency of recombination, because in each bacterium giving a burst of *λdg* there were also many vegetative infective *λ* phages. One can imagine that the presence of these infective phages affects the relation between *λdg* and bacterium, and hence the frequency of recombination. However, to gain some idea of the efficiency of this recombination, we studied also the recombination in the *gal* region of two *λdg* phages. A culture of bacteria $1^- 2^-$ was infected with a multiplicity of 3 *λdg* $1^- 2^+$, 3 *λdg* $1^+ 2^-$ and 10*λ*. The culture was chloroformed after 60 minutes at 37°C. The lysate contained per ml. 6.9×10^6 *λdg* $1^+ 2^-$, 8.2×10^6 *λdg* $1^- 2^+$, and 4.6×10^4 *λdg* $1^+ 2^+$. Since we do not know the relative efficiency of transduction of the different bacteria on which the *λdg*'s were assayed, the above numbers are only an approximate reflection of recombination frequencies. One can see, however, that the number of recombinants between *gal* 1 and *gal* 2 in two *λdg* phages is of the same order of magnitude as the number which appears in a "cross" between *λdg* and the *gal* loci of the bacterial chromosome. (In these last experiments we were careful to test the lysates at very low multiplicities of infection, to avoid simultaneous transduction by the two parental phages, which occurs in 30% of the doubly infected bacteria, giving a triplication of the *gal* loci.)

So, that is the story of *λdg*; it is also the story of the permeating influence of a questioning mind, producing in those near it another sort of questioning attitude which could be expressed this way: "What will Max think of it, if he does think about it?"

REFERENCES

Adler, J. and A. D. Kaiser. 1963. Mapping of the galactose genes of *Escherichia coli* by transduction with phage P1. Virology, *19*: 117.

Arber, W. 1958. Transduction des charactéres Gal par le bactériophage λ. Arch. Sci. (Geneva), *11*: 259.

Arber, W., G. Kellenberger, and J. Weigle. 1957. La défectuosité du phage λ transducteur. Schweiz. Z. F. Allgem. Path. U. Bakteriol., *20*: 659.

Campbell, A. 1961. Sensitive mutants of bacteriophage λ. Virology, *14*: 22.

Morse, M. L. 1954. Transduction of certain loci in *Escherichia coli* K12. Genetics, *39*: 984.

Morse, M. L., E. M. Lederberg and J. Lederberg. 1956a. Transduction in *Escherichia coli* K12. Genetics, *41*: 142.

———, ———, ———1956b. Transductional heterogenotes in *Escherichia coli*. Genetics, *41*: 758.

Starlinger, P. 1963. The formation of galactokinase in cells of *Escherichia coli* after infection with the transducing λ bacteriophage. J. Mol. Biol., *6*: 128.

Weigle, J. 1957. Transduction by coliphage λ of the galactose marker. Virology, *4*: 14.

Weigle, J., M. Meselson and K. Paigen. 1959. Density alterations associated with transducing ability in the bacteriophage λ. J. Mol. Biol., *1*: 379.

Zinder, N. and J. Lederberg. 1952. Genetic exchange in *Salmonella*. J. Bacteriol., *64*: 679.

V. *DNA*

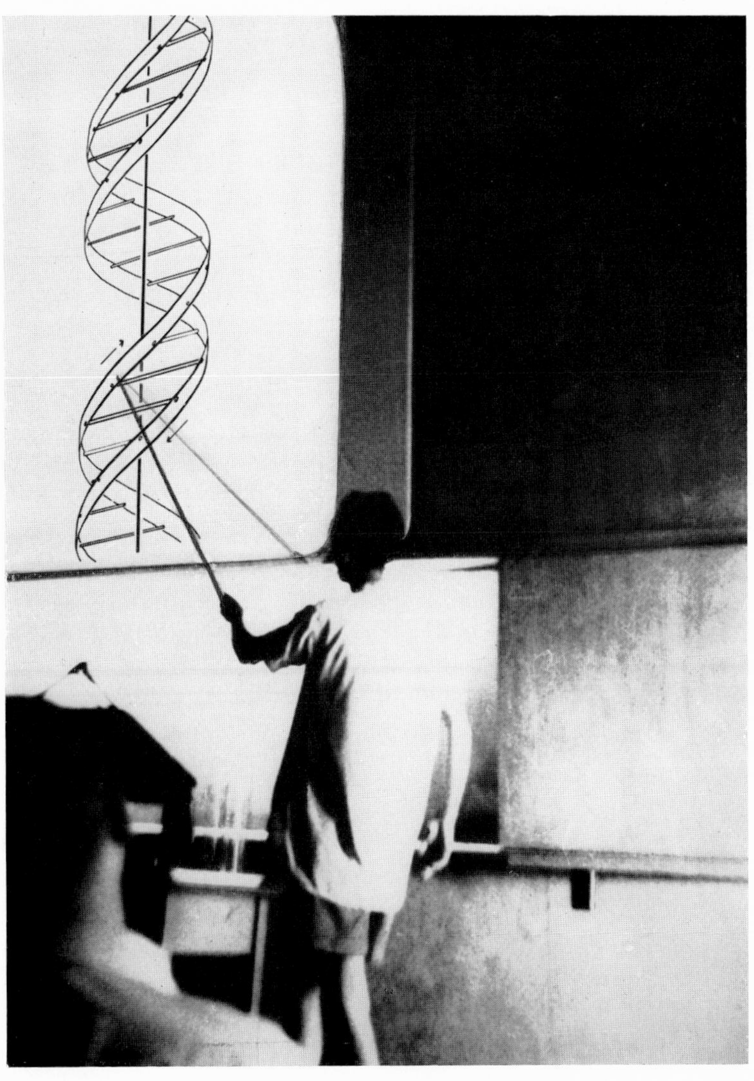

J. D. Watson, Cold Spring Harbor Symposium, 1953

J. D. WATSON
The Biological Laboratories, Harvard University, Cambridge, Massachusetts

Growing Up in the Phage Group

As an undergraduate at Chicago, I had already decided to go into genetics even though my formal training in it was negligible, with most of my course work reflecting a boyhood interest in natural history. Population genetics at first intrigued me, but from the moment I read Schrodinger's "What is Life" I became polarized toward finding out the secret of the gene. My obvious choice for graduate school was Caltech, since I was told its Biology Division was loaded with good geneticists. They, however, did not want me, nor did Harvard, to which I had applied without considering what I might find. Harvard's disinterest in me was particularly fortunate, for if I had gone there I would have found no one excited by the gene and so might have been tempted to go back into natural history. Fortunately my advisor at Chicago, the human geneticist Strandskov, also had me apply to the Indiana University in Bloomington, emphasizing that H. J. Muller was there as well as several very good younger geneticists (Sonneborn and Luria). To my relief, Indiana took a chance with me, offering a $900 fellowship for the coming 1947–48 academic year. Characteristically, Fernandus Payne, then dean of its graduate school, wanted to make sure that I knew what I was getting into. He appended a postscript to the fellowship offer saying that if I wanted to continue my interest in birds I should go elsewhere.

During my first days at Indiana, it seemed natural that I should work with Muller but I soon saw that Drosophila's better days were over and that many of the best younger geneticists, among them Sonneborn and Luria, worked with micro-organisms. The choice among the various research groups was not obvious at first, since the graduate student gossip reflected unqualified praise, if not worship, of Sonneborn. In contrast, many students were afraid of Luria who had the reputation of being arrogant toward people who were wrong. Almost from Luria's first lecture, however, I found myself much more interested in his phages than in the Paramecia of Sonneborn. Also, as the fall term wore on I saw no evidence of the rumoured inconsiderateness toward dimwits. Thus with no real reservations (except for occasional fear that I was not bright enough to move in his circle) I asked Luria whether I could do research under his direction in the spring term. He promptly said yes and gave me the task of looking to see whether phages inactivated by X rays gave any multiplicity reactivation.

The only other scientist in Luria's lab then was Renato Dulbecco, who six months previously had come from Italy to join in the experiments on the multiplicity reactivation of UV-killed phages. I was given a desk next to Dulbecco's and, when he was not doing experiments, often worked on his lab bench. Most of Luria's and Dulbecco's conversation was in Italian, and I might have felt somewhat isolated had it not been for the fact that my first experiments gave a slightly positive result. Usually, Luria never let even a few hours pass between the counting of my plaques and his knowing the answer. Also, Dulbecco's family had not yet come from Italy, and we would occasionally eat together at the Indiana Union. During one Sunday lunch, I remember asking him whether Luria's figure of 25 T2 genes should not tell us the approximate size of the gene since the molecular weight of T2 could be guessed from electron micrographs. Dulbecco, however, did not seem interested, perhaps because he already suspected that multiplicity reactivation of UV-killed phage was more messy than Luria's pretty subunit theory proposed. Then there was also the fact that despite Avery, McCarthy, and MacCleod, we were not at all sure that only the phage DNA carried genetic specificity.

Some weeks later in Luria's flat, I first saw Max Delbrück, who had briefly stopped over in Bloomington for a day. His visit excited me, for the prominent role of his ideas in "What is Life" made him a legendary figure in my mind. My decision to work under Luria had, in fact, been made so quickly because I knew that he and Delbrück had done phage experiments together and were close friends. Almost from Delbrück's first sentence, I knew I was not going to be disappointed. He did not beat around the bush and the intent of his words was always clear. But even more important to me was his youthful appearance and spirit. This surprised me, for without thinking I assumed that a German with his reputation must already be balding and overweight.

Then, as on many subsequent occasions, Delbrück talked about Bohr and his belief that a complementarity principle, perhaps like that needed for understanding quantum mechanics, would be the key to the real understanding of biology. Luria's views were less firm, but there was no doubt that on most days he too felt that the gene would not be simple and that high powered brains, like Delbrück's or that of the even more legendary Szilard, might be needed to formulate the new laws of physics (chemistry?) upon which the self replication of the gene was based. So, sometimes I worried that my inability to think mathematically might mean I could never do anything important. But in the presence of Delbrück I hoped I might someday participate just a little in some great revelation.

I looked forward greatly to the forthcoming summer (1948) when Dulbecco and I would go with the Lurias to Cold Spring Harbor. Delbrück and his wife Manny were coming for the second half while, before they arrived, there was to be the phage course given by Mark Adams. No great conceptual advances, however, came out that summer. Nonetheless, morale was high even though Luria and Dulbecco sometimes worried whether they had multiplicity reactivation all wrong. Delbrück remained confident, however, that multiplicity reactivation was the key breakthrough which soon should tell us what was what. His attention, however, was then often directed toward convincing us that an argument of alternate steady states would explain Sonneborn's data on antigenic transformations in Paramecia. The idea of cytoplasmic hereditary determinants did not appeal at all to Delbrück and he hoped we would all join together to try to bury as many of them as possible.

As the summer passed on I liked Cold Spring Harbor more and more, both for its intrinsic beauty and for the honest ways in which good and bad science got sorted out. On Thursday evenings general lectures were given in Blackford Hall by the summer visitors and generally everyone went, except for Luria who boycotted talks on extra-sensory perception by Richard Roberts and on the correlation of human body shapes with disease and personality by W. Sheldon. On those evenings, as on all others, Ernst Caspari opened and closed the talks, and we marveled at his ability to thank the speakers for their "most interesting presentations."

Most evenings we would stand in front of Blackford Hall or Hooper House hoping for some excitement, sometimes joking whether we would see Demerec going into an unused room to turn off an unnecessary light. Many times, when it became obvious that nothing unusual would happen, we would go into the village to drink beer at Neptune's Cave. On other evenings, we played baseball next to Barbara McClintock's cornfield, into which the ball all too often went.

There was also the fair possibility that we could catch Seymour Cohen and Luria each informing the other that his experiments were not only over-interpreted but off the mainstream to genuine progress. Though Cohen was spending the summer doing experiments with Doermann, a sharp gap existed between Cohen and the phage group led by Luria and Delbrück. Cohen wanted biochemistry to explain genes, while Luria and Delbrück opted for a combination of genetics and physics.

Cohen was not, however, the only biochemist about. David Shemin spent most of the summer living in Williams House while Leonor Michealis stayed for several weeks, despite his wife's complaints about the run-down condition of their apartment and of Demerec's failure to replace a toilet seat containing a large crack. When August began the Lurias went home to Bloomington because Zella Luria would soon have a child. This left Dulbecco and me even more free to swim at the sand spit or to canoe out into the harbor often in search of clams or mussels.

By the time we were back in Bloomington, all of us were again ready for serious experiments. Soon Dulbecco found photoreactivation of UV inactivated phage, thereby explaining why the plaque counts in multiplicity reactivation experiments were often annoyingly inconsistent. This discovery did not seem a pure blessing, however, for it immediately threw into doubt all previous quantitative interpretation of multiplicity reactivation. Thus much of the work of the previous eighteen months had to be repeated, both in the light and dark. When this was finally done it became clear that multiplicity reactivation was, by itself, not going to yield simple answers about the genetic organization of phage. As a corollary, my study of X-ray inactivated phage also was much less likely to yield anything very valuable. By then, however, I had begun to study the indirect as well as the direct effect of X rays, and the complexity of the inactivation curves initially kept me from worrying whether they would be very significant.

That fall I had my first extended view of Szilard, when Luria, Dulbecco and I drove up to Chicago to see him and Novick. There I first realized that most conversations with Szilard occurred during meals, which seemingly consumed half of his time awake. Briefly I tried to tell him what I was up to, but soon I was crushed by his remark that I did not know how to speak clearly. Even more to the point was that Szilard did not like to learn new facts unless they were important or might lead to something important. Szilard and Novick later came to Bloomington for a small phage meeting in the spring of 1949. Hershey, Doermann, Weigle, Putnam, Kozloff and Delbrück were also there. Doermann had just done his genetic crosses using premature lysates and guessed that the percentage of recombinants was approximately constant throughout the latent period. Stent and Wollman described how they thought T4 interacted with tryptophan. To me the most memorable aspect of the meeting, however, was the unplanned comic performances of Szilard and Novick. Neither understood the other's description of their phenotypic mixing experiments and they were constantly interrupting each other, hoping to make the matter clear to everyone else. A day later, Delbrück, Luria, Delbecco and I drove to Oak Ridge for its spring meeting where Delbrück coined the phrase "The Principle of Limited Sloppiness" in explaining how Kelner and Dulbecco came upon photoreactivation.

The following summer Manny Delbrück was expecting a child and most of the phage group congregated in Pasadena instead of Cold Spring Harbor. Several times each week, there occurred seminars dominated by Delbrück's insistence that the results logically fit into some form of pretty hypothesis. There were also innumerable camping trips occupying two to four days, long weekends often led by Carleton Gajdusek whose need for only two or three hours sleep a night allowed him to spend five or six days each week in the wilderness while maintaining the pretense that he was interested in the world between John Kirkwood and Delbrück. Because Gunther Stent shared a house in the San Gabriel foothills with Jack Dunitz, Pauling's postdoctoral student, there were frequent social contacts with the younger students who worked for Linus Pauling but, on the whole, I never got the impression that the phage group thought that Pauling's world and theirs would soon have anything in common. Occasionally, I would see Pauling drive up in his Riley and I felt very good when once he spontaneously smiled at me in the Faculty Club.

Most of the scientific arguments that summer were kinetic either about how tryptophan affected T4 adsorption or attempting to make sense of photoreactivation. Sometimes the genetic results of Hershey came into the picture but only Doermann seemed tempted to do more along that line. My experiments on X-ray phage had progressed to the point where I knew I had a thesis, and so in Pasadena I played about a little with formedaldehyde inactivated phages. Delbrück, like everyone else, was only mildly interested in my results but told me that I was lucky that I had not found anything as exciting as Dulbecco had, thereby being trapped into a rat race where people wanted you to solve everything immediately. If that had happened, he felt I would lose in the long term by not having the time either to think or to learn what other people were doing. I of course wanted something important to emerge from the masses of survival curves that filled several thick loose-leaf notebooks. Late in the evenings, I would try to imagine pretty hypotheses that tied all of radiation biology together, but so much special pleading was necessary that I almost never tried to explain them to Luria, much less to Delbrück.

In the early fall the question came up where I should go once I got my Ph.D. Europe seemed the natural place since, in the Luria-Delbrück circle, the constant reference to their early lives left me with the unmistakable feeling that Europe's slower paced traditions were more conducive to the production of first-rate ideas. I was thus urged to go to Herman Kalckar's lab in Copenhagen, since in 1946 he had taken the phage course and now professed to want to study phage reproduction. Though Kalckar was admittedly a biochemist, through his brother he knew Bohr and so should be receptive to the need of high powered theoretical reasoning. Even better, Kalckar's interest in nucleotide chemistry should immediately be applicable to the collection of nucleotides in DNA. The decision was finally settled in early November, when by accident Kalckar and Delbrück both were in Chicago on a weekend when Szilard had got the midwestern phage people together for a small meeting. Kalckar seemed excited about the possibility of using some C^{14} labeled adenine, which had just been synthesized in Copenhagen, to study phage reproduction, and he gave the impression of very much wanting phage people to come to his lab.

The midwestern phage meetings were then being held almost every month in Chicago, thanks to a small grant to Szilard from the Rockefeller Foundation, which covered some of the travel expenses and all of the food bills. Lederberg also began to come, adding a new vocal dimension, for he could give non-stop 3-4 hour orations without making a dent in the experiments he thought we should know about. By then, he and his wife had found phage λ in *E. coli* strain K12 but perhaps because of Delbrück's dislike of the possibility of lysogeny, I paid little attention to the discovery. Instead I conserved my brain for the facts about the somewhat messy partial diploids. My guess is that no one left the meetings remembering more than a small fraction of the ingenious alternative explanations that Lederberg dreamed up to explain the increasing number of paradoxes arising from his experiments.

In the spring of 1950 Luria went back to the problem of the distribution of spontaneous mutations within single bursts of infected cells, hoping he would find out whether or not the genetic material replicated exponentially. I spent a month on the first version of my thesis, but Luria did not like it and took it home for rewriting. This left me little more to do and not surprisingly the thesis was accepted without fuss at my Ph.D. exam in late May. Then I went out to Pasadena for a month, flying back East to spend a final six weeks in Cold Spring Harbor before the boat would take me to Europe. For a brief while I was afraid that the outbreak of the Korean War might keep me from sailing but without hesitation my draft board gave me permission to leave the country.

Practical jokes dominated the mood during the late summer in Cold Spring Harbor, culminating in an evening when Gordon Lark, Victor Bruce, and his sister Manny Delbrück, and I let the air out of the tires of several friends' cars parked before Neptune's Cave. Afterwards buckets of water were poured over our beds. On another evening, Visconti interrupted a staid Demerec social evening with an attack with a toy machine gun.

The growing number of phage people became noticeable at the phage meeting in late August. Some thirty people came, I being most affected by the talk of Kozloff and Putnam on their failure to observe 100% transfer of parental phage P^{32} to the progeny particles. Instead they believed that only somewhere between 20% and 40% of the parental label was transferred. While there existed loopholes in their experiments, the possibility was raised that perhaps both genetic and non-genetic phage DNA existed and that only the genetic portion was transferred. That prompted Seymour

Cohen to predict that a second generation of growth might yield 100% transfer.

These ideas I followed up as soon as I got to Europe. Gunther Stent had also chosen Copenhagen, and so Kalckar was faced with two phage people far less interested in biochemistry than he had been led to expect. At the same time, when we could follow Kalckar's words, it was apparent that he was not fixed on the problem of gene replication and seemed happier talking about nucleoside rearrangements. Luckily Kalckar's close friend, Ole Maaløe, had been bitten with the phage bug, and without ever formally acknowledging the arrangement, Stent and I began working with Maaløe in his lab at the State Serum Institut. Maaløe liked the idea of the second generation experiment and we began making labeled phage. After a few failures, we obtained the clean cut, but then disappointing result that the transfer in the second generation was the same as in the first generation. The data were quickly written up and dispatched in early February (1951) to Delbrück for his approval and possible transmission to the Proceedings of the National Academy. My turgid style was quickly rejected by Delbrück, who completely rewrote the introduction and discussion sections and then sent it on.

By then I knew that Maaløe wanted to go to Caltech the following autumn and so I had to find a place for the next year. I thus wrote to Luria of my dilemma, indicating a preference for England and mentioning Bawden and Pirie, neither of whom I had met. In Luria's reply he took me to task for laziness, saying that I should use my time to acquire the physics and chemistry necessary for a real breakthrough. Clearly my Copenhagen period was not developing the way Luria wanted it. Instead of learning anything new, Stent and I were merely transferring the phage group spirit to Denmark. The net result would be that I would end up doing routine phage work, and if that were to be the case it would make better sense for me to be in the States.

Some months later, Luria responded more warmly when I suggested that I go to Perutz's group at the Cavendish Laboratory, to work on the structures of DNA and the plant viruses. Soon after my letter came, he met John Kendrew at Ann Arbor and set into motion the events which led me to Cambridge and Francis Crick. How the DNA structure fell out I shall not tell here since the story is involved, and is soon to be published elsewhere.

MATTHEW MESELSON
Biological Laboratories, Harvard University, Cambridge, Massachusetts
FRANKLIN W. STAHL
Institute of Molecular Biology, University of Oregon, Eugene, Oregon

Demonstration of the Semiconservative Mode of DNA Duplication

We met each other as graduate students at Woods Hole in 1954, the second summer after the Watson-Crick DNA model had been announced. Our first conversations concerned the solutions to certain integrals describing cross-reactivation of UV-inactivated phages which one of us (F.W.S.) was then studying. Later we got to talking about ways to test the prediction of semiconservative duplication of DNA. Perhaps the more intriguing ideas have now been forgotten, but we remember talking about the possible use of density as a label for parental DNA molecules. The idea was to infect bacteria growing in ordinary medium with deuterium-labeled (dense) phage particles and allow one cycle of phage growth. The phage progeny would then be centrifuged in a solution whose density was adjusted to lie between that of unlabeled phages and phages which have inherited an appreciable amount of labeled parental phage DNA. The ordinary, non-labeled phage particles would rise to the meniscus, the heavy ones would sink to the bottom, and meniscus and pellet could then be assayed for their phage infectivity. Further resolution of the pattern of label-inheritance could be achieved by further centrifugations in media of appropriate densities. Jan Drake, another graduate student and veteran Woods Hole beachcomber, joined us in these conversations. A year later, in the fall of 1955, the three of us rented a house together in Pasadena. Although the human panorama of the landscape of molecular biology was enriched during that year, density label centrifugation remained a *Gedanken Experiment*, since one of us was then strengthening his character by solving the crystal structure of N, N′-dimethylmalonamide.

During this period others *were* busy attacking the question of atom transfer during DNA duplication. Those who chose T-even phage were doomed by a complication too unpleasant to think about at that time, namely that the great amount of genetic recombination occurring in T-even phage is proceeding *via* breakage of DNA. The first clear result was attained by Taylor, Wood, and Hughes (1957). Using autoradiography, they showed that the duplication of chromosomes in mitosis is semiconservative with respect to DNA. The simplest and most plausible explanation of this finding was that the chromosomes studied are uninemic and that the individual DNA molecules duplicate semiconservatively.

The structure of N, N′-dimethylmalonamide was almost solved when Rose Litman told us of her work on mutagenesis in T4 phage induced by 5-bromouracil, in which the pyrimidine analog was massively incorporated into T4 DNA. Thus we decided to use 5BU, both for testing the Watson-Crick prediction of semiconservative DNA duplication and for testing our own prediction that ionized 5BU causes forward and reverse mutation by mis-pairing with guanine. The mutation work took a rather long time to produce results (Terzaghi, Streisinger, and Stahl, 1962), but the centrifugal experiments soon began to work.

Our approach to the centrifugal separation of density labeled phages went through several stages. At first, we hoped to separate heavy, 5BU-labeled T4 from light T4 by centrifugation in a CsCl solution of uniform density. As a possible refinement of this procedure, we planned to grade the density of the CsCl solution along the length of the centrifuge cell. We hoped that in this way phages with different amounts of heavy label would be highly fractionated by centrifugation in the pre-established density gradient. Preliminary attempts to float and sediment unlabeled phages in a preparative ultracentrifuge in the laboratory of Renato Dulbecco, Jan Drake's research supervisor, seemed promising. But in order to see what the phages are doing during centrifugation, we made use of an analytical ultracentrifuge in Jerome Vinograd's laboratory. With Vinograd's assistance and advice we found that T4 phages are reasonably homogeneous with respect to their sedimentation velocity in a concentrated CsCl solution. This gave us hope that heavy and light phages could indeed be resolved, for it showed that the natural heterogeneity in density is much less than the 3% difference expected between 5BU-labeled and normal T4. It also gave us hope that phages might form a definite band when centrifuged to equilibrium in a suitable density gradient. In retrospect, it is no longer clear how we first came to realize that after only a few hours of centrifugation the necessary density gradient would be established as a result of the sedimentation of CsCl itself. Perhaps we first saw the equilibrium gradient forming in the Schlieren pattern of the light transmitted through the ultracentrifuge cell or perhaps we first read about it in a paper of Kai Pederson (1934), who reported the equilibrium sedimentation of simple salts after only a few hours in the centrifuge. In any case, we do remember being quite surprised at the speed with which such gradients approach their equilibrium state.

After several tries with a rather decomposed preparation of T4, we observed a band—actually two bands, neither of which was ever identified.

For some time after we learned how to get such bands, there was much doubt as to what the density of the banded material was, and as to whether the banded macromolecules were near their equilibrium distributions. These and other questions about density gradient sedimentation itself, and about the molecular weight of T4-DNA, led us on a rather long detour into physical chemistry and away from the biological experiments we had intended (Meselson, Stahl, and Vinograd, 1957). The detour was enforced by our consistent failure to detect phages or DNA molecules with intermediate amounts of parental 5BU. Finally, we abandoned phages and 5BU, and turned to bacteria and the heavy isotope of nitrogen (N^{15}). The second and third experiments along these lines worked beautifully; so we renumbered them 1 and 2 (see Fig. 1) and began to write the paper (Meselson and Stahl, 1958). To remove us from distractions while writing, Max Delbrück transported us to the Kerkhoff Marine Station in Corona Del Mar, confining us for several days to an upstairs room with a typewriter.

The results of the N^{15}-transfer experiments on the replication of bacterial DNA, like those of the transfer experiments of Taylor et al, had their simplest and most plausible explanation in the duplication hypothesis of Watson and Crick. However, both transfer experiments fell just short of completely substantiating that hypothesis, and they did so for the same reason. In neither case had the structure of the chromosome involved been well established. Although there were strong reasons for thinking that the extracted *Escherichia coli* DNA we had studied possessed the Watson-Crick structure, and that the conserved sub-units are indeed single DNA strands, other possibilities could not be completely ruled out. The remote possibility that the conserved sub-units are arranged end-to-end was conclusively ruled out by the experiments of Rolfe (1962). But we had lingering doubts as to whether the sub-units are simply single polynucleotide chains. Our doubts were increased by our finding that the conserved sub-units can be separated from one another by heat denaturation of the DNA, since various experiments of others and ourselves with calf and salmon DNA had led us to think that single strands are not separated by heat denaturation. But by now, the autoradiographic measurements made by John Cairns (1962a, b) have shown that the chromosomes of phage and bacteria known to duplicate semiconservatively have exactly the mass per unit length expected for the double-helical Watson-Crick structure of DNA, and a wide variety of somewhat less conclusive other evidence supports this result.

Thus Watson and Crick's prediction of semiconservative DNA duplication has been fully vindicated for phage and bacteria, exactly in the sense that *they* meant it. It would be mere petulance to insist at this stage that higher organisms are likely to be different in this regard. The several demonstrations (e.g., Simon, 1961; Sueoka, 1960) that DNAs from eukaryotes

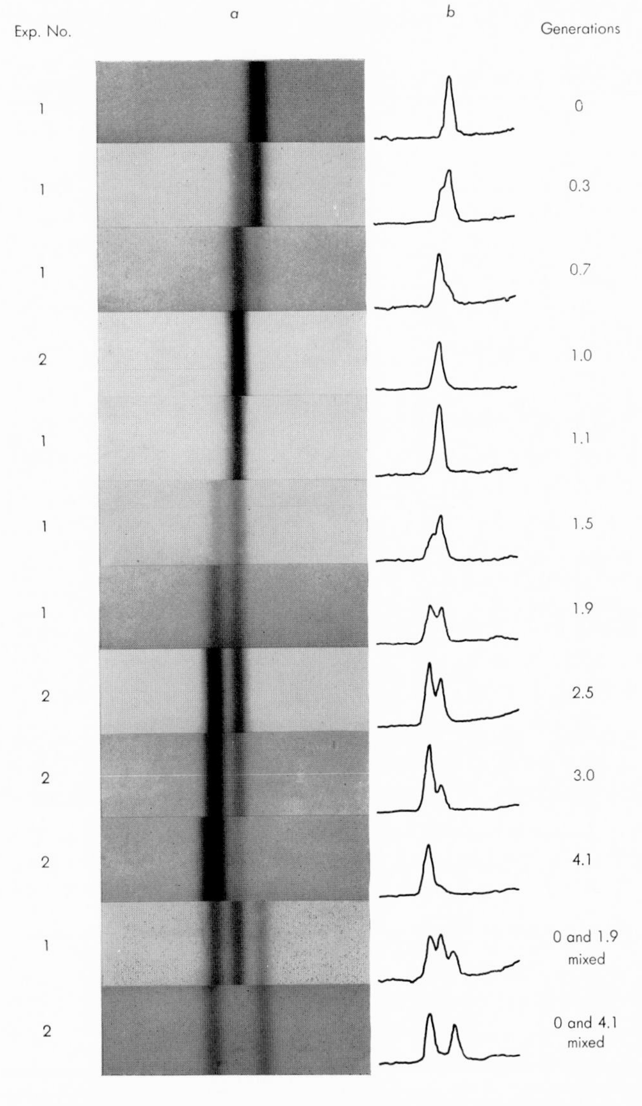

FIGURE 1. (a) Ultraviolet absorption photographs showing DNA bands resulting from density gradient centrifugation of lysates of bacteria sampled at various times after the addition of an excess of N^{14} substrates to a growing N^{15}-labeled culture. The density of the CsCl solution increases to the right. Regions of equal density occupy the same position on each photograph.

(b) Microdensitometer tracings of the DNA bands shown in the adjacent photographs. The microdensitometer pen displacement above the base line is directly proportional to the concentration of DNA. The degree of labeling of a species of DNA corresponds to the relative position of its bands between the bands of fully labeled and unlabeled DNA in the lowermost frame, which serves as a density reference. This figure appeared in Proc. Natl. Acad. Sci., *44* (1958), p. 675.

generate density-isotope transfer patterns analogous to those of *E. coli* and phage are relevant for both philosophical and anectodal reasons. Philosophically, it seems sound to assume that, until proven otherwise, DNAs from different organisms showing the same transfer pattern do so by virtue of a basic structural similarity in the mechanism of their replication. Anecdotally, we can point out that Edward Simon, in a burst of well-aimed optimism, conducted successful density-label transfer experiments using 5BU in *human* cells while we were still in the process of generating publishable data for bacteria.

REFERENCES

CAIRNS, J. 1962a. A Minimum Estimate for the Length of the DNA of *Escherichia Coli* obtained by Autoradiography. J. Mol. Biol., *4*: 407–409.

———1962b. Proof that the Replication of DNA involves Separation of the Strands. Nature, *194*: 1274.

MESELSON, M. and F.W. STAHL. 1958. The Replication of DNA in *Escherichia Coli.* Proc. Natl. Acad. Sci., *44*: 671–682.

MESELSON, M., F. W. STAHL, and J. VINOGRAD. 1957. Equilibrium Sedimentation of Macromolecules in Density Gradients. Proc. Natl. Acad. Sci., *43*: 581–583.

PEDERSON, K. O. 1934. Über das Sedimentationsgleichgewicht von anorganischen Salzen in der Ultrazentrifuge. Zeitschrift für Physikalische Chemie, *170*: 41–61.

ROLFE, R. 1962. The Molecular Arrangement of the Conserved Subunits of DNA. J. Mol. Biol., *4*: 22–30.

SIMON, E. H. 1961. Transfer of DNA from Parent to Progeny in a Tissue Culture Line of Human Carcinoma of the Cervix (strain HeLa). J. Mol. Biol., *3*: 101–109.

SUEOKA, N. 1960. Mitotic Replication of Deoxyribonucleic Acid in *Chlamydomonas Reinhardi*, Proc. Natl. Acad. Sci., *46*: 83–91.

TAYLOR, J. H., P. S. WOODS, and W. L. HUGHES. 1957. The Organization and Duplication of Chromosomes as Revealed by Autoradiographic Studies Using Tritium-Labeled Thymidine. Proc. Natl. Acad. Sci., *43*: 122–128.

TERZAGHI, B. E., G. STREISINGER, and F. W. STAHL. 1962. The Mechanism of 5-Bromouracil Mutagenesis in the Bacteriophage T4. Proc. Natl. Acad. Sci., *48*: 1519–1524.

JOHN CAIRNS
The Cold Spring Harbor Laboratory of Quantitative Biology,
Cold Spring Harbor, New York

The Autoradiography of DNA

Having just retraced in my mind the train of events that led me to apply the technique of autoradiography to the problem of DNA replication, I am surprised and disillusioned to find how, on most occasions of conscious choice, I made the wrong decision and how it was only by default, as it were, that I returned to more fruitful things. Because opportunities for retrospective analysis rarely arise, I shall try to tell my allotted story in such a way that the irrelevancies, the alarums and excursions are given, for once, their proper place.

I had been engaged, in Australia, with the affairs of influenza virus for some years when, in 1957, a fellowship from the Rockefeller Foundation allowed me to spend four months at the California Institute of Technology. The purpose of this trip was, among other things, for me to learn the techniques being used so successfully by Dulbecco in the study of animal viruses in tissue culture. For it was felt, rightly I am sure, that even the humble hen's egg, with which I had been occupied, was too complex a system in which to pursue the life history of any virus. (I can remember that I was obsessed with complexity at that stage—against it, that is; the obsession was soon to lead me into a time-consuming and profitless excursion.)

Australia was in the grip of an outbreak of Asian influenza when I left Canberra. So I was not unduly disturbed when the influenza I brought with me changed into pneumonia—a disease for which, when it comes, one is only too pleased to have some identifiable cause. As a result, therefore, I spent my first days in California in the Caltech faculty club, treating myself with sulfaguanidine.

This seemingly trivial episode was, however, to play a decisive role. For it prevented me from finding out, until too late, that I could not afford even the cost of lying in bed at the faculty club, and this in turn forced me to find somewhere to live that did not demand an advance payment. I sought refuge therefore in a nearby house, which was being run, at that time, by Jan Drake, Matthew Meselson and Howard Temin.

This house had several unusual qualities, one being that it was then in the throes of the N^{15}-N^{14} transfer experiment that was to settle the semiconservative nature of the replication of bacterial DNA. Although the principles of the experiment now seem to me straightforward enough, repeated explanations and numerous diagrams on the paper napkins at mealtimes and the construction of two models of DNA (one open, one space-filling) were needed before the entire issue became clear to me, as I had no background of genetics and, until then, little call to remember any chemistry.

At this point I should make some attempt at describing the feelings of an outsider plunged into that extraordinary world. It was 1957, as I have said. So the Delphic pronouncements of Watson and Crick had had four years in which to be assimilated by the inhabitants—who were already equipped with exactly the right household gods (Beadle and Delbrück, to name but two) for supervising receipt of the message. Grand issues were in the air. But the debate over these issues had not extended, as it has now, beyond a limited coterie. Understandably, therefore, I floundered—my flounderings including a magnificent camping trip with the Delbrücks on which, as I waited through the night for dawn to come, I thought I would freeze to death.

The transfer experiment (Meselson and Stahl, 1958) is dealt with elsewhere in this book, but I should recapitulate the results here because my understanding of them (by dint of repetition) was to determine much of what I did later. The experiment showed that the DNA of *Escherichia coli* is duplicated semiconservatively and, equally important, that this process of duplication is strictly ordered, in that none of the DNA is duplicated twice until all has been duplicated once. (This bare summary does not do justice to the great elegance and economy of the experiment, or to the unusual precision with which the results were finally presented; for these you must return to the original and to its precursor by Delbrück and Stent [1957].)

Semiconservative replication had, of course, been one of the predictions made by Watson and Crick for the DNA double-helix (Watson and Crick, 1953), so it was tempting to regard the Meselson-Stahl experiment as giving uncontestable proof for the separation of the two polynucleotide chains; after all, just such a picture had already graced the cover of a most respectable book, the 1956 McCollum Pratt Symposium on the "Chemical Basis of Heredity." Yet so strong was the prevailing wish for precision that I can remember Meselson spending almost as much time explaining to me exactly what the results of the transfer experiment did *not* necessarily mean. In fact, neither his transfer experiment nor an earlier but somewhat similar autoradiographic experiment (Taylor, Woods and Hughes, 1957) showed that the units of DNA, that separate during duplication, are the individual polynucleotide chains, rather than each of a pair of DNA double helices, for example.

It remained to be shown therefore that the DNA molecules which duplicate semiconservatively contain only two polynucleotide chains. One method would have been to measure the width of the "hybrid" DNA molecules under the electronmicroscope; however the technique of electronmicroscopy had not then advanced to the point where this could be done. The other method was to relate the length of these molecules, as measured by electronmiscoscopy, to their mass; however, attempts to do this (Hall, Meselson and Stahl, unpublished) failed for want of some method for measuring the mass of DNA molecules that was known not to contain a possible two-fold error. And there the matter rested for some years.

At this point I returned to Australia, determined to devote the next few years to dissecting the affairs of some RNA-containing virus in much the way the DNA-containing bacteriophages were then being exploited. Since metazoan cells, even under conditions of continuous cultivation, are too complex in themselves and also too demanding of complex media (I said to myself), and since all bacterial viruses were then thought to contain DNA, not RNA, I embarked on a search for RNA viruses infecting certain soil amoebae that have simple nutritional requirements. At the end of a year of futile search, I felt in the need for some project that was sufficiently straightforward to restore my dwindling reserves of *amour propre*; so I deserted the boundlessly healthy amoebae and their elusive RNA viruses for an autoradiographic study of the multiplication of the DNA virus, vaccinia, and was thereby happily occupied for the next year.

In 1960, the Australian National University's custom of granting sabbatical leave every five years gave me almost a year in Hershey's laboratory at Cold Spring Harbor. It was a fortunate moment for such a visit, because he had just elaborated the methods for extracting and handling the "chromosome" of the larger bacteriophages (in particular, that of T2). In addition, the New England Nuclear Corporation had just succeeded in producing tritiated thymine of such high specific activity that one could consider visualizing individual DNA molecules by autoradiography and, since tritium gives rise to a β particle of very low energy and hence very short range, therefore getting some idea of their shape. (I believe that many others had made the same calculation, but none of them had access to large enough molecules of DNA to make the exercise seem at all easy, or to molecules of such precisely determined molecular weight that the exercise seemed worthwhile.)

The discovery that even the most complex bacteriophages contain their DNA in single molecules (Rubenstein, Thomas and Hershey, 1961; Davison, Freifelder, Hede and Levinthal, 1961) somehow constituted a turning point in my thinking about the matter. For it seemed to take the business of DNA out of the realm of orthodox biochemistry (to which I, for one, am irrevocably antipathetic) and back into the everyday world of morphology.

Nevertheless, I spent the first three months at Cold Spring Harbor devising some remarkably complicated experiments to do with the distribution of T4 bacteriophage DNA among its progeny. Luckily, these experiments were so complicated that even the preliminary stages defeated me, and so I retreated into the world of autoradiography that had provided a refuge once before.

Getting an estimate of the length of the DNA molecule of T2 bacteriophage proved a fairly simple task, although the necessary period of exposure under film (not less than two months, for the most highly active tritiated thymine) turned out to be longer than I had hoped. As a result, it was readily established that the T2 DNA molecule, or "chromosome," has only two polynucleotide chains over most, if not all, of its length (Cairns, 1961).

At this point, I should digress to discuss the relative merits of autoradiography and electronmicroscopy for displaying DNA molecules, since the two methods have been to some extent in competition. Autoradiography has one special advantage, in that it only shows labeled molecules; this fact allows one either to look at a particular species of molecule in the presence of excess homologous non-labeled carrier, and so avoid the risk of being misled by aggregation, or to look at the particular part of a molecule that has been synthesized during a limited period of labeling, and so get some notion of the history of the molecule. Electronmicroscopy, on the other hand, has the advantage of much finer resolution; for this reason it can show the form of much smaller molecules, or of those larger molecules that are too tangled for autoradiography. There are, however, DNA molecules which are so large that they can only fall entirely within the electronmicroscope grid boundaries if they are very tangled; these are still studied most easily by autoradiography. Ultimately, of course, the electronmicroscope will reign supreme, since it is only a matter of time before even this last technical difficulty is overcome.

To return to the chronicle: A bacteriophage such as T2 indulges in such extensive recombination that any semiconservative replication it may undergo is largely masked. Showing that its DNA is two-stranded had not, therefore, proved that the units of semiconservative replication are the individual chains—an ambition which had become implanted in my mind during my indoctrination period in California. No such difficulty applied to bacteriophage λ, since this phage undergoes much less recombination and therefore often succeeds in handing over one half of its DNA to one of its 100 or so progeny (Meselson and Weigle, 1961). It was necessary, in short, to repeat for phage λ what had been done for T2.

I had by this time returned to Australia (though, happily, not also to the soil amoeba), and the λ project was completed uneventfully. Since the molecular weight of the λ chromosome had been determined by Hershey and others, it was possible to say that its length observed by autoradiography demanded that it is only two-stranded (Cairns, 1962). Any doubts about the nature of the units of semiconservative replication were now settled. (An entirely different kind of proof was published at the same time by Wake and Baldwin [1962], and other varieties of proof have emerged since then.)

Soon after starting on the autoradiography of T2 DNA, I had resolved that if all went well I should turn as quickly as possible to bacterial DNA. There were two reasons for considering this the ultimate subject for autoradiography. Both Hershey and Levinthal had shown that large molecules of DNA are very fragile; it seemed likely therefore that if the bacterial chromosome also were a single molecule it would be so fragile that its size could not be determined by conventional methods (all of which demand moving the molecule in some manner or other) and so would have to be measured by a static method (i.e., by some form of microscopy). Secondly, it was known that certain bacteria synthesize DNA continuously when they are multiplying rapidly; it followed therefore that if the bacterial chromosome could be isolated intact it should, unlike the DNA of bacteriophage particles, show the form of replicating DNA.

The image of the act of DNA duplication that most people had in mind since the authoritative exposition of the problem by Delbrück and Stent had become somewhat clouded by Kornberg's failure to isolate any polymerase that can add nucleotides to chains ending in 5′-triphosphate. So there was the feeling either that only one of the new DNA chains is being synthesized in each region of duplication or that the Kornberg enzyme, which can only add nucleotides to the 3′ end of a chain, plays no part in replication.

The exercise with bacterial DNA occupied me for most of the two years that turned out to be the end of my time in Australia. During that period I maintained a steady correspondence with Hershey and Delbrück; for, like so many before me, I had elected them to be my moral guardians in matters of fact and fantasy.

Although at first the autoradiographic appearance of bacterial DNA was extremely confusing, the problem gradually straightened itself out with the result that at least the form of the replicating chromosome is now reasonably clear: Replication occurs at a single locus that traverses the entire molecule from one end to the other; at this locus, both new chains are being synthesized; during replication (and perhaps at other times too) the ends of the chromosome are somehow joined together to make the whole structure into a closed circle (Cairns, 1963 a, b).

Fortunately, this description of DNA replication by bacteria does not depend solely upon the results of autoradiography (a technique which, more than most, speaks to the reader through the primary beholder), but has received crucial contributions from the experiments of Bonhoeffer and Gierer, Forro, Lark, Maaløe, Nagata and Sueoka, and of course from the pioneer work on the bacterial chromosome by Hayes, Jacob, Monod and Wollman (the references are too numerous to list). Indeed, I now believe that autoradiography—at least in this form—has made its main contribution. The problems that lie ahead are not static ones, resolvable by static techniques, but concern the "machinery" of replication. This machinery, which is responsible for the rapid and accurate synthesis of a single enormous polymer, promises to have a physical complexity that matches the scale of the operation. But I see that I have wandered into mixing prophecy with history. It is therefore time to stop.

REFERENCES

CAIRNS, J. 1961. An estimate of the length of the DNA molecule of T2 bacteriophage by autoradiography. J. Mol. Biol., *3*: 756–761.

———1962. A proof that the replication of DNA involves separation of the strands. Nature, *194*: 1274.

———1963a. The bacterial chromosome and its manner of replication as seen by autoradiography. J. Mol. Biol., *6*: 208–213.

———1963b. The chromosome of *Escherichia coli*. Cold Spring Harbor Symp. Quant. Biol., *28*: 43–45.

DAVISON, P. F., D. FREIFELDER, R. HEDE, and C. LEVINTHAL. 1961. The structural unity of the DNA of T2 bacteriophage. Proc. Natl. Acad. Sci., *47*: 1123–1129.

DELBRÜCK, M., and G. S. STENT. 1957. On the mechanism of DNA replication. p. 699–736. *In* W. D. McElroy and B. Glass [Ed.] The Chemical Basis of Heredity. Johns Hopkins Press. Baltimore.

MESELSON, M., and F. W. STAHL. 1958. The replication of DNA in *Escherichia coli*. Proc. Natl. Acad. Sci., *44*: 671–682.

MESELSON, M., and J. J. WEIGLE. 1961. Chromosome breakage accompanying genetic recombination in bacteriophage. Proc. Natl. Acad. Sci., *47*: 857–868.

RUBENSTEIN, I., C. A. THOMAS, and A. D. HERSHEY. 1961. The molecular weights of T2 bacteriophage DNA and its first and second breakage products. Proc. Natl. Acad. Sci., *47*: 1113–1122.

TAYLOR, J. H., P. S. WOODS, and W. L. HUGHES. 1957. The organization and duplication of chromosomes as revealed by autoradiographic studies using tritium-labeled thymidine. Proc. Natl. Acad. Sci., *43*: 122–128.

WAKE, R. G., and R. L. BALDWIN, 1962. Physical studies on the replication of DNA *in vitro*. J. Mol. Biol., *5*: 201–216.

WATSON, J. D., and F. H. C. CRICK. 1953. The structure of DNA. Cold Spring Harbor Symp. Quant. Biol., *18*: 123–131.

ROBERT L. SINSHEIMER
Division of Biology, California Institute of Technology, Pasadena, California

ϕx: Multum in Parvo

WHY ϕX?

Like many accounts of modern bacterial virology my story begins on a quiet morning in Max Delbrück's office. Out of a conversation with Delbrück on "whither virology" in about May 1953 there came the idea that it could be useful to explore the characteristics of very small phages, which might be "simpler"—physically, chemically, genetically, physiologically—than the T phages which then were at the focus of attention and whose structure and biological behavior was turning out to be more and more complex.

Only two "small" phages were then known—"small" as defined by passage through membrane filters and cross section to inactivation by ionizing radiation. These were S13 (Elford, 1936) and *ϕx*174 (Bonet-Maury and Bulgakov, 1944). Neither phage had been successfully purified, nor had the growth of either been studied in any detail.

After preliminary reconnaissance of culture conditions for these two phages and selection of a suitable host (*E. coli*, strain C), I decided to concentrate on *ϕx*, for the prosaic but highly heuristic reasons that it could be grown to higher titer and appeared to be more stable than S13. (In retrospect, both of these apparent virtues of *ϕx* probably derived from its much less extensive adsorption to bacterial debris than S13 in our culture media.)

A SINGLE STRAND IS SUFFICIENT

I considered the quantitative determination of the physical and chemical parameters of the virus to be a necessary prelude to any biophysical analysis of its mode of replication. But to determine these parameters required the preparation of reasonable amounts of pure virus. Because of the limited techniques then available, the culture of large volumes of *ϕx* lysate and development of procedures for concentration and purification of the virus were primitive and empirical enterprises. Progress was laborious but, happily, not imperceptible. Fortunately, the technique of density gradient centrifugation (Meselson, Stahl and Vinograd, 1957) was developed just in time for the last step of purification, the separation of infective virus from the nearly empty "70 S" virus shells that are abundantly present in every lysate (Sinsheimer, 1959).

It transpired that the *φx* virus really *is* small. A low sedimentation coefficient (114 S compared to about 1000 S for the T-even phages) suggested a molecular weight in the neighborhood of $4\text{–}8 \times 10^6$ and early electron micrographs indicated a "spherical" particle of diameter about 25 mμ. It seemed difficult to believe that a "conventional" double-stranded, stiff DNA molecule with enough genetic information could be packaged into such a small volume. The heretical notion that *φx* might be an RNA virus appeared to us as a bright, but brief vision that was quickly erased by chemical analysis; *φx* contains 29% DNA.

The molecular weight of the DNA extracted from the virus was found to be 1.7×10^6, or 27% of the particle weight of the virus. Thus it followed that this DNA-containing virus carries only *one* single DNA molecule, an inference that was entirely novel at the time, though now considered to be commonplace.

Further work on *φx* DNA demonstrated it to be a rather different molecule from the "conventional" types of DNA studied until then. I was first led to this inference by considering the surprising finding that its sedimentation coefficient (24 S) is much too high (Doty, McGill and Rice, 1958) for a typical DNA of its molecular weight. Then, measurements of the dependence of the UV absorption of *φx* DNA on temperature and on ionic strength made it quite clear that the extracted *φx* DNA resembles "denatured" rather than native DNA.

It then seemed necessary to prove that these physico-chemical properties really represented the state of the DNA within the virus—that a denaturation had not somehow occurred during extraction. The ready reaction of *φx* DNA with formaldehyde, an agent known not to react with native double-stranded DNA (Fraenkel-Conrat, 1954), could be demonstrated to occur even within the intact virus particles, showing that the viral DNA is really "denatured" naturally.

Chemical analysis of the nucleotide composition of *φx* DNA showed that the purine and pyrimidine bases are not present in the complementary frequencies demanded by the Watson-Crick structure. This finding eliminated any possibility that *φx* DNA is a "naturally" denatured double-stranded DNA and provided the final link in the chain of evidence which proved that in *φx* a single-stranded DNA carries the genetic information. Where, then, *is* the complementary strand? It was conceivable that each of two different classes of virus particles—present in the lysate in unequal proportions—carries one of the complementary strands. This possibility was later eliminated by negative results of self-annealing experiments with the extracted viral DNA and, in any case, would not have altered the inescapable conclusion that a single virus particle carrying a single-stranded DNA molecule can initiate infection.

THE REPLICATIVE FORM—COMPLEMENTATION AND ALL THAT

Just about that time, replication of double-stranded DNA was shown by Meselson and Stahl (1958) to proceed via the semiconservative process, based upon double strand unwinding and purine-pyrimidine base pairing, as first postulated by Watson and Crick (1953). How, then, does a single-stranded DNA molecule replicate? Is a new principle of replication involved?

Answering these questions requires that one can follow the fate of parental viral DNA inside the infected cells. To this end, the parental DNA could readily be labeled in various ways—with radioactive and with stable isotopes—but before being able to do the experiments, means had to be developed to obtain and analyze DNA extracts from *φx*-infected cells without degradative artifacts.

By the use of gentle (enzymatic) lysis of the infected cells to avoid molecular breakage by shear, low temperature, high pH, and versene to minimize nuclease digestion in the lysate, and rapid treatment with phenol to remove and destroy nucleases, a viral DNA preparation of apparent macromolecular character could be obtained from infected cells. Since just at this time it finally proved possible to develop an assay system (Guthrie and Sinsheimer, 1960) by which the infectivity of free viral DNA—initially DNA isolated from phage particles and later DNA extracted from infected cells—could be demonstrated, we were justified in concluding that the viral DNA of our extracts was not only macromolecular but also a reasonable facsimile of the in vivo structure.

Density gradient ultracentrifugal analysis of such extracts of cells infected with isotopically labeled virus revealed that soon after infection the parental viral DNA is converted into a *replicative form* (RF) (Sinsheimer, Starman, Nagler and Guthrie, 1962). The physical and chemical properties of this replicative form are those of a "conventional" complementary double-stranded DNA molecule and it is also infective in the assay system. The intracellular amount of this RF increases several-fold during infection and only at late stages of the phage growth cycle does the single-stranded DNA destined for incorporation into the progeny particles appear in the infected cell.

Thus replication of the single-stranded *φx* DNA does involve the "conventional" principle of complementary base pairing, but through conversion of the infecting single strand into a complementary "replicative form." Subsequent studies on other viruses endowed with single-stranded nucleic acids [M13 and the RNA viruses (Kelly and Sinsheimer, 1964; Montagnier and Sanders, 1963; Weissmann, Borst, Burdon, Billeter and Ochoa, 1964)] have shown that formation of a complementary, double-stranded replicative intermediate is a general feature of their reproduction.

At the time, the discovery of the capacity of *E. coli* to convert an infecting single strand of viral DNA into a double strand came quite unexpectedly. It revealed an entirely new facet of DNA metabolism, whose significance for the mechanisms of DNA repair and recombination is only now becoming clear (Boyce and Howard-Flanders, 1964; Setlow and Carrier, 1964; Howard-Flanders, 1965).

A DNA RING—THE SHAPE OF THINGS TO COME

In the early studies of *φx* DNA, a most curious and at first uninterpreted phenomenon was observed. When the DNA was sedimented under conditions which favored polymer chain extension—in low ionic strength, in alkali, or after treatment with formaldehyde—there appeared two discrete sedimentation boundaries representing molecular species differing by about 10% in their sedimentation rate. This observation remained unexplained, but not forgotten, for nearly three years. Then Walter Fiers began a series of experiments intended to explore the effect upon the infectivity of *φx* DNA of the removal of terminal nucleotides by exonuclease action (Fiers and Sinsheimer, 1962a). He found to his surprise that the exonuclease would not remove any terminal nucleotides, at least not from the infective DNA molecules. Further study of this anomalous behavior removed or excluded various possible trivial explanations for the immunity of *φx* DNA against exonuclease attack, and some fundamental explanation had to be sought. One such explanation was the concept that the infective *φx* DNA molecule has no ends, which would also neatly explain the double boundary previously observed in the ultracentrifugal studies. For here the faster sedimenting of the two components could be infective DNA rings and the other, more slowly sedimenting component could be linear DNA molecules derived from rings that had somehow been opened by a single cleavage. These linear molecules would most likely be noninfective if the rings from which they derived were opened at random. Any second cleavage of a DNA chain would produce fragments of random size, and thus no material sedimenting with a discrete boundary.

This concept of *φx* DNA rings was tested by introducing deoxyribonuclease scissions deliberately into a preparation of viral DNA, initially composed of mainly the faster sedimenting component. Samples of this DNA exposed to deoxyribonuclease action for various lengths of time were assayed for their residual infectivity and analyzed for their sedimentation characteristics in the ultracentrifuge. Now, since the scissions introduced into DNA by deoxyribonuclease occur at random, the proportion of rings having sustained no scission (and hence retained their infectivity), one scission, and more than one scission, ought to vary as the corresponding terms of a Poisson distribution. Since these proportions did vary in just this way the results of this test confirmed the ring hypothesis in a most gratifying way (Fiers and Sinsheimer, 1962b).

It seemed plausible that such a DNA ring would possess some distinguishable feature, possibly marking the site at which replication or transcription is initiated. An inquiry into the action of exonuclease upon opened *ϕx* DNA rings then revealed the existence of such a feature, a single exonuclease-resistant "discontinuity" in the rings. But the nature of this "discontinuity" remains a problem for the future.

Not only the *ϕx* viral DNA but also the *ϕx* replicative form is a ring (Kleinschmidt, Burton and Sinsheimer, 1963; Chandler, Hayashi, Hayashi and Spiegelman, 1964). Since the time of the discovery of the *ϕx* DNA ring, many other viral DNAs have been shown either to be rings or to pass through a ring stage during their replication (Dulbecco and Vogt, 1963; Weil and Vinograd, 1963; Kleinschmidt, Kass, Williams and Knight, 1965; Young and Sinsheimer, 1964). It does not seem too far-fetched to suppose now that circularization of viral DNA is a most general circumstance. But replication of such rings would seem to require the existence of specific ring-opening and ring-closing enzymes that remain to be discovered.

AND AGAIN, WHY *ϕ*X?

A decade after that quiet morning talk and in possession of a few, hard-won facts, it is reasonable to inquire: what may we hope to learn from future *ϕx* research? The original hope of a structural and chemical simplicity was not disappointed. The small size of the viral DNA has permitted physico-chemical analysis with relative ease, although the technology available *now* is capable of coping with much larger DNA molecules. The small DNA size proved of value also in that it permitted development of a bioassay system for the free DNA of exceptionally high efficiency.

As a single-stranded DNA of non-complementary composition, *ϕx* DNA has been of considerable value in the analysis of the modes of action of DNA- and RNA-polymerases (Kornberg, 1961; Chamberlin and Berg, 1962; Chamberlin and Berg, 1964; Sinsheimer and Lawrence, 1964; Hayashi, Hayashi and Spiegelman, 1963). As it is available in both single- and double-stranded, circular forms, both infective, we may expect that *ϕx* DNA will continue to be useful for in vitro analyses of varied aspects of DNA enzymology.

For less evident reasons, *ϕx* virus has proven to be an exceptionally useful and easily assayed antigen (Uhr and Finkelstein, 1963).

A corollary of the original hope of structural simplicity was the expectation that the limited amount of DNA in such a small virus can correspond only to a very limited number of viral cistrons, and thus viral functions. By now, a good start has been made toward enumeration and analysis of these cistrons through isolation of many conditional lethal mutants of ϕx and through tests of these mutants for functional complementation and genetic recombination. The necessity for certain viral functions, such as invasion, replication, coat protein formation, and cell lysis, was evident *a priori*, but the necessity for other, less obvious functions, such as control of and intervention in cellular metabolism may soon become apparent.

Having become more sophisticated with the years, we might now suggest, however, that the quest for a "simple" viral infection may be, in part, illusory. For, if the virus provides fewer functions, the cell must provide more. In that sense, ϕx is undoubtedly a more parasitic virus than T2. Thus the network of reactions proceeding within the infected complex of a "simple virus" actually may be not much less intricate than the network obtaining within the infected complex of a large and complicated virus. But even this, at first sight depressing thought gives cause for hope, because study of ϕx infection may well provide a useful key to more detailed analysis of the role of the *host* in viral infection.

Being a single-stranded DNA phage, ϕx provides an example of a rather specialized parasitism: a virus which makes use of an agency most likely "intended" by the host cell for repair of lesions in its own DNA. This way of life appears to be a novel situation in virology, and further research on ϕx reproduction may well illuminate also the still obscure processes of DNA repair and recombination.

More and more, the participation of ring structures in DNA replication —cellular or viral—is seen to be widespread. The teleonomic reasons for this device, and the details of the replication process, are not yet well understood. And here study of the reproduction of ϕx DNA rings may serve as a miniature, more readily analysable model for DNA ring replication in general, an amenable experimental system with which to explore many still rather mysterious facets of DNA synthesis.

REFERENCES

Bonet-Maury, P. and N. Bulgakov. 1944. Comptes Rendus, Paris, *138*: 499.

Boyce, R. P. and P. Howard-Flanders. 1964. Release of ultraviolet light-induced thymine dimers from DNA in *E. coli* K-12. Proc. Natl. Acad. Sci., *51*: 293.

Chamberlin, M. and P. Berg. 1962. Deoxyribonucleic acid-directed synthesis of ribonucleic acid by an enzyme from *Escherichia coli*. Proc. Natl. Acad. Sci., *48*: 81.

———, ———1964. Mechanism of RNA polymerase action: formation of DNA-RNA hybrids with single-stranded templates. J. Mol. Biol., *8*: 297.

Chandler, B., M. Hayashi, M. N. Hayashi and S. Spiegelman. 1964. Circularity of the replicating forms of a single-stranded DNA virus. Science, *143*: 47.

DOTY, P., B. B. McGILL, and S. A. RICE. 1958. The properties of sonic fragments of deoxyribonucleic acid. Proc. Natl. Acad. Sci., *44*: 432.

DULBECCO, R. and M. VOGT. 1963. Evidence for a ring structure of Polyoma virus DNA. Proc. Natl. Acad. Sci., *50*: 236.

ELFORD, W. J. 1936. Centrifugation studies. Critical examination of a new method as applied to the sedimentation, bacteria, bacteriophages, and proteins. Brit. J. Exp. Path., *17*: 399.

FIERS, W. and R. L. SINSHEIMER. 1962a. The structure of the DNA of Bacteriophage *ϕx* 174. I. The action of exopoly nucleotidases. J. Mol. Biol., *5*: 408.

———, ———1962b. The structure of the DNA of Bacteriophage *ϕx* 174. III. Ultracentrifugal evidence for a ring structure. J. Mol. Biol., *5*: 424.

FRAENKEL-CONRAT, H. 1954. Reaction of nucleic acid with formaldehyde. Biochim. Biophys. Acta, *15*: 307.

GUTHRIE, G. D. and R. L. SINSHEIMER. 1960. Infection of protoplasts of *Escherichia coli* by subviral particles of Bacteriophage *ϕx* 174. J. Mol. Biol., *2*: 297.

HAYASHI, M., M. N. HAYASHI, and S. SPIEGELMAN. 1963. Replicating form of a single-stranded DNA virus: Isolation and properties. Science, *140*: 1313.

HOWARD-FLANDERS, P. 1965. Molecular mechanisms in the repair of irradiated DNA. Japan. J. Genetics, *40* suppl. 256.

KELLY, R. B. and R. L. SINSHEIMER. 1964. A new RNA component in MS2-infected cells. J. Mol. Biol., *8*: 602.

KLEINSCHMIDT, A. K., A. BURTON, and R. L. SINSHEIMER. 1963. Electron microscopy of the replicative form of the DNA of the Bacteriophage *ϕx* 174. Science, *142*: 961.

KLEINSCHMIDT, A. K., S. J. KASS, R. C. WILLIAMS, and C. A. KNIGHT. 1965. Cyclic DNA of Shope papilloma virus. J. Mol. Biol., *13*: 749.

KORNBERG, A. 1961. Enzymatic Synthesis of DNA. John Wiley and Sons, New York. 103 pp.

MESELSON, M. and F. W. STAHL. 1958. The replication of DNA in *Escherichia coli*. Proc. Natl. Acad. Sci., *44*: 671.

MESELSON, M., F. STAHL, and J. VINOGRAD. 1957. Equilibrium sedimentation of macromolecules in density gradients. Proc. Natl. Acad. Sci., *43*: 581.

MONTAGNIER, L. and F. K. SANDERS. 1963. Replicative form of Encephalomyocarditus virus ribonucleic acid. Nature, *199*: 664.

SETLOW, R. B. and W. L. CARRIER. 1964. The disappearance of thymine dimers from DNA: an error-correcting mechanism. Proc. Natl. Acad. Sci., *51*: 226.

SINSHEIMER, R. L. 1959. Purification and properties of bacteriophage *ϕx* 174. J. Mol. Biol., *1*: 37.

SINSHEIMER, R. L. and M. LAWRENCE. 1964. In vitro synthesis and properties of a *ϕx* DNA-RNA hybrid. J. Mol. Biol., *8*: 289.

SINSHEIMER, R. L., B. STARMAN, C. NAGLER and S. GUTHRIE. 1962. The process of infection with Bacteriophage *ϕx* 174. J. Mol. Biol., *4*: 142.

UHR, J. W. and M. S. FINKELSTEIN. 1963. Antibody formation. J. Exp. Med., *117*: 457.

WATSON, J. D. and F. H. C. CRICK. 1953. Genetical implications of the structure of deoxyribonucleic acid. Nature, *171*: 964.

WEIL, R. and J. VINOGRAD. 1963. The cyclic helix and cyclic coil forms of Polyoma viral DNA. Proc. Natl. Acad. Sci., *50*: 730.

WEISSMANN, C., P. BORST, R. H. BURDON, M. A. BILLETER, and S. OCHOA. 1964. Replication of viral RNA, III. Double-stranded replicative forms of MS2 phage RNA. Proc. Natl. Acad. Sci., *51*: 682.

YOUNG, E. T., II and R. L. SINSHEIMER. 1964. Novel intra-cellular forms of lambda DNA. J. Mol. Biol., *10*: 562.

OLE MAALØE
University Institute of Microbiology, Copenhagen, Denmark

The Relation between Nuclear and Cellular Division in Escherichia coli

"*Nur das Schaffen erhält jung.*"—K. Delbrück

Twenty years ago nearly all senior bacteriologists and virologists were trained either as chemists or as physicians. The reason was that most of the jobs available to them were in technological institutes (like the famous Laboratory of Microbiology in Delft) or in institutes of medical bacteriology (such as the Pasteur Institute in Paris). Few scientists with a background in general biology ever established themselves in microbiology and those that did were mostly taxonomists.

The dramatic change through which microorganisms, including viruses, became objects of choice for the study of a variety of biological phenomena occurred when physicists and geneticists were first attracted to the field. The phage courses catalyzed this process in a decisive way.

A very small "class of 49," consisting of one physicist (Jean Weigle) and one M.D. (myself) graduated in Pasadena under Max Delbrück's supervision, and subsequently took part in a phage meeting devoted largely to multiplicity- and photo-reactivation. This was very exciting, but the individual, startling discoveries, so characteristic of that period, probably meant less to me than the simple lesson that processes as complex as those involved in phage reproduction could be initiated throughout a bacterial population *at a precisely defined time*, thus permitting a study of the natural sequence of events during the synthesis of new particles. This growth cycle is of course a perfectly "normal" biological event, and although the infected cell eventually lyses, the one-step growth experiment is very different from the altogether pathological cases in which the death of bacteria results from chemical or physical treatments. In fact, one of the few features common to such experiments is that nearly all the cells behave alike.

The normal growth cycle of a bacterial cell could be studied in the same way as phage growth if the ideal conditions of a well-timed phage experiment were created in an uninfected culture; i.e., if synchronous growth and division could be obtained. Being very interested in the normal behavior of the bacterial cell, I decided to try to establish such conditions and, together with Gordon Lark, found a procedure involving temperature shifts that produced reasonably good division synchrony.

When the time came to investigate the division cycle of the bacterial cell we were disappointed, however, because it turned out that the technique used to obtain synchronous division seriously disturbed the normal sequence of events in growing cells; i.e., the object we wanted to study was no longer "normal" (see Maaløe, 1960). For this reason synchronization was abandoned in the laboratory until recently when a new and better technique became available (see below).

However, the experiments which demonstrated the inadequacy of the temperature shift method revealed new possibilities. It became clear that many features of the growth physiology of bacteria could be studied without synchronization just by following the sequence of events by which cells adjust to a new medium known to support a higher growth rate; i.e., to a shift-up. In such an experiment the entire cell population responds as a unit, much as does a phage infected culture, and it is a curious fact that this extremely simple approach was developed in such a roundabout way.

In a recent monograph we have described in detail how this primitive start led to an extensive study of the control mechanisms by means of which bacteria adjust the relative rates of DNA, RNA, and protein synthesis (Maaløe and Kjeldgaard, 1966). In this work we have tried to keep as close as possible to "normal" growth conditions, and the simple shift-up has been as important in our limited area of study as the one-step growth experiment was for the development of phage research in general.

It is not uncommon that certain striking observations made during the early development of a new line of research remain unexplained for a long time (in phage work such conspicuous phenomena as "mutual exclusion" and "lysis inhibition" remain mysterious despite all the advances made since they were first observed). Our work on control mechanisms will be illustrated by an example showing how one of the first and most intriguing observations, which for several years was quite incomprehensible, can now be analyzed in terms of a reasonable model. I refer to the fact that, after a shift-up, the preshift rate of cell division is rigorously maintained for about one hour (at 37°).

ANALYSIS OF THE "RATE MAINTENANCE EFFECT"

The more we study bacterial growth the stronger becomes the impression of order. All the major synthetic activities of the cells seem to be effectively regulated, and it is natural to assume, as is usually done, that nuclear and cellular division are linked in such a way as to ensure that each daughter cell can receive at least one full complement of DNA (i.e., one genome) at the time of cell division.

The bacterial nucleus is a very small, rather ill-defined cytological object and mitotic configurations are not observed. We shall consider nuclear division to be accomplished in a bacterial cell when, at some point after the completion of a round of replication, two identical DNA units separate (see Fig. 1). It is not clear whether, at this point, two distinct nuclear bodies would be observable in the microscope.

In uninucleate cells of higher organisms, nuclear and cellular division certainly are coupled, but it is not obvious that the same is true in the more primitive procaryotic cells. In bacteria, nuclear and cellular division could be imagined to be independent events occurring with similar frequencies, and it is difficult to rule out this notion by direct observations. However, in the course of many generations of growth such independence would lead to a steady state in which no definite time relation would exist between nuclear and cellular division; i.e., knowledge of the time a cell had divided would not indicate *when* during the cycle nuclear division had taken place.

Experiments by D. J. Clark (see Maaløe and Kjeldgaard, 1966) indicate that a more or less precise time relation actually exists between nuclear and cellular division in *E. coli* (strain B/r), and we therefore conclude that these events in fact are linked. This seems reasonable, and it is therefore surprising to find that the bond between the two division processes apparently is broken during the transition period following a shift-up. Thus, when a bacterial culture is transferred from minimal medium to broth, the pre-shift rate of cell division is maintained for 60–70 min. (at 37°), after which time the new and higher division rate is established; in contrast, the rate of DNA synthesis increases to its new value within 15–25 min. after the shift (i.e., about 40 min. *before* the cell division rate changes). As a result of the temporary dissociation between these rates, the average number of nuclei per cell increases during the first hour after the shift (Kjeldgaard, Maaløe, and Schaechter, 1958).

For years this observation has been a puzzle; in particular, no attractive teleological explanation has suggested itself. The purpose of this paper is to present a model according to which the maintenance of the pre-shift rate of cell division (and with that the increase in number of nuclei per cell) is a necessary consequence of the pattern of replication characteristic of rapidly growing cells.

The model is based on the following three assumptions: (1) that the increase in the *rate* of DNA synthesis observed between 15 and 25 min. after a shift-up is due to the introduction of "extra" growing points (see Fig. 1 B); (2) that cell division *normally* occurs about 20 min. after the initiation of a new round of replication; and (3) that the processes leading to cell division are *not* triggered unless two separate DNA units exist in the cell at the time replication is initiated. The experimental evidence for (1) and (2) will be briefly discussed; the last assumption is arbitrary, but seems plausible.

The establishment of several growing points per nucleus (as illustrated in Fig. 1 B) seems to be the means by which rapidly growing bacteria manage to double their DNA content in a time shorter than that required for one round of replication (defined as the more or less fixed time it takes for *one* growing point to travel through the whole length of the genome). This notion is due to Sueoka and his colleagues, and it is based on transformation studies with *B. subtilis* by Yoshikawa and Sueoka (1963); Yoshikawa, O'Sullivan, and Sueoka (1964); and Oishi, Yoshikawa, and Sueoka (1964). With a technique that combines density and tracer labeling, it has been shown

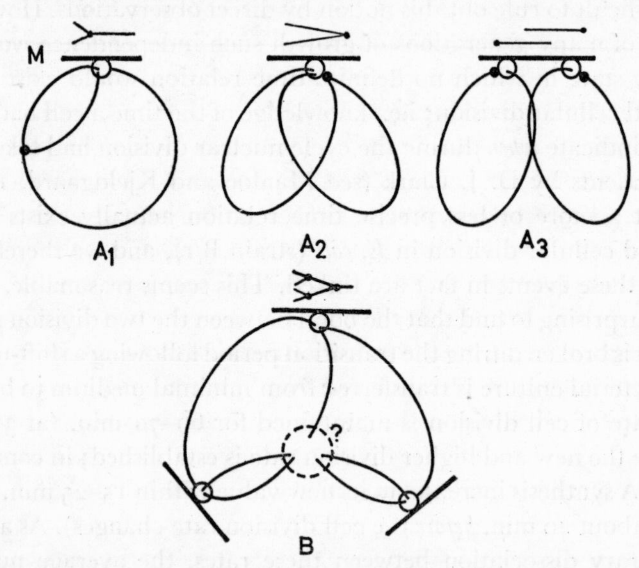

FIGURE 1. The bacterial nucleus in various states of replication. A_1 shows a growing point, imagined to be attached to the cell membrane (M), through which the circularized genome is threaded during replication. A_2 shows the situation at the end of a round of replication. The next steps are thought to be 1) the actual separation of the two genomes (nuclear division) and 2) the initiation of a new round of replication involving both (A_3). According to our model, this is the sequence of events during growth in glucose-minimal medium.

B illustrates a nucleus with three growing points, all attached to sites on the membrane. Replication is thought to follow this pattern in rapidly growing cells. The spatial arrangement in Figure 1 B is arbitrary, and the same is true of the way in which circularization has been retained by allowing the five free ends to meet and be held together in a hypothetical structure.

Each of the four states of replication (A_1–A_3 and B) is also illustrated by a linear symbol. This simplified notation, in which a black dot represents a growing point, is used in Figure 2.

that "extra" growing points also can be established in the nuclei of *E. coli* (Pritchard and Lark, 1964). The increase in the overall rate of DNA synthesis observed around 20 min. after a shift-up is therefore interpreted to mean that the replication pattern changes from that illustrated in Fig. 1 A to that of Fig. 1 B.

The second assumption is that cell division normally occurs some 20 min. after the initiation of a new round of replication. This notion is derived from experiments with synchronously dividing cultures of *E. coli*, strain B/r (see Maaløe and Kjeldgaard, 1966). The technique of Helmstetter and Cummings (1963) was used to select "new-born" cells from populations of bacteria representing steady states of growth in media producing different growth rates. Throughout the first two division cycles the *residual capacity for DNA synthesis* was determined in samples to which chloramphenicol had been added. It is believed that replication already in progress continues under these conditions, whereas new rounds of replication cannot be initiated. The relative quantities of DNA produced in the presence of chloramphenicol should therefore be related to the replication cycle in the following way: samples receiving the drug immediately after the initiation of a new round should produce maximum amounts of DNA, and subsequent samples should produce gradually decreasing quantities. Thus, if initiation occurs at a defined time in the division cycle, a more or less abrupt increase in the residual capacity should be observed. In Clark's experiments such an increase occurred regularly some 20 min. before cell division, irrespective of the growth rate. This in turn indicates that at least *two* separate genomes must be involved in the act of initiation, since 20 min. presumably is only about half the time it takes to complete a round of replication. (The rather abrupt increase in residual DNA synthesis in strain B/r of *E. coli* suggests that replication of *both* genomes is initiated more or less simultaneously, and that this is the case even in slow-growing cells. A different model has been proposed by Lark and Lark [1965]. With one of the $15T^-$ strains of *E. coli* they show that DNA synthesis *always* takes place during most of the division cycle, and further experiments lead them to the interesting hypothesis that, during *slow* growth, the two genomes of a cell replicate *sequentially* rather than simultaneously. At the moment it is not easy to reconcile their results and those of Clark, except by assuming that the strains used actually behave differently.)

The third and last assumption was that cell division can be triggered only if *two* separate DNA units are present at the time replication is initiated. In a cell of the type shown in Fig. 2 A, the *separation* of the two genomes, which immediately precedes initiation, is therefore thought to condition the cell division that takes place about 20 min. later. In the terminology used here this means that nuclear division *must* precede cell division by about 20 min., and it implies that an act of initiation *not* preceded by nuclear division cannot start the chain of events that lead to cell division.

This last point is important when analyzing the relation between replication and cell division during the period following a shift-up. Consider a culture which, at time zero (t=0), is transferred from a glucose-minimal medium to broth, and let the doubling times characteristic of balanced growth in the two media be 40 and 20 min., respectively. Figure 2 shows the time course of replication and of cell division in bacteria that are transferred at the time when replication is initiated. The new, or extra growing points are thought to be established 20 min. after the shift, i.e., when the two genomes are half replicated and the cell is about to divide (Fig. 2 includes a schematic representation of the replication patterns before and after the shift).

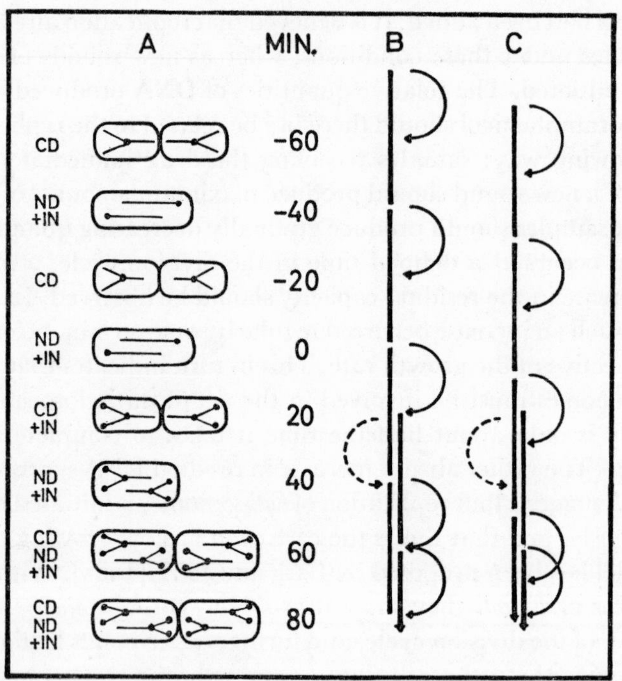

FIGURE 2. The time course of replication and of cell division before and after a shift-up of the type discussed in the text. CD, ND, and IN stand for cellular division, nuclear division, and initiation of replication, respectively. The symbols used to indicate the state of replication in a cell (Column A) are explained in Figure 1. In Columns B and C the vertical bars indicate time of onset and the duration of individual rounds of replication. Solid arrows indicate the time elapsing between a nuclear division and the corresponding cell division; the cases in which replication is initiated *without* being accompanied by nuclear division (at 20 min.) and in which, according to our model, cell division cannot be triggered, are indicated by stippled arrows.

The first round of replication to be initiated after the shift (at 20 min.) is not preceded by nuclear division and, according to our rules, it cannot trigger cell division. This will not occur until nuclear division takes place at 40 min., and at 60 min. when the cell divides it will have 4 genomes, each of which is half replicated (see Fig. 2). This rigid scheme obviously produces the transition pattern observed in a shift-up; i.e., maintenance of the pre-shift rate of cell division for 60 min. and a doubling of the number of nuclei per cell. Notice that the maintenance period is composed of two elements: the 20 min. lag preceding the establishment of extra growing points, plus the 40 min. it takes for the first of these new points to complete a round of replication. As of this time (60 min.) a round of replication will be completed every 20 min.; i.e., the new rhythm of nuclear *and* of cellular division can now be established.

The case just considered is simple because the time of the shift was chosen so as to coincide with nuclear division. Column C of Fig. 2 illustrates a shift taking place 10 min. before nuclear division, and if the "extra" growing points again are assumed to be established at 20 min., the overlap between the replication periods will be asymmetric. This would mean that eventually the cells should divide at intervals alternating between 10 and 30 min. Probably, in such cases the "extra" growing points are established relatively later, say, at 25 min.; in this way the initial asymmetry will be reduced and it may disappear gradually during subsequent growth. Conversely, if the shift takes place soon *after* nuclear division, the "extra" growing points may be introduced earlier than 20 min. This assumption is consistent with the fact that the overall rate of DNA synthesis does not increase abruptly at 20 min. as indicated by the formal scheme of Fig. 2, but changes gradually over a period of at least 10 min. It can be shown graphically that cell division in the culture as a whole will follow the same time course as in the special case shown in Fig. 2 (Columns A and B).

The shift from glucose-minimal medium to broth has been extensively studied both with respect to DNA synthesis and to cell division, and the model just described is based on these data. Shifts from media "slower" than glucose-minimal into broth have also been done, and again cell division has been observed to continue at the pre-shift rate for about 60 min. However, DNA synthesis was not followed in these experiments, and since, furthermore, the replication pattern in slow-growing cultures needs to be studied more carefully, shifts involving such cultures will not be discussed further at this time.

The effects of a shift-up on the rates of DNA and on protein synthesis are very similar. In both cases the pre-shift rates are maintained initially and they reach the higher post-shift values more or less simultaneously. For this and other reasons it is tempting to speculate that somehow the absolute rate of protein synthesis in a cell determines the *frequency* with which new rounds of replication are initiated (see Maaløe and Kjeldgaard, 1966).

EPILOGUE

The problem discussed above is rather special, but nevertheless of considerable significance as part of the general study of bacterial growth carried out in our laboratory during the past ten years. Actually, the "rate maintenance effect," as applying to cell division, was the first recorded characteristic of a shift-up. The delayed but rather sudden change from the pre- to the post-shift division rate suggested that more or less discrete steps were involved in the transition. Further experiments revealed that the immediate response to a shift-up is a dramatic, but transient dissociation between RNA and protein synthesis. This observation led to a more systematic study of different steady states of growth. In fact, the shift-pattern—with its abrupt increase in the rate of RNA synthesis and the initial maintenance of the pre-shift rates of DNA and of protein synthesis—has incited much of our later work on control mechanisms (Maaløe and Kjeldgaard, 1966).

During the past six to eight years the cell division problem has frequently been re-examined, always without success. The model presented here could not have been built up gradually because it is based on new and unsuspected information about DNA replication in rapidly growing cells; it may not contain the whole truth (nor nothing but the truth) but it looks promising.

REFERENCES

Helmstetter, C. E., and D. J. Cummings. 1963. Bacterial synchronization by selection of cells at division. Proc. Natl. Acad. Sci., *50*: 767.

Kjeldgaard, N. O., O. Maaløe, and M. Schaechter. 1958. The transition between different physiological states during balanced growth of *Salmonella typhimurium*. J. Gen. Microbiol., *19*: 607.

Lark, K. G., and C. Lark. 1965. Regulation of chromosome replication in *Escherichia coli*: Alternate replication of two chromosomes at slow growth rates. J. Mol. Biol., *13*: 105.

Maaløe, O. 1960. The nucleic acids and the control of bacterial growth. Symp. Soc. Gen. Microbiol., *10*: 272.

Maaløe, O., and N. O. Kjeldgaard. 1966. Chapter 6 in Control of macromolecular synthesis. W. A. Benjamin, Inc., New York.

Oishi, M., H. Yoshikawa, and N. Sueoka. 1964. Synchronous and dichotomous replications of the *Bacillus subtilis* chromosome during spore germination. Nature, *204*: 1069.

Pritchard, R. H., and K. G. Lark. 1964. Induction of replication by thymine starvation at the chromosome origin in *Escherichia coli*. J. Mol. Biol., *9*: 288.

Yoshikawa, H., A. O'Sullivan, and N. Sueoka. 1964. Sequential replication of the *Bacillus subtilis* chromosome. III. Regulation of initiation. Proc. Natl. Acad. Sci., *52*: 973.

Yoshikawa, H., and N. Sueoka. 1963. Sequential replication of *Bacillus subtilis* chromosome. I. Comparison of marker frequencies in exponential and stationary growth phases. Proc. Natl. Acad. Sci., *49*: 559.

VI. *Ramifications of Molecular Biology*

California Institute of Technology, Pasadena

T. T. PUCK
The Eleanor Roosevelt Institute for Cancer Research and the Florence R. Sabin Laboratories of the Department of Biophysics, University of Colorado Medical Center, Denver, Colorado

The Mammalian Cell

The articles in this book all bear testimony to the deep and extensive impact Max Delbrück produced on biology. His development of the bacteriophage system into a device, capable of rapid and exquisitely precise answers to specific questions dealing with biological replication and its associated processes, furnished an experimental tool whose usefulness continues to expand after more than 25 years. In addition, he introduced into biology a different kind of physical thinking from that of several other physicists who had previously attempted to approach biological phenomena. These people often asked of the biological system questions of great generality and of the kind that is often asked of inanimate matter such as: "Can living organisms manufacture negative entropy?" "or "What are the implications of the uncertainty principle for living systems?" Delbrück instead focused on uncovering the intimate and basic phenomenology of the replicative process, and on expressing these as precisely as possible in terms of specific, time-ordered transformations of definite entities. His passionate rejection of vagueness in the building and testing of conceptual models has helped to change radically the entire philosophy of biological research. Finally, he combined the results of quantitative experimentation, and highly specific model-building approaches in the bacteriophage system, with the newly emerging conceptual advances of biochemical genetics. The results of his innovations together with those of his colleagues, students, and their students, and of scientists who were attracted by the excitement so generated have taken the form of what is now called Molecular Biology—one of the great human intellectual achievements.

Like many of Delbrück's students and co-workers, I became enchanted with the possibilities of the bacteriophage system. On coming to Colorado, I engaged with Alan Garen, the first graduate student in the Department of Biophysics of this University, to study details of the attachment and penetration reactions initiated by bacteriophages invading their host cells. Later, I conceived the possibility of attempting to apply the powerful methods of genetic-biochemical analysis, which microbial techniques offer, to mammalian cell systems. Such studies began in 1955 and have involved contributions by and collaboration with the following persons, most of whom have been graduate students or post-doctoral fellows in the Department: Mrs. J. R. Christianson, P. Marcus, H. Fisher, S. Cieciura, D. Morkovin, G. Sato, J. Tjio, L. Tolmach, H. Lee, A. Robinson, J. Engelberg, M. Oda, R. Ham, M. Yamada, R. Gamewell, J. Daniel, P. Sanders, R. Tobey, D. Peterson, E. Anderson, and F. Kao.

DEVELOPMENT OF QUANTITATIVE METHODS FOR THE STUDY OF GROWTH AND GENETICS OF MAMMALIAN CELLS

In 1955 Philip Marcus and I devised a simple method for growing single mammalian cells into clones in vitro in a quantitative fashion analogous to that of the standard plating technique of quantitative microbiology. In further steps the composition of the nutrient solution employed was improved so that it was possible to eliminate the feeder layer which we had devised as a means for providing the necessary conditioning of the earlier tissue culture media. Then, methods were developed making possible reliable initiation of cultures from cell samples taken from a large variety of tissues of many different mammals, and long-term cultivation of cell strains without the karyotype disorganization which previously had characterized most, if not all, of the standard tissue culture strains (Puck and Marcus, 1955; Puck, Marcus, and Cieciura, 1956; Ham and Puck, 1962).

In this way it became possible to quantitate the number of cells capable of reproduction in normal or specially treated cell cultures; to select mutants from such cultures; and to establish long term cell cultures from individual animals of particular genetic constitutions.

CHROMOSOMAL ANALYSIS AND ITS APPLICATION TO HUMAN GENETICS

It now seems almost unbelievable that for thirty years an erroneous value was accepted for the number of the human chromosomes. Tjio and Levan first challenged the number of 48 with their finding of only 46 chromosomes in cells of human fetal origin. Other reports indicated that a variety of different chromosomal numbers are commonly to be found in man (Kodani, 1957). Since we had developed methods of reliably cultivating cells from the skin of any person, we invited Tjio to join our laboratory and to devise a method for examination of the chromosomes of any person. This made it possible to demonstrate that only 46 chromosomes were to be found in repeated sampling of the cells of normal persons. The first complete analysis of the human karyotype resulted from these studies and, shortly thereafter, an International Conference held at Denver resulted in the adoption of a uniform classification system for human chromosomes. Study of mammalian chromosomes in vitro has become an exceedingly active scientific field, in which many new techniques contributed by a variety of laboratories have steadily provided more simple and rapid chromosomal delineation. One of the unexpected results of these studies has been the demonstration of the amazingly high incidence of human diseases due to gross chromosomal anomalies, particularly non-disjunction. At least ½% of all human live births display deviations in chromosomal number or structure and in virtually every case these are attended with metabolic diseases, most of which include mental deficiency as a symptom. Medical cytogenetics has suddenly become an extremely active branch of medicine. In a recent study, we have shown that the frequency of anomalies of the sex chromosomes in human newborns displays a pattern which strongly indicates that these aberrations do not occur spontaneously but are the result of external factors, a conclusion which offers hopes that the specific causes may be identifiable and controllable (Tjio and Puck, 1958a, b; Puck, Ciecura, and Robinson, 1958; Tjio, Puck, and Robinson, 1959; Puck, Robinson, and Tjio, 1960; Tjio, Puck, and Robinson, 1960; Böök et al, 1960; Puck and Robinson, 1966; Robinson and Puck, 1965).

NUTRITIONAL REQUIREMENTS FOR MAMMALIAN CELL GROWTH; CELL DIVISION PROTEIN—FETUIN A

Early in these studies, we began a long term program to achieve a medium of completely defined constitution in which single cells would routinely grow to form colonies. Although our first media required both whole mammalian sera and a feeder layer of X-rayed cells in order to provide the molecular environment needed for self-sustaining growth from single cells, further studies on medium enrichment made possible elimination of the feeder layers. However, the continued need for mammalian sera with its large numbers of unknown and highly variable constituents in both the macro- and the micro-molecular fractions made it difficult to secure well controlled conditions for quantitative, reproducible experiments, and for nutritional analysis.

In these studies we began with results of work from other laboratories which had defined nutritional requirements of mammalian cells grown in massive inocula. It soon became apparent, however, that the requirements for single cell growth are much more demanding. Eventually a medium was devised in which single cells of some of our standard clonal strains would grow with 100% efficiency, and which contained a synthetic micromolecular part plus two highly purified proteins. These latter consisted of serum albumin, and an α-globulin protein which could be isolated from a variety of different mammalian serum but which was found in highest concentration in the α-globulin component of fetal calf serum that had previously been named fetuin by Pedersen who first characterized this protein fraction.

The availability of this medium of highly defined composition greatly increased the reproducibility of cell plating experiments. It also raised the question about the role of these two proteins in promoting cell division and whether a completely synthetic medium might be possible in which single mammalian cells of at least some mammalian strains could reproduce to form colonies.

In attempting to answer these questions, we changed over from the S3 HeLa cell clone to clones derived from the Chinese hamster, because the latter were more nearly diploid in contrast to the highly aneuploid and polyploid HeLa cells. Analysis of the role of albumin revealed that it could be replaced by the addition of judicious amounts of linoleic acid. Dr. Ham then demonstrated that linoleic acid, although a nutritional growth requirement of these cells, inhibits growth in concentrations only slightly greater than the optimum amount. Apparently albumin, which binds fatty acids strongly, provides a reservoir of linoleic acid adequate for the cell needs without allowing the free concentration in solution ever to become great enough to exercise a toxic effect.

The metabolic role of the α-globulin protein has proved more intricate. The need for this α-globulin component was announced almost simultaneously by Lieberman and Ove (1957) and ourselves (Sato, Fisher, and Puck, 1957; Fisher, Puck, and Sato, 1958, 1959), and was identified by the former group as a glycoprotein. We confirmed its glycoprotein nature and demonstrated that the fetuin fraction of fetal calf serum contains this activity in amounts greater than any other serum yet described. The fetuin fraction has been highly purified so as to exhibit maximal biologic activity and uniform electrophoretic and sedimentational patterns. It appears to be a typical glycoprotein with 15% of its mass consisting of sugar residues. Its biological activity can easily be destroyed by relatively simple procedures. In addition to making cell division possible in an otherwise inadequate medium, fetuin also has the property of producing a striking cytological reaction: When added in concentrations of approximately one microgram per milimeter to a cell suspension, in a completely synthetic medium, it causes the rounded cells to become attached to glass or plastic surfaces and to stretch out in highly characteristic, spindle-shaped forms (Fisher, Puck, and Sato, 1958, 1959; Fisher, O'Brien, and Puck, 1962).

While other laboratories have confirmed these findings that mammalian sera contain a α-globulin protein which confer upon cells the ability to stretch out on a glass surface and to reproduce in an otherwise inadequate medium, it has been questioned whether fetuin (or its active component if not all of the constituents of this protein fraction should contain activity) is indeed the same substance obtained from adult mammalian sera which produces similar effects. Moreover, the fact that certain mammalian cell cultures were discovered which appeared to reproduce at least in massive culture with no protein supplement whatever raised the question as to whether fetuin is indeed a significant substance in the reproductive process.

In an effort to answer both of these questions, antisera were prepared by injection into sheep of fetuin prepared from fetal calf serum. Gamma-globulin was prepared from the resulting serum and as a control, also from the sera of other sheep which had been injected in a similar way with bovine serum albumin. It was found that the anti-fetuin γ-globulin would neutralize the cell growth and stretching activities of all mammalian sera tested including human and bovine, adult and fetal sera. The γ-globulin obtained from sheep injected with bovine albumin was inactive. Moreover, γ-globulin from fetuin-injected sheep prevented the growth and stretching of all mammalian cells tested including even those that were capable of growth in the absence of any macromolecular constituents. The activity of the anti-fetuin γ-globulin could be completely neutralized by the addition of fetuin. Finally, it was shown that even cells capable of growth and stretching without any protein supplement will increase their activity in these respects when given purified fetuin. We have provisionally interpreted these experiments as indicating that fetuin appears to be required in the reproductive process of all mammalian cells tested but that some cells may manufacture a supply of this substance adequate to promote a low level of reproduction without an external source. Recent experiments have furnished evidence that these cells can synthesize proteins with properties similar to that of fetuin. Table I exhibits the results of a typical effect of the fetuin antibody preparation on cell stretching.

Table 1

Demonstration of action of anti fetuin gamma globulin on the glass attachment and stretching of mouse L-cells. Gamma globulin prepared from sheep injected a) with fetuin and b) with bovine albumin was added to mouse L-cells, which are capable of growth in massive culture in a completely synthetic medium.

Concentration of gamma globulin (μgm/ml) a) Anti fetuin	b) Anti albumin	Per cent of cells remaining attached to glass	Per cent of glass-attached cells which are stretched
0	0	100	92.0
100	0	100	1.0
1000	0	42	0.2
5000	0	0	—
0	5000	72	95.0

In a further attempt to define the growth requirements of single mammalian cells, a standard strain of Chinese hamster cells was investigated. A medium was achieved in which, by the addition of both linoleic acid and

1,2 diamino butane, single cell growth with high efficiency was obtained in the absence of any macromolecular supplement whatever. Although cell growth is more rapid when fetuin is furnished, the completely synthetic medium can be used on a routine basis (Fisher, Puck, and Sato, 1959; Ham, 1963, 1965).

GENETIC BIOCHEMICAL STUDIES

One of the main purposes underlying development of microbiological techniques for use with mammalian cells was to permit biochemical and genetic studies like those that had been so enormously successful for microorganisms like *E. coli*. However, great difficulty was encountered in obtaining good genetic markers for such studies. This difficulty arose from two facts: It is more difficult to obtain mutants displaying phenotypic differences in diploid cells, particularly in the absence of methods for achieving genetic recombination. In addition, however, it proved exceedingly difficult to find genetic markers that are sufficiently stable and possess a low enough background reversion rate (either phenotypic or genotypic) to be used in mutational studies. Attempts were made to find isolatable markers involving resistance to viruses, antimetabolites or specific antibodies. In early experiments we were unable to find any such genetic markers that were sufficiently clean to permit definitive studies. However, when the completely synthetic medium, F12, was developed for one strain of our Chinese hamster cell lines, it was tested on our other Chinese hamster cell stocks in an effort to determine whether some of these might exhibit nutritional differences useful in genetic experiments. And this revealed that one of our standard Chinese hamster cells strains exhibits a specific requirement for the amino acid proline (Ham, 1963, 1965).

In the standard medium without proline, or in proline concentrations as high as 10^{-5}M, no growth whatever results. In 10^{-3}M proline maximum plating efficiency and growth rate are achieved. Moreover, if ornithine is added, the proline-deficient mutant multiplies with high plating efficiency but reduced growth rate. When ornithine plus 10^{-5}M proline are simultaneously present, maximum plating efficiency and growth rate result. It would appear then that, as shown in Figure 1, the conversion of glutamic acid to its semi-aldehyde may be blocked in this mutant while the pathway from ornithine through glutamic γ-semi-aldehyde to proline is probably intact (Kao and Puck, 1966).

Cytologic examination showed that this strain contains only 21 chromosomes and that the missing chromosome is the large sub-metacentric No. 2. In this strain, arginine will not replace proline even in high concentrations but it will stimulate growth in the presence of sub-optimal amounts of proline. Hence, the pathway from arginine through ornithine to proline may be intact but it is not a major source of proline in these cells. Citrulline does not overcome the need for proline under any conditions.

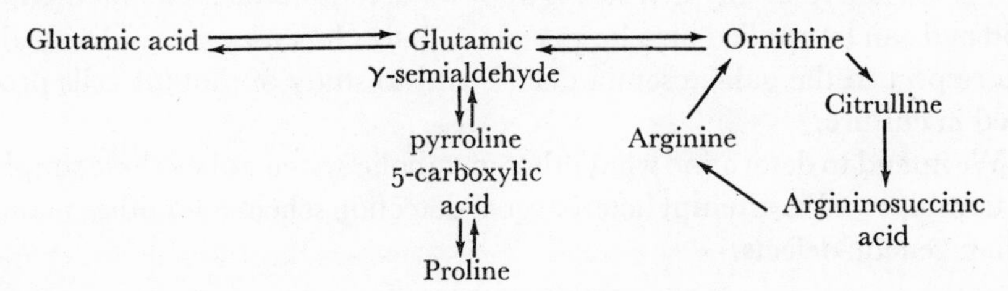

FIGURE 1. Metabolic relationships between glutamic acid, proline, and ornithine. (After Vogel and Bonner, Proc. Natl. Acad. Sci., *40*: 688, 1954.)

The spontaneous rate of reversion of this mutant to proline independence was carefully measured and found to be equal to or less than 1.8×10^{-6} per generation. Therefore, the marker appears to be stable and to possess a sufficiently low background for genetic experiments. The action of a variety of mutagenic agents on this marker is currently under study.

In principle these developments appear to be generalizable. Hence, search is now under way for cell strains which are monosomic for other chromosomes in the hope that these will provide additional genetic markers and at the same time make possible the localization of specific genes on the various mammalian chromosomes.

DETECTION OF HETEROZYGOTES

One of the most important problems in human genetics is the detection of recessive heterozygous gene defects in man, which would identify affected persons and permit them, if they wished, to avoid producing children with the homozygous disease. Moreover, such detection methods could provide estimates of the rates of various mutational events in man; evaluation of possible subclinical effects of heterozygous situations involving recessive genes; and estimation of selection pressures exercised by certain genes in various situations. Using the enzymes of the galactose metabolic chain as a representative system, a method was developed for rapid measurement of the cellular reaction rate constants for mammalian cells. The procedure has been applied to the sequential conversion of galactose to galactose-1-phosphate and uridine-diphosphogalactose, in red cells, leucocytes, and cells grown in culture (Robinson, 1963). Estimates of the specific reaction rate constants per cell have been made for normal cells homozygous for the genes of the galactose oxidation pathway; for cells heterozygous for the transferase gene, which are obtainable from the parents of galactosemia patients; and for tissue culture cells which appear to be euploid tetraploids. While the results are still in an early stage, in each case the specific reaction rate constant for the transferase enzyme shows the expected variation corresponding to the number of active genes present. Hence, the number of these genes active in any cell strain appears determinable. The method is rapid and can be applied on a large scale to detect heterozygous individuals with respect to the galactosemia defect, and to study of mutant cells produced in culture.

We intend to determine what other metabolic systems obey these simple relationships and to attempt heterozygote detection schemes for other mammalian genetic defects.

LIFE CYCLE ANALYSIS OF MAMMALIAN CELLS

A method for analysis of the biochemical events in the life cycle of mammalian cells has been devised. It involves addition of a specific inhibitor to a random cell population so that passage of the cells around their life cycle is blocked at one specific point. The accumulation of cells behind the block is scored by microscopic or other means and the results are expressed as a function (called the accumulation or collection function) which is linear for random cultures and which deviates in clearly recognizable fashion at each point where the cell population deviates from the random distribution. The method can be used to measure the duration of each of the standard segments (mitosis, G1, DNA synthesis and G2) of the life cycle for any cell population; the point of synthesis of any macromolecule for which a labeled precurser can be furnished; and the point in the life cycle at which specific inhibitors block passage of the cells around the cycle. The life cycle patterns of a variety of cells grown in tissue culture have been compared and differences have been found between cells that exhibit no differentiation characteristics and those exhibiting specific differentiation processes. Metabolic events can be mapped with a resolving power of approximately 15 minutes in the total life cycle of ten or more hours. The points of action in the life cycle for a variety of agents including colcemide, velban, an excess of thymidine, actidione, actinomycin, streptonigrin, flurophenylalanine, and doses of X-irradiation as small as 9 rads have been accurately located. A variety of specific blocking points have been defined including arrests in anaphase, the G2/mitosis boundary, the middle of G2, a point slightly beyond the middle of S and points in G1. One example of the use of this method has been the demonstration that the reversible G2 lag induced in the mammalian cells by X-irradiation is like the irreversible killing process in that chromosomal damage is involved. Experiments to test this point were carried out by measurement of the effect on the G2 period of a variety of agents. Only streptonigrin, which produces chromosome breaks strongly resembling those obtained from X-irradiation, yielded a reversible G2 lag like that characteristic of X-irradiation. In another experiment, cells were labeled with bromo-deoxy-uridine which, like thymidine, is incorporated only into DNA. Such labeled cells show a completely normal G2 period. However, after exposure to visible light, which produces localized photochemical action in the DNA, these cells, but not unlabeled cells, also exhibit a G2 lag like that resulting from X-irradiation.

This life cycle method has also been employed to demonstrate that in the part of the life cycle of Chinese hamster cells, just preceding the onset of mitosis, the time between completion of messenger RNA formation and its associated protein synthesis (which possibly may involve some component of the spindle protein) lies in the neighborhood of 1.3 hours (Puck and Steffen, 1963; Puck, Sanders, and Petersen, 1964; Tobey, Petersen, Anderson, and Puck, 1966; Puck, 1964c).

MAMMALIAN RADIOBIOLOGY: SINGLE CELL SURVIVAL CURVES; THE RADIATION SYNDROME; CELL TURNOVER TIMES IN TISSUES

The availability of a simple, precise method for measuring the ability of single cells to reproduce and form colonies made it possible to determine survival curves following exposure to a wide variety of physical, chemical, and biological agents. X-irradiation survival curves for mammalian cells were constructed and these demonstrated that the mean lethal dose for cell killing (defined as loss of the ability of a cell to form a macroscopic colony) was only about 100 rads in contrast to previous estimates which, in general, were much higher and often reached values beyond 50,000 rads. The mean lethal dose was found to be very similar for cells taken from a variety of different mammalian species and from different tissues. Several lines of evidence have demonstrated beyond reasonable doubt that the damaging process responsible for death is located in the chromosomes and is probably associated with the production of chromosomal breaks and abnormal restitutions, some of which are microscopically visible in subsequent mitoses.

On the basis of the finding that mammalian cells have an extraordinarily high sensitivity to killing by X-irradiation and that the cells, whose reproductive careers have been so ended, continue to carry out many metabolic functions including macromolecular synthesis at a very high rate, it was proposed that cell reproductive death, which displays a mean lethal dose of about 100 rads, is the principal immediate action resulting from exposure of mammals to ionizing radiation, at least in doses less than 1,000 rads. Since the mean lethal dose for killing is approximately the same for all mammalian cells so far tested, this idea implies that the basic lesion is similar in the somatic cells of the various tissues. It differs, therefore, from the classical interpretation of the effects of radiation on mammals which supposed that the basic radiation-induced lesion is different in the cells of different tissues and involves cell disintegration or other accelerated removal processes depending on the particular interaction of radiation with the cells of different tissues. Both theories can account for the fact that some tissues, like bone marrow, exhibit a high rate of cell depletion immediately after X-irradiation of the animal while others, like the brain, reveal little change even after exceedingly high doses.

Experiments were designed to distinguish between these two theories. It was shown that the rate of cell depletion should display a different kind of dose dependence if the action of radiation is limited only to the suppression of cell reproduction or if, in addition, radiation triggers other processes causing disintegration or removal of cells.

The pattern of cell depletion in the bone marrow, spleen and thymus of young mice following total and partial body X-irradiation was studied as a function of time. In every case, the results followed the predictions of the hypothesis that cell reproductive death alone is the major effect of X-irradiation. Thus, it was demonstrated that the initial rate of nucleated cell depletion at first increases with dose and then reaches a plateau at about 250 rads which remains constant even for doses beyond 1,000 rads. The dose at which the plateau is reached is the same even for tissues with significantly different cell turnover times. The same limiting depletion rate is achieved regardless of whether the radiation is applied to the whole body or only to the part under study. The depletion achieved in the bone marrow and spleen after administration of saturating amounts of drugs like colcemide and vinblastin, which inhibit cell reproduction only, is the same as the maximal rate due to X-irradiation. Finally, a reproductive survival curve can be calculated from the cell depletion following X-irradiation which agrees quantitatively with survival curves measured by single cell cloning experiments (Puck and Marcus, 1956; Puck, Morkovin, Marcus, and Cieciura, 1957; Puck, 1959; Puck, 1964a, b; Puck, 1965).

FORECAST FOR THE FUTURE

The results of these experiments indicate that the depletion constant observed in these tissues, after administration of doses of X-irradiation equal to or greater than the limiting dose, afford a rapid and convenient measurement of the cell turnover time in these particular tissues since the depletion measured under carefully specified conditions is approximately the normal rate of cell removal in the animal. The half time for depletion has been found to be 10.8, 6.5 and 4.9 hours, respectively, for the bone marrow, spleen and thymus of the young adult mouse. Definition of these constants plus understanding of the differences in action between cancer-therapeutic agents like X-irradiation, which kills cells throughout their life cycle, and velban, which kills cells only as they enter mitosis, increase the power of the therapist in attempting selectively to kill malignant cells.

To those of us working in this field, it appears that cellular and molecular biology will also revolutionize understanding of mammalian systems and ultimately provide a continuum of understanding from bacteriophage to man. Progress is still somewhat slow because the basic tools are still being fashioned. However the rate and therefore the interest of the field appear to be accelerating continuously.

REFERENCES

Böök, J. A., et al. 1960. A proposed standard system of nomenclature of human mitotic chromosomes. J. Amer. Med. Assoc., *174*: 159–162.

Fisher, H. W., D. O'Brien, and T. T. Puck. 1962. The hydrolytic products of fetuin. Arch. Biochem. and Biophys., *99*: 241–248.

Fisher, H. W., T. T. Puck, and G. Sato. 1958. Molecular growth requirements of single mammalian cells: The action of fetuin in promoting cell attachment to glass. Proc. Natl. Acad. Sci., *44*: 4–10.

Fisher, H. W., T. T. Puck, and G. Sato. 1959. Molecular growth requirements of single mammalian cells. III. Quantitative colonial growth of single S3 cells in a medium containing synthetic small molecular constituents and two purified protein fractions. J. Exp. Med., *109*: 649–660.

Ham, R. G. 1963. Albumin replacement by fatty acids in clonal growth of mammalian cells. Science, *140*: 802–803.

——— 1965. Clonal growth of mammalian cells in a chemically defined, synthetic medium. Proc. Natl. Acad. Sci., *53*: 288–293.

Ham, R. G., and T. T. Puck. 1962. Quantitative clonal growth of isolated mammalian cells, pp. 90–119. *In* S. P. Colowick and N. D. Kaplan [Ed.] Methods in Enzymology, Vol. 15. Academic Press, New York.

Kao, F. T., and T. T. Puck. 1966. Genetic-biochemical studies on mammalian cells with nutritional markers. In press.

Kodani, M. 1957. Three diploid chromosome numbers of man. Proc. Natl. Acad. Sci., *43*: 285–292.

Lieberman, I., and P. Ove. 1957. Purification of a serum protein required by a mammalian cell in tissue culture. Biochim. et Biophys. Acta, *25*: 449–450.

Puck, T. T. 1959. Quantitative studies on mammalian cells *in vitro*. Revs. Modern Physics, *31*: 433–448.

——— 1964a. Cellular interpretation of aspects of the acute mammalian radiation syndrome. *In* R. J. C. Harris [Ed.] Symp. of the Intern. Soc. for Cell Biology, *3*: 63. Academic Press, New York and London.

——— 1964b. Cellular aspects of the mammalian radiation syndrome. I. Nucleated cell depletion in the bone marrow. Proc. Natl. Acad. Sci., *52*: 152–160.

——— 1964c. Studies of the life cycle of mammalian cells. Cold Spring Harbor Symp. Quant. Biol., *29*: 167–176.

——— 1965. Cell turnover in mammalian tissues. I. Use of cell depletion measurements to estimate x-ray reproductive survival curves. Proc. Natl. Acad. Sci., *54*: 1797–1803.

Puck, T. T., and P. I. Marcus. 1955. A rapid method for viable cell titration and clone production with HeLa cells in tissue culture. (The use of x-irradiated cells to supply conditioning factors.) Proc. Natl. Acad. Sci., *41*: 432–437.

———, ——— 1956. Action of x-rays on mammalian cells. J. Exp. Med., *103*: 653–666.

Puck, T. T., and A. Robinson. 1966. Some prospectives in human cytogenetics. *In* Robert E. Cooke [Ed.] Biologic Basis of Pediatric Practice—Infancy, Childhood and Adolescence. McGraw Hill, New York. In press.

PUCK, T. T., and J. STEFFEN. 1963. Life cycle analysis of mammalian cells. I. A method for localizing metabolic events within the life cycle, and its application to the action of colcemide and sub-lethal doses of x-irradiation. Biophys. J., *3*: 379–397.

PUCK, T. T., S. J. CIECIURA, and A. ROBINSON. 1958. Genetics of somatic mammalian cells. III. Long-term cultivation of euploid cells from human and animal subjects. J. Exptl. Med., *108*: 945–956.

PUCK, T. T., P. I. MARCUS, and S. J. CIECIURA. 1956. Clonal growth of mammalian cells *in vitro*. Growth characteristics of colonies from single HeLa cells with and without a "feeder" layer. J. Exp. Med., *103*: 273–284.

PUCK, T. T., D. MORKOVIN, P. I. MARCUS, and S. J. CIECIURA. 1957. Action of x-rays on mammalian cells. II. Survival curves of cells from normal human tissues. J. Exp. Med., *106*: 485–500.

PUCK, T. T., A. ROBINSON, and J. H. TJIO. 1960. Familial primary amenorrhea due to testicular feminization: Evidence for a human sex-linked gene affecting sex differentiation. Proc. Soc. Exp. Biol. and Med., *103*: 192–196.

PUCK, T. T., P. SANDERS, and D. PETERSEN. 1964. Life cycle analysis of mammalian cells. II. The Chinese hamster ovary grown in suspension culture. Biophys. J., *4*: 441–450.

ROBINSON, A. 1963. The assay of galactokinase and uridyl diphosphogalactose enzyme activity in human erythrocytes: A presumed test for heterozygous carriers of the galactosemic defect. J. Exp. Med., *118*: 359–379.

ROBINSON, A., and T. T. PUCK. 1965. Sex chromatin studies in newborns: Presumptive evidence for external factors in human nondisjunction. Science, *148*: 83–85.

SATO, G., H. W. FISHER, and T. T. PUCK. 1957. Molecular growth requirements of single mammalian cells. Science, *126*: 961–964.

TJIO, J. H., and T. T. PUCK. 1958a. Genetics of somatic mammalian cells. II. Chromosomal constitution of cells in tissue culture. J. Exp. Med., *108*: 259–268.

———, ———1958b. The somatic chromosomes of man. Proc. Natl. Acad. Sci., *44*: 1229–1237.

TJIO, J. H., T. T. PUCK, and A. ROBINSON. 1959. The somatic chromosomal constitution of some human subjects with genetic defects. Proc. Natl. Acad. Sci., *45*: 1008–1016.

———, ———, ———1960. The human chromosomal satellites in normal persons and in two patients with Marfan's syndrome. Proc. Natl. Acad. Sci., *46*: 532–539.

TOBEY, R. A., D. F. PETERSON, E. C. ANDERSON, and T. T. PUCK. 1966. Life cycle analysis of mammalian cells. III. The inhibition of division in Chinese hamster cells by puromycin and actinomycin. Biophys. J. In press.

RENATO DULBECCO
The Salk Institute for Biological Studies, San Diego, California

The Plaque Technique and the Development of Quantitative Animal Virology

In the late Forties a prominent citizen of one of the small cities which lie east of Los Angeles became ill with shingles (herpes zoster), a viral disease that produces considerable discomfort and for which there is no effective therapy. As the patient expressed his surprise at this state of affairs, he was reminded by his physician that very little was known about viruses affecting man. Hope had arisen, however, from the results that were being obtained from studying viruses that affect bacteria; results of no immediate practical consequence, but of great theoretical interest. The physician was prompt to add that a great center for such work was nearby, at that famous institution, The California Institute of Technology. Would the patient help develop a center for animal virus work at the same place so that the knowledge already gleaned from bacterial viruses might be used for a more direct human purpose? The patient agreed. In this way Caltech found itself with a welcome endowment and a problem: how to use it.

Virology at Caltech was Max Delbrück; so while the endowment was wisely invested for future use, the problem was tossed into his lap. Since Delbrück knew very little about animal viruses, he thought it wise to arrange a meeting of virologists of all kinds to provide inspiration. The meeting was held at Caltech; the year was 1950; the result was a memorable little book, *Viruses 1950*. The book was closed by a brief statement on bacteriophagology, which contained all the available wisdom on the subject and in due time became the excuse for Gunther Stent to write his wonderful *Molecular Biology of Bacterial Viruses*.

No important revelation came from this meeting, however, and Delbrück turned to a more pragmatic approach. One day Seymour Benzer and I were called to his office: Delbrück pointed out that animal virology appeared to be ready for major advances. Would either of us be interested in trying his hand at it? To me it sounded wonderful. I had been thinking, perhaps with nostalgia, of my work with tissue cultures, years before, in Giuseppe Levi's laboratory in Torino; so I immediately expressed my interest, before Benzer could say anything. Benzer, on the other hand, was not interested, so everything was settled without delay. At that moment I became an animal virologist. It was agreed that I should visit a few laboratories, to see what was going on and to learn a few tricks.

The several months spent in the various laboratories were sad and discouraging. For one thing, it was the winter of the Eastern Seaboard, and my mind would constantly wander to the thought of the warm, sunny California desert, where Caltech's virologists were wont to spend their weekends; then the reading of innumerable articles full of new information, often irrelevant to my purposes, was tiring; and finally the approaches, methods and goals of animal virology seemed unconvincing. I learned that the state of tissue cultivation was, except for a few bright spots, scarcely different from what it had been when I was working in Torino. What seemed very exciting, however, was the success Earle was having in his attempts to cultivate animal cells without plasma, using liquid medium and sheets of perforated cellophane; and Gey's obtaining a moderate growth of certain cells in suspension in rapidly rotated bottles.

I returned to Pasadena, where I found Delbrück and many of his group at the railroad station ready to carry me off to a camping trip to the desert: extrasensory perception?

Before setting to work, I spent some time collecting my thoughts and I wrote a report to the Biology Division at Caltech. In it I pointed out that the main deficiency in animal virology lay in the lack of a proper quantitative approach. Assay was conducted by laborious end-point methods of low statistical efficiency. Since at least some animal viruses kill cells, a method of plaque assay could possibly be developed for them.

The next problem was how to proceed. After a few attempts to use explants of cells growing out of tissue fragments, according to the usual tissue culture procedure of the times, it was obvious that this would lead to nowhere. By then it was late summer and I decided to plant a new lawn in my back yard. A favorite ground cover in the Los Angeles area is dichondra, which is sold in flats of turf, pieces of which are planted in the soil about a foot apart. The dichondra quickly spreads from the patches to cover the entire lawn area. It occurred to me that the same approach ought to give the answer to the problem of producing a good cell sheet. I started experimenting with mechanical gadgets, such as stacks of razor blades thinly spaced, in order to reduce a chicken embryo to relatively uniform small bits; the system was cumbersome but produced encouraging results.

One day, Earle sent me a photograph of something he called, in the accompanying letter, a "steak": a wonderful layer of cells growing at the bottom of a flask. He knew of my attempts to grow cells in a layer without cellophane, and since he had succeeded in obtaining that result he sent me the recipe. According to his procedure, the chicken embryos were broken into small bits, by centrifuging them through wire gauzes of various mesh, beginning coarse and ending fine. The method worked.

The final problem was to grow the cultures in petri dishes. I had decided that future work would require many manipulations of the infected cells, which would prove cumbersome if they were grown in a flask. A CO_2-flushed incubator had to be built, but in the meantime I started experimenting with petri dishes enclosed in properly gassed and sealed plastic bags.

Everything worked as expected. Soon afterwards, one evening, I removed a dish which had been infected with diluted western equine Encephalomyelities virus. Upon examination, I found small, unusual areas. Under the microscope, through the agar layer, I could see a number of small, perfect plaques. It was a wonderful sight. I left the basement lab and walked home thinking it was a good beginning.

Within a few days I could proudly walk to Delbrück's office and invite him downstairs. He asked no questions, just followed me. I produced a plate with wonderful white plaques for his admiration. He asked: "What date is it? We should remember it." The date, I have forgotten.

It seemed that a whole world of experiments was open in front of me, as I set to working out the growth curve of the virus, adapting tissue culture techniques to the new task. After a month or so something distressing started to happen. The plaques were less and less evident, sometimes none would form, and finally a whole series of experiments ended in total failure. How was it possible? I started wondering whether I had ever really seen a plaque. Maybe it was a mistake and I would have to retract the paper that Delbrück had sent to the *Proceedings of the National Academy of Sciences*.

I returned to my lawn. It was winter: the dichondra was there, healthy and green, but for the first time I noticed a few leaves of contaminant grass turning slightly yellow. By looking more carefully, I noticed that the grass had invaded the dichondra, forming patches here and there, and was almost indistinguishable from the dichondra itself. In a day or two, however, the invader rapidly turned from green to yellow, from yellow to red, its patches becoming very noticeable against the green background. This sight was a revelation. Of course, the plaques had been forming all the time in my cultures; they were invisible and required only a stain. A. G. Strickland, who collaborated with me at that time, suggested the use of a vital stain. After a few tests we found that neutral red worked well and we adopted that. With that our technical problems were over.

The final step in the improvement of the method was to adopt trypsin digestion to prepare a cell suspension, rather than using cell fragments. This came as a consequence of a visit I made to John Enders in Boston. Enders had been using trypsin to favor the attachment of fragments of tissue to glass in his work, now classical, with poliovirus. With this last improvement, a system for studying animal viruses very similar to that used for bacteriophage work became available.

The system lacked the supreme simplicity of that used for bacteriophage: it required the CO_2-gassed incubator, the agar we used was sloppier, the cells were much more labile and fussy. I was doomed to envy Jean Weigle on some Friday afternoon, as he prepared to go camping in the desert. He never had the dilemma of whether to postpone an important experiment or to postpone an attractive camping trip. Just before leaving, he would prepare the necessary plates, tuck them inside his shirt and set off. At night he put the experiment inside his sleeping bag and the next morning, when the sun cast long shadows of Joshua trees on the bright sand, Weigle would read the results.

The approach that these developments permitted in work with animal viruses was rapidly accepted, but not without argument. The test came in the summer of 1953, at the Cold Spring Harbor Symposium dedicated to viruses. For instance, it was not an easy matter to convince everybody that the linearity of the dose response of the plaque assay means that a plaque derives from a single virus particle and, therefore, contains a virus clone. I relived the predicament of d'Herelle and I resorted to his ultimate weapon, namely citing Einstein on my behalf.

During the following years, in collaborating with many investigators attracted by the possibilities of the new approach, we developed a complete system for quantitative work with animal viruses. It included the infection of cells in suspension, which could then be plated for infective center formation, thus allowing the direct count of the numbers of infected cells; the study of single cycle multiplication; the analysis of single yields both by the statistical method currently used in bacteriophage work, and by the microdrop method, which is more suitable for animal cells since their large size makes them easily transferable. We were also able to prepare genetically pure virus stocks.

With this arsenal at our disposal, we started to investigate many problems of animal virology from a variety of aspects of their multiplication to genetics.

Looking from the vantage point of present achievement, the consequences of these early results seem large. The plaque method was progressively extended in many laboratories to a large proportion of animal viruses. It was not too difficult to extend it to cytopathic viruses, except for problems sometimes deriving from the presence of a sulphated polysaccharide in the agar which adsorbs the virus particles: this problem was eliminated by the substitution of methylcellulose for agar. Although it was more difficult to extend the plaque method to non-cytopathic viruses, it was often possible to do so. For instance, the plaque-forming ability of one influenza virus strain (NWS) was introduced, by Hirst, into other strains by recombination; plaques of other non-cytopathic viruses could be obtained by hemadsorption, by interference, and by locating the viral antigens in the cells by means of fluorescent antibodies. A very important development was the extension of a similar approach to tumor viruses by Harry Rubin, who was in my laboratory at that time.

The widespread use of these techniques, and other parallel developments, afforded not only efficient assay of animal viruses, but also experimentation on a quantitative level hitherto unknown.

It would be futile to try to assess the results brought about by the adoption of this quantitative approach, since it permeates essentially all present-day animal virology. The biochemical investigations which have clarified many molecular aspects of animal virology would be meaningless without a precise reference to the multiplication phase determined by biological assay. The genetic investigations have been extended to a number of viruses, a notable one being poliovirus, whose genetic mechanisms are now fairly well known. These genetic investigations, in turn, allowed the development of live poliovirus vaccines, which would have been impossible without them. The complex interactions of viruses with antibodies have been investigated and largely clarified. Much progress in tumor virology was made possible by the quantitative approach.

In going back to the early times in the work, it is interesting to remark how a single development, such as that of a plaque assay, could have great consequences, and this happened because the time was ripe. It is also interesting that this development was an obvious necessity to investigators coming from the phage field, but not equally evident to most professional animal virologists of the time. The lesson to be drawn is that scientific progress requires cross-fertilization of different fields and the willingness of investigators to jump from one field to another as the opportunity or necessity arises.

H. RUBIN
Department of Molecular Biology and Virus Laboratory,
University of California, Berkeley, California

Quantitative Tumor Virology

THE BACKGROUND

There has been so much publicity in recent years about viruses and cancer that the casual reader might get the impression that the discovery of tumor viruses is a recent development. Actually, viruses which can cause cancer were discovered shortly after the discovery of viruses that cause infectious disease. The first unequivocal demonstration of a tumor virus was made in 1911 by Peyton Rous at the Rockefeller Institute (Rous, 1911). Although the initial discovery of Rous' virus was followed in the next few years by that of other tumor viruses, all isolations of such viruses were made from chickens, and all early attempts failed to show the existence of anologous tumor viruses in mammals. It was not until the 1930's that tumor viruses were first isolated from rabbits (Shope, 1932, 1933) and that a virus was shown to be responsible for the mammary cancer of mice (Bittner, 1936).

The current popular excitement about tumor viruses had its origin in the discovery that mouse leukemia can be induced by a virus (Gross, 1951). The excitement engendered by this discovery was heightened by the isolation of the polyoma virus from mice (Stewart, Eddy, Gochenour, Borgese and Grubbs, 1957). This virus, true to its name, caused a variety of tumors in a variety of mammalian species, and caused the final breach in the wall of indifference to the possible role of viruses in human cancer.

I think it is safe to say, however, that there was little increase in our understanding of how viruses cause cancer until quantitative methods were introduced for studying virus-cell interactions. From the time of their discovery until the 1950's it was the common practice to assay tumor viruses by inoculating experimental animals with serial dilutions of the viruses to determine the highest dilution which produces tumors. This endpoint dilution technique had serious limitations when used for the assay of ordinary animal viruses, but it was even less satisfactory for the assay of tumor viruses. First, the infective titer of many tumor virus stocks was very low, thus greatly restricting the range over which serial dilutions could be made. Second, many potential test animals were highly resistant to infection by tumor viruses. This resistance is probably connected with the ubiquity of tumor viruses in nature. Since many of the tumor viruses commonly found in field and laboratory animals are antigenically related to viruses under investigation, a high proportion of experimental animals have acquired immunity to the test virus prior to their use in its assay. Third, the pathogenicity of many tumor viruses is extremely low; only a rare cell among those infected undergoes an observable pathological alteration, and the manifestation of the alteration is sometimes delayed for months, and even years.

Because of the limitations of the assay method, tumor virology remained an essentially descriptive science during this era. Some very provocative observations were made, and it was realized that the explanation of these observations, if they could only be fathomed, would be likely to provide important insights. But because of the difficulty in carrying out quantitative studies, such an analysis was impossible.

THE METAMORPHOSIS

The advent of intensive quantitative research with bacteriophages, heralded by the work of Max Delbrück and his associates in the 1940's, began to change the orientation of animal virus research. The animal virologist, who was previously able to rationalize his position of relative ignorance of fundamental processes by invoking methodological limitations, now watched knowledge about the nature of bacteriophages accumulate at an impressive rate. But although the plaque assay technique, which was a prerequisite for the rapid progress in bacteriophage work, had been known since its invention by d'Herelle in the 1920's, no animal virologist made an attempt to adapt it to animal viruses. It was left to a phage worker, Renato Dulbecco, to grasp the initiative and develop the appropriate techniques for quantitative animal virology. The outcome was the tissue culture plaque assay for cell-killing animal viruses, now used in laboratories everywhere, and described in another chapter of this book.

The success of Dulbecco's efforts inspired attempts to adapt the plaque assay technique to tumor viruses. One very great difficulty had to be faced in these attempts. Since there was no evidence that the multiplication of tumor viruses causes the death of cells in the animal, it seemed unlikely that it would do so in tissue culture. Hence the necrotic areas of host cells on the tissue culture monolayer that form the plaques of Dulbecco's method, could not be expected to be produced by tumor viruses. An alternative to plaque production had to be found. The obvious alternative was to attempt to induce and recognize the transformation of normal cells into malignant cells in tissue culture.

When I joined Dulbecco's laboratory at the California Institute of Technology in 1953, some two years after his invention of the animal virus plaque assay, this was precisely what I set out to do. I chose to work with the transformation of chicken cells in tissue culture by the Rous sarcoma virus (RSV) because RSV was known to produce highly malignant tumors in chickens within a few days after its inoculation. Unfortunately, the art of culture of cell monolayers was still in a fairly primitive state, inasmuch as the media were inadequate, and cell growth poor. Furthermore, I had no idea of how the malignant transformation, were it to occur in tissue culture, would manifest itself in the appearance of the tumor cells.

I struggled with the problem for a few months, infecting chicken cells with RSV. With a lively sense of fantasy, it was easy to imagine seeing the hoped-for cell transformation quite frequently, but it could be neither quantitated nor reproduced. In the meantime, I resuscitated an assay method for RSV devised by Keogh in 1938 but neglected since that time. The assay depended upon the production of discrete ectodermal tumors on the choriallantoic membrane of the developing chick embryo, and represented a great improvement over the endpoint assay in chickens.

In this method, as I could now show, each tumor is produced by a single virus particle, and each virus particle has the potential for producing either ectodermal or mesodermal tumors, depending only on the germ layer of the chick embryo from which the infected cell is derived (Rubin, 1955). The assay proved to be sufficiently quantitative to allow study of virus release during the culture of cells derived from sarcomas induced in chickens by RSV. This study showed that a fairly high proportion of such sarcoma cells produce RSV in small amounts continuously while the cells continue to divide. Finding a continuous release of virus was, of course, a marked departure from the burst-like release of bacteriophage by lysis of the host bacterium culminating in cell death.

FOCUS FORMATION AT LAST

After working with it for a few years, it became evident that the chorioallantoic membrane assay had really outlived its usefulness. There was great variation from embryo to embryo in the number of tumors induced by the same RSV stock, and some embryos were completely resistant to infection. It was impossible to wash the membranes to free them of unadsorbed virus or to manipulate the infected cells in other ways, and it was clear that using the available techniques, the level of understanding attained with bacteriophages would never be reached with RSV. In 1956, however, a note by Manaker and Groupé appeared in *Virology* which reported the occurrence of foci of transformed cells in cultures of chick embryo cells infected with RSV. This renewed my hopes for developing a workable and reproducible assay for RSV. Howard Temin, then a graduate student in embryology, expressed an interest in adapting Manaker and Groupé's technique for this purpose. Now we had the psychological advantage of knowing that observable Rous sarcoma foci really occur in tissue culture, and we were not so likely to be discouraged by failures. Furthermore, the techniques for tissue culture had been markedly improved in the intervening few years. Medium 199 and Eagle's medium supported cell growth more effectively than did the earlier concoctions, although the addition of tryptose phosphate broth to the medium was perhaps the most crucial improvement. It became possible to start chick embryo cell cultures at relatively low cell densities. This allowed extensive growth in the cultures and was essential for the assay.

After a considerable period of trial, error, and frustration, a reproducible assay for RSV and for RSV-infected cells was finally developed (Temin and Rubin, 1958). It differs from the animal virus plaque assay in that it requires transformation and multiplication, rather than killing, of infected cells. The foci of transformed cells are easily recognized because the cells round up, become refractile and tend to heap up in layers.

The development of this assay made possible an analysis of the dynamics of RSV infection (Temin and Rubin, 1959). These studies showed that RSV has a relatively long latent period. The first progeny virus particles in a large population of heavily infected cells appear twelve hours after infection, and the peak of virus production is reached in about three days. The results confirmed the previous finding that cells continuously release the virus and yet multiply indefinitely. The kinetic data indicated that the virus is completed at the cell surface, and then released into the medium.

THE LEUKEMIA DIVERSION

Despite its modest success for study of the growth cycle of RSV, the assay technique still had some unexplained complications. Occasionally, a group of chicken embryos yielded cells which were highly resistant to RSV infection and thus whole experiments were ruined. It seemed unlikely that this resistance had a genetic basis, since about ten embryos were generally pooled at a time to prepare the assay cell cultures, and the pool was either fully sensitive or highly resistant. Investigation of this annoying complication showed that an occasional embryo is congenitally infected with an avian leukosis virus which induces cellular resistance to infection with RSV (Rubin, 1960).

The infection of the one donor embryo then spreads to cells from the other embryos in the pool which were originally sensitive to RSV infection, and renders the whole culture resistant. The leukosis virus produces no detectable change in the cells other than to make them resistant to RSV infection. The resistance does not extend to other viruses, nor even to all strains of RSV. Resistance, as shall be seen presently, is conferred only against RSV particles whose surface antigens resemble those of the resident avian leukosis virus.

The induced resistance to RSV was now used as the basis for assaying avian leukosis viruses. It might be pointed out here that these viruses cause the widespread leukemia of chickens, which is of great economic importance to the poultry industry and of great theoretical importance in attempts to control leukemia in all species.

Previous assay methods of the leukosis viruses in chickens were inaccurate, could only be carried out in an isolated virus-free chicken flock maintained in East Lansing, Michigan, and required nine months for their completion. By contrast, the induced RSV resistance assay can be carried out within two weeks using individual embryos from any flock, and provides data of a fairly high degree of accuracy. This assay became known as the RIF assay, because the first leukosis virus uncovered in this was designated Resistance-Inducing Factor. The RIF assay was used to analyze the natural history of the avian leukosis viruses in field flocks (Rubin, Fanshier, Cornelius and Hughes, 1962). The main results of this analysis can be summarized as follows. Leukosis viruses can be transmitted to the embryo through the egg, but not through the sperm. An adult chicken infected at its embryonic stage is immunologically tolerant to the virus, and produces virus for its lifetime in large amounts in a high proportion of its cells. Despite the heavy infections in such congenitally infected chickens, there are, in most cases, no overt manifestations of pathology. Some such chickens do develop leukemia, of course, but these are usually in the minority. The virus spreads to uninfected chicks after hatching, but again only a minority become sick. The economic importance arises from the fact that almost all chickens are ultimately infected, and although only a minority of infected birds become sick, the absolute number of chickens dying from leukemia is much higher than the number dying from any highly fatal virus infections which attack only a tiny segment of the poultry population.

Congenitally infected females transmit the virus to almost all their offspring, indicating that a high proportion of their ova are infected. Congenitally infected males do not transmit the virus through the spermatozoa, although many cells of their testes produce virus. This suggests that the virus genome, which is composed of RNA, is jettisoned along with the cytoplasm and most of the cellular RNA during spermiogenesis.

These results indicate that long-term maintenance of leukosis virus infection in a cell lineage does not depend on attachment of the viral genome to the chromosomes of the host cell, as has sometimes been proposed by enthusiasts of the prophage notion derived from the study of lysogenic bacteria. It is somewhat bizarre that, despite all the reductionist advances that have been made in analyzing cell structure and function by means of fancy experimental techniques, barnyard experiments should have provided the best insight to date into the cellular localization of a viral genome.

DEFECTIVENESS AND ITS CONSEQUENCES

The RIF assay was useful in unravelling other aspects of the epidemiology of avian leukemia, but its most important role was in disclosing the defectiveness of RSV. Upon diluting stocks of RSV beyond their focus-forming endpoint, it was found that these stocks also contain a leukosis virus which is present in somewhat higher concentration than RSV itself and which, like RIF, interferes with the RSV focus-forming assay (Rubin and Vogt, 1962). This leukosis virus was called Rous Associated Virus or RAV. All attempts to free the RSV stock of RAV failed, although a stock of RAV which is free of RSV can be easily obtained. Not only is RAV indistinguishable from RSV in all physical characteristics studied, but it is identical to RSV in antigenic composition, as measured by serum neutralization tests.

One of these unsuccessful attempts to free the RSV stock of RAV consisted of isolating individual foci of RSV transformed cells, and maintaining them in culture through many passages. Foci were found which produce both RSV and RAV, and others which produced neither virus, but no foci were found which produced only RSV (Hanafusa, Hanafusa and Rubin, 1963). Some of the foci which produced no virus when first isolated, began to produce RSV after several cell passages. In such cases the production of RSV was always preceded or accompanied by the production of RAV. Many other foci of transformed cells could, however, be maintained indefinitely without virus production. But whenever RAV was added to such transformed cells production of both RAV and RSV would begin within a day.

This observation led us to conclude that RSV is *defective* and requires a *helper virus*, such as RAV, to achieve the production of infectious RSV. Any virus of the avian leukosis complex can serve as a helper virus for RSV production, but viruses such as the Newcastle disease virus, which belong to another structurally similar group of avian viruses, cannot. Further studies now showed that the antigenicity of RSV is always identical to that of its helper virus (Hanafusa, Hanafusa, and Rubin, 1964) and that transformed cells which produce no RSV do not contain any RSV coat antigen. These findings indicated that RSV is defective in the production of its coat, and depends on a helper virus to carry out this function.

There are several functions which RSV *can* carry out independently of helper virus. Most importantly, RSV can, by itself initiate and perpetuate the malignant transformation of cells. Not only does the RSV genome control the transformation as such, but it also controls the precise form the transformation will take (Rubin, 1964). There is no indication that the virus coat influences the transformation in any way. The RSV genome is also capable of taking care of its own replication, since cells infected and transformed by RSV alone can be maintained for many generations in the absence of production of infectious virus. Yet, at the end of this time a helper virus can be added to the transformed cells, and all of the cells respond by producing RSV. The RSV genome is, therefore, present in all the cells and must have been multiplying continuously through the lineage of the transformed cells.

The defectiveness of RSV also made possible an analysis of those viral properties which are controlled by the coat of the infectious particle. One such property, obviously, is antigenicity. Another property is the growth rate of infectious RSV. In the presence of a slow growing helper, RSV is produced slowly, and in a presence of a fast growing helper, it is produced rapidly. This correlation may seem obvious, but it does indicate that virus maturation, i.e., acquisition of the coat, is the limiting step in the production of infectious virus, an inference that is not certain *a priori*.

A third virus property controlled by the coat is the specificity of interference (Hanafusa, 1965). Avian leukosis viruses interfere only with infection by RSV particles that carry an antigenically related coat. Such specificity suggested that interference with RSV infection by leukosis viruses derives from a block of some early step of RSV infection, such as adsorption or penetration. This hypothesis seemed eminently reasonable, because the coat protein of leukosis viruses is produced in copious amounts at the cell surface and might be expected to occupy receptor sites for an exogenous superinfecting virus with a similar coat. Recent experiments have borne out this concept of interference. In particular, they have shown that the interference represents a great reduction of the rate at which superinfecting RSV reaches a cellular position where it is no longer susceptible to neutralization by added antibody. In normal, i.e., previously uninfected cells, this step presumably occurs as soon as the RSV particle has made a firm fit with a specific surface receptor.

The coat of RSV has also been found to determine its host range (Hanafusa, 1965). Cells derived from certain type of embryos are resistant to RSV with a certain type of coat. But the same types of embryo can be infected with high efficiency by RSV endowed with an antigenically unrelated coat provided by a different helper virus. Analysis of chicken pedigrees showed that this differential susceptibility of embryos to RSV with different coats is controlled by a single autosomal gene of the chicken genome. The allele for susceptibility is fully dominant over that for resistance. From these findings we conclude that susceptibility or resistance of a host cell to RSV is determined by the firmness of the union formed between a specific surface receptor of the cell and the coat of the virus.

These considerations lead one to suppose that the recently observed ability of certain strains of RSV to produce tumors in mammals (Ählstrom and Forsby, 1962) derives from the particular antigenic structure of the coat of these RSV strains. Indeed, those strains of RSV which are most effective in causing mammalian tumors are antigenically distinct from the strains which are least effective, a fact that is at least consistent with the premise that the determination of the host range of RSV even outside avian species is determined by the virus coat.

SPECULATIONS AND EXTRAPOLATIONS

The fundamental question now arises whether the defectiveness of RSV is connected with its extremely high carcinogenic capacity. Unlike many other tumor viruses, RSV induces the malignant transformation, at least in tissue culture, in every cell it infects. Its effect is thus sharply distinguished from that of polyoma virus, for instance, which kills many of the cells it infects, or from that of the leukosis viruses, which cause no significant change in most cells. Rous sarcoma virus is also the only tumor virus known to be intrinsically defective in synthesis of its coat.

One can invent plausible models to account for the high rate of malignant transformation by RSV in terms of its defectiveness. One such model envisages that since the virus coat is produced in the cell membrane a defective coat makes a defective cell membrane and thereby releases the cell from contact inhibition and other membrane-controlled regulatory processes.

Another such model would suppose that "late" virus functions regulate "early" functions, as has sometimes been thought to be the case in bacteriophage multiplication. In this case defectiveness in coat production could lead to an overproduction of materials concerned with early virus functions, and thereby disrupt the regulatory processes of the cell.

In any case, the hypothesis that the malignant transformation and defectiveness of RSV are causally connected leaves us with some clear alternatives, and methods are now being sought for testing the alternatives. Some of the methods will require the use of intact viral RNA, which has now been obtained from mixtures of RSV and RAV, and from RAV alone (Robinson, Pitkanen, and Rubin, 1965). The RNA of these viruses was found to have the surprisingly high sedimentation coefficient of about 70 S, indicating that it is very large—of estimated molecular weight over 10^7—and that it exists in the infectious particle as one, single-stranded molecule. Efforts are now being made to find the viral RNA in the infected cell, particularly to determine whether a complementary viral RNA strand arises there, and to ascertain the state of the viral RNA in transformed cells having no helper virus and hence producing no infectious RSV. We shall also look for virus-specific enzymes in these cells and ask what all this has to do with cell transformation. At this stage it should be evident to the reader of this book that unlike bacterial virology, most of whose glories now lie in the past, tumor virology has most of its road still ahead. For what else, if not tumor virology, can lead to an unravelling of the molecular basis for the malignant behavior of cells?

REFERENCES

Åhlstrom, C. G., and N. Forsby. 1962. Sarcomas in hamsters after infection with Rous chicken tumor material. J. Exptl. Med., *115*: 839–852.

Bittner, J. J. 1936. Some possible effects of nursing on the mammary gland tumor incidence in mice. Science, *84*: 162.

Gross, L. 1951. "Spontaneous" leukemia developing in C3H mice following inoculation, in infancy, with AK-leukemic extracts, or Ak-embryos. Proc. Soc. Exptl. Biol. Med., *76*: 27–32.

Hanafusa, H. 1965. Analysis of the defectiveness of Rous sarcoma virus III. Determining influence of a new helper virus on the host range and susceptibility to interference of RSV. Virology, *25*: 248–255.

Hanafusa, H., T. Hanafusa, and H. Rubin. 1963. The defectiveness of Rous sarcoma virus. Proc. Natl. Acad. Sci., *49*: 572–580.

———, ———, ———1964. Analysis of the defectiveness of Rous sarcoma virus II. Specification of RSV antigenicity by helper virus. Proc. Natl. Acad. Sci., *51*: 41–48.

Keogh, E. V. 1938. Ectodermal lesions produced by the virus of Rous sarcoma. Brit. J. Exptl. Pathol., *19*: 1–8.

Manaker, R. A., and V. Groupé. 1956. Discrete foci of altered chicken embryo cells associated with Rous sarcoma virus in tissue culture. Virology, *2*: 838–840.

Robinson, W. S., A. Pitkanen, and H. Rubin. 1965. The nucleic acid of the Bryan strain of Rous sarcoma virus. Purification of the virus and isolation of the nucleic acid. Proc. Natl. Acad. Sci., *54*: 137–144.

Rous, P. 1911. Transmission of a malignant new growth by means of a cell-free filtrate. J. Amer. Med. Assoc., *56*: 198.

Rubin, H. 1955. Quantitative relations between causative virus and cell in the Rous No. 1 chicken sarcoma. Virology, *1*: 445–473.

———1960. A virus in chick embryos which induces resistance *in vitro* to infection with Rous sarcoma virus. Proc. Natl. Acad. Sci., *46*: 1105–1109.

———1964. Virus defectiveness and cell transformation in the Rous sarcoma. J. Cell. Comp. Physiol., *64*: sup. 1, 173–180.

Rubin, H., L. Fanshier, A. Cornelius, and W. F. Hughes. 1962. Tolerance and immunity in chickens after congenital and contact infection with an avian leukosis virus. Virology, *17*: 143–156.

Rubin, H., and P. K. Vogt. 1962. An avian leukosis virus associated with stocks of Rous sarcoma virus. Virology, *17*: 184–194.

Temin, H. M., and H. Rubin. 1958. Characteristics of an assay for Rous sarcoma virus and Rous sarcoma cells in tissue culture. Virology, *6*: 669–688.

———, ———1959. A kinetic study of infection of chick embryo cells *in vitro* by Rous sarcoma virus. Virology, *8*: 209–222.

Shope, R. E. 1932. A filterable virus causing tumor-like condition in rabbits and its relationship to virus myxomatosum. J. Exptl. Med., *56*: 803–822.

———1933. Infectious papillomatosis of rabbits. J. Exptl. Med., *58*: 607–624.

Stewart, S. E., B. E. Eddy, A. M. Gochenour, N. G. Borgese, and G. E. Grubbs. 1957. The induction of neoplasms with a substance released from mouse tumors by tissue culture. Virology, *3*: 380–400.

NIELS K. JERNE
Paul Ehrlich Institute, Frankfurt am Main, Germany

The Natural Selection Theory of Antibody Formation; Ten Years Later

"Can the truth (*the capability to synthesize an antibody*) be learned? If so, it must be assumed not to pre-exist; to be learned, it must be acquired. We are thus confronted with the difficulty to which Socrates calls attention in *Meno* (Socrates, 375 B.C.), namely that it makes as little sense to search for what one does not know as to search for what one knows; what one knows one cannot search for, since one knows it already, and what one does not know one cannot search for, since one does not even know what to search for. Socrates resolves this difficulty by postulating that learning is nothing but recollection. The truth (*the capability to synthesize an antibody*) cannot be brought in, but was already inherent."

The above paragraph is a translation of the first lines of Søren Kierkegaard's "Philosophical Bits or a Bit of Philosophy" (Kierkegaard, 1844). By replacing the word "truth" by the italicized words, the statement can be made to present the logical basis of the selective theories of antibody formation. Or, in the parlance of Molecular Biology: synthetic potentialities cannot be imposed upon nucleic acid, but must pre-exist.

I do not know whether reverberations of Kierkegaard contributed to the idea of a selective mechanism of antibody formation that occurred to me one evening in March 1954, as I was walking home in Copenhagen from the Danish State Serum Institute to Amaliegade. The train of thought went like this: the only property that all antigens share is that they can attach to the combining site of an appropriate antibody molecule; this attachment must, therefore, be a crucial step in the sequences of events by which the introduction of an antigen into an animal leads to antibody formation; a million structurally different antibody-combining sites would suffice to explain serological specificity; if all 10^{17} gamma-globulin molecules per ml of blood are antibodies, they must include a vast number of different combining sites, because otherwise normal serum would show a high titer against all usual antigens; three mechanisms must be assumed: (1) a random mechanism for ensuring the limited synthesis of antibody molecules possessing all possible combining sites, in the absence of antigen, (2) a purging mechanism for repressing the synthesis of such antibody molecules that happen to fit to auto-antigens, and (3) a selective mechanism for promoting the synthesis of those antibody molecules that make the best fit to any antigen entering the animal. The framework of the theory was complete before I had crossed Knippelsbridge. I decided to let it mature and to preserve it for a first discussion with Max Delbrück on our freighter trip to the U.S.A., planned for that summer.

During the preceding five years, a succession of molecular biologists had made their impact upon Ole Maaløe's laboratory where I worked. The first to come was Hans Noll, and soon after Gunther Stent and James Watson descended upon us. The air was filled with the phage particles that Delbrück had picked out as one of the weakest spots in the armour behind which Nature guards her secrets. Most experiments then began with mixing phage with bacteria and ended with a plaque assay. Since it turned out that there are amazingly many ways in which this can be done, Nature winced under the onslaught of young men counting specks in Petri dishes. Noll carried up to our lab the bricks of lead that were to shield our first Geiger-Müller tube. One of Stent's contributions was the flick of the fingers that produces whirling uniformity in a dilution tube. Watson never trusted anybody to make his phage dilutions for him; with arachnid movements he would minister to series of 5 ml volumes of medium, and it is probable that nobody's plaque counts ever approached the bare Poisson error more closely. Maaløe, exercising his bacteria to synchronize, developed virtuosity in shocking them from hot into cold medium and vice versa, whilst smoking a cigar and making dilutions and platings on the dot of the stop watch. Meanwhile, in the same small laboratory room, I injected mixtures of diphtheria toxin and antitoxin into shaven rabbits, in order to study an esoteric property of antibodies that went under the name of "avidity." I admire the friendly stoicism with which the molecular biologists bore this incongruous activity.

Over it all hovered the spirit of Max Delbrück who was shepherding his handpicked band along the last stretch of the narrow path to the central fortress of biology. He made a few triumphant visits to Copenhagen, both before and after Lwoff assembled the court at Royaumont in 1952. In the spring of 1954, Delbrück gave a lecture on phycomyces at the Serum Institute. I remember waiting at the gate for Niels Bohr, who arrived a little late. With long strides he hastened in front of me, saying: "I do not want to miss a word of what Delbrück says; I so enjoy to listen to this man." Delbrück spoke in Danish about the "lille stilk" that grew up out of a "lille urtepotte," and refused to discuss the chemistry of the pigments involved until the growth response to light-dark programs had been properly described.

Immunology was not then an "in" subject, and I had to apply antibodies to bacteriophage in order to hang on to the fringe. My avidity observations strengthened my faith in the truth of antibody selection. Antibodies produced by an animal against one antigen appeared to increase in "goodness of fit" during the course of immunization. This was true both for antitoxin and for anti-T4 antibodies. The phenomenon had Darwinian overtones. Since that time Herman Eisen (1964) has demonstrated that lymphoid cells explanted from an animal at increasing time intervals after antigenic stimulation with a hapten produce antibodies in vitro that show a progressive increase in average association constant of the hapten-antibody complex.

This heterogeneity in combining power among antibodies of the same "specificity" has received less attention during the past ten years than other types of heterogeneity in the molecular structure of antibodies, elucidated by Gerald Edelman (Edelman and Gally, 1964) and many others. Each of the two polypeptide chains, a light chain of about 200 amino acid residues and a heavy chain about twice as long, that make up each subunit of an antibody molecule has a "variable" stretch of about 100 amino acid residues at its NH_3^+ – terminal end and a remaining "constant" stretch at the COO^- – terminal end. The specific combining site of each subunit is thought to be formed by the "variable" stretches, whereas part of the "constant" stretches of two heavy chains constitute a molecular fragment that can be removed enzymatically and is associated with properties that antibodies of different specificities have in common, such as complement fixation, and the ability to cross the placenta from mother to fetus. On the basis of antigenic differences and differences in peptide composition it has been established that there are at least seven types of heavy chains and at least two types of light chains in the antibodies produced by one individual. Since a single antibody molecule contains only one type of heavy chain and one type of light chain in its subunits, this heterogeneity permits at least 14 types of antibody molecules that can have combining sites of the same "specificity." Mice and men seem to be quite similar in this respect. Furthermore, in different individuals of the same species the same types of heavy chains and light chains can present different antigenic determinants controlled by different alleles at certain genetic loci. Thus, in the rabbit, three alleles at each of two unlinked loci determine heavy chains of three "allotypes" and light chains of three allotypes. It would seem that at any given time a plasma cell generally produces antibody molecules of only one specificity and composed of only one type of heavy chain and one type of light chain. The lymphoid tissue of an animal that is heterozygous with respect to allotypes appears to be a mosaic. Some of its cells produce antibody molecules of one allotype and others produce antibody molecules of another allotype, though both types may have the same antibody specificity. It would seem therefore that only one of two homologous chromosomes of a diploid plasma cell is functioning in antibody formation, like the case of the two *X*-chromosomes in female cells.

There is also reason to believe that plasma cells engaged in the production of antibody molecules containing a certain type of heavy chain can "switch" to the production of antibody molecules having a different type of heavy chain, though preserving both allotype and antibody specificity. This would suggest that function of a gene for the structure of the heavy chain can be partly shut off, allowing continued function of only those parts that code for the "variable" stretch of the heavy chain and for the allotype of its "constant" stretch, while replacing the rest of the chain with a part whose structure is coded by another gene. Crucial in all this complexity is the question of the origin of the immense variability of the nucleic acids that determined the "variable" parts of the polypeptide chains, permitting the construction of antibodies of a vast number of different specificities. It would seem most likely that this diversity arises from somatic rearrangements of a few prototype genes present in the zygote. There is still no reliable estimate of the number of different antibody combining sites that an animal can produce. In my 1955 paper, I roughly estimated the number of specificities to be one million and, at the moment, there seems no reason to revise this estimate. If the "variable" parts of both a heavy and a light chain contribute to the structure of an antibody combining site, a thousand different chains of both kinds might suffice to construct a million different sites. Evidence produced by David Pressman and his associates (Roholt, Radzimski and Pressman, 1965) suggests, however, that this matter is more complicated. Heavy chains and light chains isolated from the same antibody can, when mixed, recombine with good recovery of antibody activity, and this recovery is reduced by only about 50% when a nine-fold excess of non-specific light chains from the same serum is added to the specific light chains before presentation to the heavy chains. The heavy chains thus seem to be able to pick out a class of better fitting light chains from the mixture.

None of the features described in the two preceding paragraphs were known in 1954, when I embarked upon the Dutch freighter that brought the Delbrücks and us to New Orleans from Antwerp. We played chess and ping-pong. We placed a sealed bottle containing a piece of paper bearing Arabic script into the bucket of seawater that was hauled up along the shipside every day at noon on its way to the bridge, for measurement of the ocean temperature. The captain developed a fanciful theory to account for the origin of the bottle and, having made a fool of himself, retaliated by ordering the radio-telegraphist to produce items in his daily newsletter that he hoped would upset Delbrück. Thus, the day after Delbrück had demonstrated at the dinner table why the government of Mendès-France was the last hope for Europe the newsletter mercilessly announced that the French cabinet had been forced to resign. Unfortunately, the atmosphere did not seem to permit rather far-fetched theories of antibody formation to get more than scant attention.

Pasadena was a peculiar place, with endless rows of bungalows neatly arranged in streets and perpendicular avenues that were as boring as they were long. Pedestrians had long since ceased to exist. Not having a car, I always risked arrest for loitering. As an escape, Delbrück arranged frequent expeditions to the desert, which was even more desolate than Pasadena. By sheer will-power, he would make everyone climb nameless mountains strewn about in the hot sand. Here I gained a deeper insight into pure randomness. Inadvertently, one day, I displayed some enthusiasm for a provocative rock, the size of the Empire State Building, that was littering the Sea of Tranquility. Immediately, the veteran member of the Swiss Alpine Club, Jean Weigle, got out his rope and led Seymour Benzer and me in a perilous ascent half-way up, which was as far as we dared. He climbed like a cat, guarding from untimely death both the selective theory of antibody formation and the elucidation of genetic fine structure, whose survival depended on the clinging of benumbed fingers and toes to tiny irregularities carelessly provided for us by Nature on the face of the rock.

One of the tasks that Delbrück assigned to me in Pasadena was to help him correct the grammar in American scientific texts. We had a system by which a manuscript would finally bear my *non obstat* followed by his *imprimatur*. Attired in shorts, as if we had just been expelled from a tennis court, we attended frequent jam-sessions in the lab at which Delbrück would relentlessly pursue kinetic clues to the action of tryptophan, indole and specific antibody on T4 phage infectivity. Gordon Sato and Ping Yao Cheng provided helpful data on the influence of urea and temperature. Nobody quite understood why Delbrück continued to hold this problem in such great esteem, even after he had made Gunther Stent and Eli Wollman practically exhaust the subject. It is now clear that he was on the earliest track of the allosteric effect. Renato Dulbecco produced polio-plaques, and André Lwoff came along and, with rubber gloves, counted the number of poliovirus that are released by a single cell. Werner Reichardt and Delbrück had a workshop cloaked in black curtains in which an impressive apparatus focussed light beams onto phycomyces. Harry Rubin drilled holes into eggs through which he forced Rous' virus to disclose its peculiarities. Rubin and I shared a cellar laboratory that bore the notice: "In this room Max Delbrück discovered bacteriophage, and André Lwoff lysogeny; it is therefore placed under the protection of the public."

Meanwhile, George Streisinger, and indeed all of us, busily created heaps of genetic and kinetic data that would probably have remained forever uninterpretable, had not Charley Steinberg been around to solve everybody's mathematical and logical difficulties. Riding in a taxi out to a seminar in Los Angeles one day, Delbrück and I discussed the calculation of the number of fruitful collisions in a mixture of phage particles and antibody molecules. My objection to the use of the concept of an antibody concentration gradient forming around each phage particle was that since only a single hit is required to inactivate a particle, thereby putting that particle out of the picture, no gradient could ever be established around surviving particles. Delbrück had an argument against this, which struck me by its brilliancy but which has since escaped me. Anyway, it seemed likely that virtually each collision of an antibody molecule with a T4 particle has to be fruitful to account for the observed inactivation rates, but, at the same time, it also seemed likely that only an antibody molecule attaching near the tip of the phage tail (fibers were not yet discovered) could cause inactivation. This led Delbrück to the concept of the walking antibody molecule. From later electron micrographs taken by T. F. Anderson, showing that an antibody molecule can form a loop, attaching with both combining sites to two adjacent antigenic determinants on the same particle, it would now appear that reversible attachment could indeed permit an antibody to walk along a surface, if the relevant determinants are suitably spaced.

One morning, Delbrück had arranged for me an audience with Linus Pauling. The *maestro* kindly listened to my selective idea for antibody formation; he understood and rejected the thing, probably within five seconds from the start of my exposition, perhaps recalling his own instructive exercises in this field. I had no better luck with James Watson, whom I accosted one late night in an all-night restaurant; in his characteristic way of producing a succinct, unambiguous answer to any question, he said: "It stinks."

In the face of these blows, it was difficult to maintain faith in the selective idea. It seemed unrewarding to present the case to real immunologists, as the instructive theory of antibody formation was firmly entrenched in those days. To all immunologists that I knew, instruction by the antigen seemed the only idea compatible with the fact that antibodies are formed in response even to synthetic haptens, such as arsanilic or tartranilic acid, and with the widespread belief that the number of antigens, and therefore the number of antibody specificities, is potentially infinite and thus greater than the possible content of genetic information pre-existing in the responding animal. Antigen was thought to be present in antibody producing cells and to leave its steric imprint on every molecule of antibody globulin produced, by union of the antigenic determinants with the peptides of the globulin *in statu nascendi*. To molecular biologists, furthermore, the Natural Selection Theory seemed appalling, because its assumption that protein molecules can be reproduced apparently violated their Central Dogma. It has always escaped my comprehension why it should be more reasonable to believe that a cell can be stimulated by an antigenic protein molecule to produce antibodies carrying sites that are *complementary* to this protein, than to believe that a cell can be stimulated by an antibody molecule to produce antibodies carrying *identical* sites. As soon as one assumes that the structural genes requisite for antibody synthesis pre-exist in the cell, the immune response becomes a matter of control of gene function, which could be accomplished one way as well as the other without violation of sacred Dogma.

My only hope now was my friend Gunther Stent, whom I knew to be an insatiable consumer and critic of theories. In Berkeley, late that spring, we spent a happy month together. We conducted incredibly involved experiments passing P^{32} atoms through several phage generations, and spent the remainder of our energies detecting the flaws in daily rejuvenated theories concerning the manner in which the phage DNA is fragmented in T4-reproduction. Unfortunately, the selective theory of antibody formation somehow only got some spare moments of undivided attention. Each time it came up, there happened to be some more important matter at hand. I still regret this since the proposal would have gained in depth if we had found occasion to discuss it more seriously at that time.

After Berkeley, I crossed the American continent in an old Studebaker driven by Werner Reichardt. We paid our way out of fees for seminars at selected seats of learning. Reichardt had a spectacular lecture about the eyes of a beetle and I talked about antibodies. We slept alternately one night in the grass and the next in a motel, and discovered that the bedbug threshold in the United States lay at six dollars for a double room. Reichardt, being a specialist of insect behavior, was very deft at detecting them. Being from Berlin, furthermore, he was popular at gas stations. There would always be an attendant who had been stationed during his military service at Oberniederstein, or some such place in Germany. Denmark, in contrast, had failed to make much impact on the ex-servicemen. In a place in New Mexico where I answered the standard question "Where are you from?" the man asked "Denmark, Texas?" During our 600 miles per day, we entertained each other with immunology, semi-conductors (which was Reichardt's other specialty), and with observations concerning the difference between the U.S. and Europe, the last being an inexhaustible subject. On that trip, the only persons who commented favorably on my selective ideas of antibody formation were Colin McLeod and Wendell Stanley. This finally encouraged me to write up the manuscript during my last two weeks in Pasadena in August 1955. Delbrück was away, but had already promised me to send it to the *Proceedings of the National Academy of Sciences*. I left it on his desk, with the words: "Do not make any changes in this text, but please leave it as it now stands." A month later, he told me on a post-card to Copenhagen that he had indeed dispatched the manuscript after adding (hoping that I would not mind) a comma.

This was the end of my career in the heroic era of molecular biology. The following year I joined the staff of the World Health Organization in Geneva, and the Natural Selection Theory of Antibody Formation would probably have been forgotten had it not caught the fancy of Sir Macfarlane Burnet in 1957. He proceeded to modify the theory by shifting the object of selection from antibody molecules to cells pre-committed to the synthesis of antibody molecules of only one specificity, and by identifying the random element of the theory with somatic mutation, thus placing his Clonal Selection Theory in the mainstream of current biological thought. Burnet's paper marked the decline of the instructive theories of antibody formation that have since received a number of mortal blows, such as the evidence that the specific structure of the combining site of an antibody molecule does indeed derive from a particular amino acid sequence of the polypeptide chains, and the demonstration that an antibody secreting plasma cell does not contain any antigen, although it synthesizes antibody molecules simultaneously at a hundred thousand ribosomic sites. Plasma cells explanted from the lymph nodes or spleen of an immunized animal can continue their antibody synthesis for many hours. Synthesis is halted by the addition of puromycin or cycloheximide, but not by actinomycin D. Hence plasma cells appear to possess many thousand copies of long lived antibody messenger RNA. It has been suggested that nucleic acid carrying the correct sequential information is not part of the cellular chromosomes but is brought into the plasma cells from without, by analogy to infection with a virus. All the same, nucleic acids possessing the correct nucleotide sequences for antibody specificity must be derived from the DNA of the immunized animal, and ultimately from the DNA of the zygote from which it developed, since antigens can hardly be endowed with the capacity for arranging DNA or RNA nucleotide sequences that code for proteins having combining sites fitting to these antigens.

During the days following a primary antigenic stimulus, antibody forming plasma cells multiply and undergo a sequence of morphological changes called maturation. It is not known what type of cell is the ancestor of these plasma cells nor whether the cells that first receive the antigenic stimulus are plasma cells or the ancestor cells. A likely candidate for the ancestor is the small lymphocyte which is by far the most numerous cell in lymphoid tissue. It is evident that most of the antigen that arrives in the spleen or lymph nodes becomes phagocytized by macrophages. Some of these cells line the lymphoid parenchyma of these organs and move in among the lymphocytes after having taken up antigen. Histological studies suggest that a localized area is then established in which the macrophages, extending a reticulum of pseudopodia, remain in close contact with the mass of lymphocytes. It is not yet known, however, whether the antigenic stimulus is transmitted to the lymphocytic cells by a macrophage that has either processed the antigen in some way or merely holds the antigen *in situ* for presentation to the lymphocytes, or whether the stimulus proceeds directly from an encounter between non-phagocytized antigen and a lymphocyte or a plasma cell. In the latter case, the macrophage would be merely a scavenger, whereas in the former case, the macrophage may have a crucial function in permitting or denying the antigen access to the site of antibody formation. In this respect it is interesting to consider the evidence presented by Nossal (Nossal et al, 1964a, b) showing that gamma globulins are the only circulating protein molecules of the animal itself that are taken up by the relevant macrophages. This suggests that the macrophages recognize the gamma-globulins of the animal itself, leaving to these gamma-globulins, or "normal" antibodies, the task of recognizing the presence of foreign antigen in the circulation.

After my return to experimental work, at the University of Pittsburg in 1962, Albert Nordin, Claudia Henry and I worked out a practical method for recognizing individual antibody-forming cells by a plaque method in agar (Jerne et al, 1963a, b) that is similar to the plaque method which has rendered such good service in phage research. We have demonstrated by this method that an injection of a large dose of foreign red blood cells into a mouse leads to the proliferation of plasma cells producing 19 S antibodies, followed four days later by the appearance of a large number of cells producing 7 S antibody molecules. The latter have a different heavy polypeptide chain, and it seems that during the immune response cells or cell clones switch to production of antibodies of the same specificity but made with different heavy chains. It would seem, moreover, that the first 7 S antibody molecules to replace the 19 S are themselves replaced, within a day or so, by a different type of 7 S antibody molecules. A small quantity of this last type of antibody, when injected into a normal mouse, specifically inhibits this mouse from initiating a primary response to an injection of the corresponding antigen, i.e. from producing 19 S antibody.

Little has become known otherwise about homeostatic mechanisms, except that excessive antigen doses can, under certain conditions, provoke unresponsiveness (immunological paralysis or tolerance), and that the thymus gland has some regulatory function, as already suggested prematurely in my 1955 paper.

The agar plaque technique has also revealed that all normal animals, before exposure to an antigen, already contain a small number of plasma cells producing specific 19 S antibody. Thus, among 10^8 spleen cells of a normal mouse the number of plasma cells producing anti sheep red cell antibody is of the order of 100. Two important observations concerning the relation of antigen dose to host response were made, furthermore. One is that the rate at which plasma cells proliferate is dependent on the dose of antigenic red cells given. After one intravenous dose of 4×10^4 red cells, antibody producing plasma cells appear to multiply with a doubling time of about 36 hours, whereas a rapid proliferation with a doubling time of 7 hours follows a dose of 4×10^7 red cells. Intermediate antigen doses are followed by intermediate rates of plasma cell proliferation, but the peak number of 19 S antibody producing plasma cells reached is proportional to the antigen dose. The second observation is that a further increase of the single antigenic dose from 4×10^7 to 4×10^9 red cells does not lead to a further increase in response. Some factor involved in the response must become saturated or exhausted at a dose level of 4×10^7 red cells. This factor is specific, because the animal responds normally to a small dose of a second kind of antigen given simultaneously with an exhausting dose of the first antigen. Since this pre-existing factor is specific to the antigen, it is almost by definition a preformed antibody. This suggests one of two interpretations, or both. Either the mouse contains a small number of pre-committed cells that are the only cells that can respond to a given antigen, or the mouse contains a small number of pre-formed antibody molecules of a certain type that have to attach to the antigen as a first step toward antibody formation. This leads back to the crucial unresolved question of the nature of the antigenic stimulus. Does it consist of selection by the antigen of appropriate pre-committed uni-competent cells for multiplication, as proposed by Burnet (1957, 1959, 1962), or are the plasma cell precursors multi-competent and their appropriate antibody genes selected by antigen performing intracellular derepression, as proposed by Leo Szilard (1960); or did Burnet perhaps strip the Natural Selection Theory of one true basic feature, namely that recognition of a foreign antigen is accomplished by pre-formed antibody molecules, attaching to the antigen and crucial for the transmission of the stimulus? After all, when the combining site of a pre-formed antibody molecule has attached to an antigenic determinant, the complex carries, as it were, the appropriate information concerning the desired structure of the antibody combining site *in duplo*, namely in the antibody molecule as well as in the antigen. It still seems as likely to me today, as it did ten years ago, that this antibody molecule is used for transmitting the antigenic stimulus, because it is more simple to allocate this task to a uniform class of pre-formed antibody molecules than to a heterogeneous class of antigenic determinants that may be parts of proteins, polysaccharides, nucleic acids, or inorganic haptens. The transmission of one appropriate molecule, or a small number of such molecules, to the surface of a cell can be imagined to induce profound changes in cellular function, by analogy to the action of colicine on bacterial cells, described by Masayasu Nomura (Nomura and Maeda, 1965) and Salvador Luria (1964).

The antibody problem now has become an "in" subject, since it has caught the attention of molecular biologists looking for the simplest road toward understanding differentiation.

In the foregoing, I have not conveyed Max Delbrück's crucial significance as a teacher, as I do not recall any striking occasion or incident with which to illustrate this. I shall therefore continue my translation from the first paragraphs of the Philosophical Bits, in order to let Kierkegaard (1844) describe the intangible Socratic aspect of Delbrück's influence:

"Examining the difficulty of seeking the truth, we showed how Socrates concluded that the truth must be inherent in every human being. He therefore entered into the role of midwife, because he perceived that a teacher can serve only accidentally, as it were, to *remind* the learner. The occasion at which this occurs is immaterial; for as soon as the learner is reminded that he has known the truth all along without knowing it, from that same moment this accidental instant is hidden in eternity, and, so to speak, can no more be found, because it is neither here nor there but *ubique et nusquam*."

REFERENCES

Burnet, F. M. 1957. A Modification of Jerne's Theory of Antibody Production using the Concept of Clonal Selection. Austral. J. Sci., *20*: 67.

———1959. The Clonal Selection Theory of Acquired Immunity, University Press, Cambridge, England.

———1962. The Integrity of the Body. Harvard University Press, Cambridge, Mass.

Edelman, G. M., and J. A. Gally. 1964. A Model for the 7 S Antibody Molecule. Proc. Natl. Acad. Sci., *51*: 846.

Eisen, H. N. 1964. The Immune Response to a Simple Antigenic Determinant. The Harvey Lectures, September 1964, New York.

Jerne, N. K. 1955. The Natural Selection Theory of Antibody Formation. Proc. Natl. Acad. Sci., *41*: 849.

———1960. Immunological Speculations. Ann. Rev. Microbiol., *14*: 341.

Jerne, N. K., A. A. Nordin, and C. Henry. 1963a. Plaque Formation in Agar by Single Antibody Producing Cells. Science, *140*: 405.

———, ———, ———1963b. The Agar Plaque Technique for Recognizing Antibody Producing Cells, p. 109. *In* B. Amos and H. Koprowski [Ed.] Cell-bound Antibodies. The Wistar Institute Press, Philadelphia.

Kierkegaard, Søren Aa. 1844. Philosophiske Smuler eller En Smule Philosophi. C. A. Reitzel, Copenhagen.

Luria, S. E. 1964. On the Mechanisms of Action of Colicines. Ann Inst. Pasteur, *107*: 5 Suppl., p. 67.

Nomura, M., and A. Maeda. 1965. Mechanism of Action of Colicines. Zentralbl. für Bact., Parasit., Infektionskr. und Hyg., Abt. I, *196*: 216.

NOSSAL, G. J. V., A. SZENBERG, G. L. ADA, and C. A. AUSTIN. 1964a. Antigens in Immunity. Austral. J. Exp. Biol. and Med. Sci., *42*: 311, 331.

———, ———, ———, ———1964b. Single Cell Studies on 19 S Antibody Production. J. Exp. Med., *119*: 485.

ROHOLT, O. A., G. RADZIMSKI, and D. PRESSMAN. 1965. Preferential Recombination of Antibody Chains to Form Effective Binding Sites. J. Exp. Med., *122*: 785.

SOCRATES. 375 B.C. "Meno" *80* (Plato, editor), Athens.

SZILARD, L. 1960. The Molecular Basis of Antibody Formation. Proc. Natl. Acad. Sci., *46*: 293.

WERNER E. REICHARDT
Max-Planck-Institut für Biologie, Tübingen, Germany

Cybernetics of the Insect Optomotor Response

A. INTRODUCTION

Investigations in sensory physiology are characterized mainly by the application of two methods: electrophysiological microprobing of the electrical activity in sense and nerve cells, and study and quantitative evaluation of evoked behavioral responses. Hence present day sensory physiology is not yet a domain of molecular biology; it still belongs to the phenomenological sciences.

Both methods of study have their advantages and their limitations. Electrophysiological studies enable us to trace the information flux, especially in those parts of the nervous system which are isotropic in function, i.e. whose interactions depend entirely on the distance between two or more interacting neurons rather than on their locations. But the technological difficulties of this method become formidable whenever the subsystem to be studied is nonisotropic in function. Behavioral studies enable us to analyze the functional principles of an input-output relation, regardless of the structural complexities or physical parameters of the components of the system. This method however does not give us information on how these functional principles are realized physically and what components of the nervous structure are actually involved.

Despite a vast number of investigations based on the application of these two experimental methods, it is not yet known whether the functional principles of the nervous system are related to the circuitry of its nets by one-to-one correspondence or whether these principles are realized by classes of nerve networks differing in structure and synaptic function.

The account I shall present here concerns an application of the second method to the optomotor response of insects, i.e. their evoked response to movements relative to themselves in their visual surroundings. Optomotor responses are highly reproducible and allow a quantitative system analysis that makes possible the detection and specification of some fundamental functional principles of the insect central nervous system. However the system analysis, at least as far as it has been carried until now, has not told us where in the mass of nerve cells connected to the insect eye the mechanism responsible for the optomotor response is actually located.

I began these studies at the Fritz-Haber-Institut der Max-Planck-Gesellschaft in 1950, in collaboration with the zoologist Bernhard Hassenstein, now at the University of Freiburg. But in 1954, Max Delbrück invited me to join him in an effort to make a system analysis of the light response of the fungus Phycomyces. And so, being persuaded that the behavior of a fungus ought to be simpler to analyze than that of an insect, I interrupted these studies and illuminated Phycomyces for a couple of years. When I finally resumed work on the optomotor response, its analysis seemed easy compared with trying to understand the influence of light on the dynamics of fungus growth.

B. OPTOMOTOR RESPONSE MEASUREMENTS

1. Methods

In a typical optomotor experiment an animal is placed on the axis of a hollow cylinder whose inside is lined with perpendicular black and white stripes or other contrast patterns (Hassenstein, 1951, 1958a, b; Mittelstaedt, 1951). When the cylinder rotates, the animal tends to turn with the movement, so that the displacement of the surroundings it observes is reduced. Thus animal and moving pattern together form a feedback loop. Such a cylinder arrangement used for eliciting optomotor responses is shown schematically in Fig. 1.

In the "classical" experiment, the beetle Chlorophanus is held fixed on the axis of the patterned cylinder (Hassenstein, 1951). This procedure opens the feedback loop, as it prevents the animal from actually changing its position relative to the stimulus. Nevertheless, the beetle is allowed to manifest "walking" along a lightweight spherical maze which it holds with its legs, as shown in Fig. 2. This maze is so constructed that the animal continually reaches Y-shaped branches in its path. These branches obviously force a series of right-left choices on the part of the animal which provide a sensitive measure of its turning response induced by the rotation of its surroundings. The strength of the response is measured by the ratio $R=(W-A)/(W+A)$, where W is the number of choices made with and A the number made against the direction of rotation of the patterned cylinder. Statistical considerations have shown that if R represents a linear measure of the strength of the response it does not exceed 0.7. Hence measurements of the optomotor response have been limited to the range $-0.7<R<+0.7$.

During the last few years a new technique for measuring the optomotor response has been developed in our laboratory (Kunze, 1961; Fermi and Reichardt, 1963; Götz, 1964) which records the torque exerted by fixed, flying insects (Drosophila and Musca). The principal components of the apparatus are shown schematically in Fig. 3. By means of a cardboard tab attached to its thorax and head the fly is clamped in a horizontal attitude to a rigid double coil. The double coil is suspended from a fixed frame by tension wires that allow the double coil to turn about an equilibrium position. The upper coil lies between the poles of a permanent magnet, while the lower coil lies in an alternating magnetic field produced by the high-frequency core in a coil fed with current from an AC generator. The

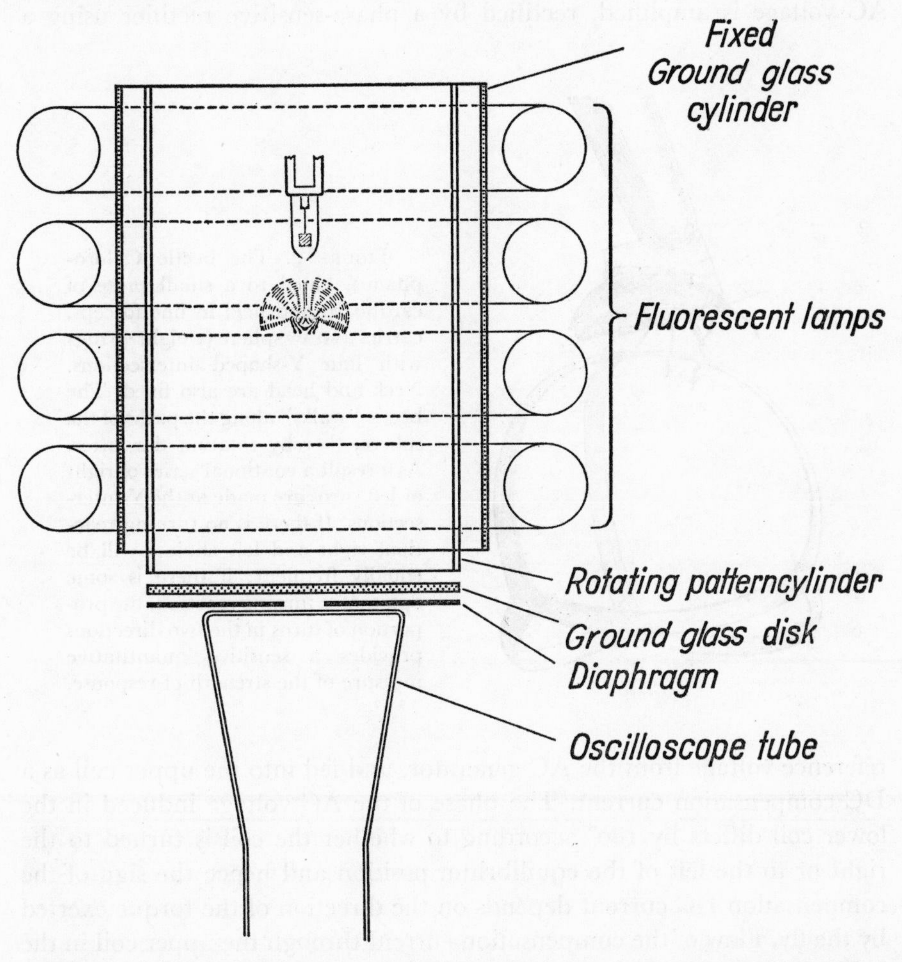

FIGURE 1. Schematic representation of a typical arrangement for eliciting optomotor responses of insects. The cylinder arrangement presented here consists of two concentric cylinders. The outer ground-glass cylinder is stationary, whereas the inner cylinder, which carries the pattern, can be rotated by an AC motor through a planetary gear system. Outer illumination of patterned cylinder is provided by annular fluorescent lamps. Inner illumination with constant or pulsed light is provided in this arrangement by means of an oscilloscope tube. The oscilloscope screen is covered by a diaphragm containing an optical window; the light from the screen is scattered in a ground-glass disk.

Many other cylinder arrangements and kinds of illumination also have been used in the various experimental tests.

double coil system is at equilibrium when the lower coil is so oriented in the alternating magnetic field that no voltage is induced in it. When producing the optomotor response, the fly exerts a torque on the double coil system and displaces it from its equilibrium position, inducing thereby an AC voltage in the lower coil. For small displacements the magnitude of the induced voltage is proportional to the angular displacement. The induced AC voltage is amplified, rectified by a phase-sensitive rectifier using a

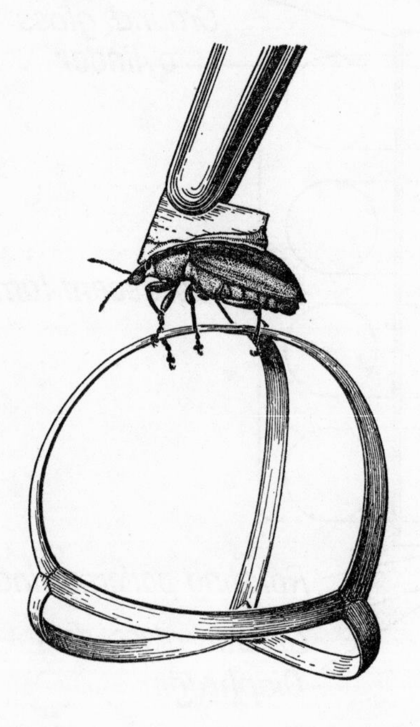

FIGURE 2. The beetle Chlorophanus, glued to a small piece of cardboard and held in fine forceps, carries a straw sphere (weight 30 mg) with four Y-shaped intersections. Neck and head are also fixed. The beetle "walks" along the paths of the sphere, thereby rotating the latter. As a result a continual series of right or left turns are made at the Y-intersections. If there is no turning reaction, right and left choices will be equally frequent; if there is some optomotor turning reaction, the proportion of turns in the two directions provides a sensitive quantitative measure of the strength of response.

reference voltage from the AC generator, and fed into the upper coil as a DC compensation current. The phase of the AC voltage induced in the lower coil differs by 180° according to whether the coil is turned to the right or to the left of the equilibrium position and hence the sign of the compensation DC current depends on the direction of the torque exerted by the fly. Flow of the compensation current through the upper coil in the field of the permanent magnet then produces a torque on the double coil. The entire device has been so mounted that this torque always opposes that exerted by the fly. The magnitude and sign of the compensation current is recorded and thus provides a linear measure of the torque exerted by the fly. As long as the amplification of the voltage induced in the lower coil is sufficiently large, any torques exerted by the flying insect result in only very small angular displacements (1/100 of a degree) of double coil and attached fly from the equilibrium position.

2. Summary of Main Results

The results of a large number of experiments lead to some general conclusions about the optomotor response of the beetle Chlorophanus and the flies Drosophila and Musca (Reichardt, 1961a, b, 1962, 1965). The following brief summary of these conclusions will provide the background necessary for understanding the theoretical analysis of how these insects perceive movement.

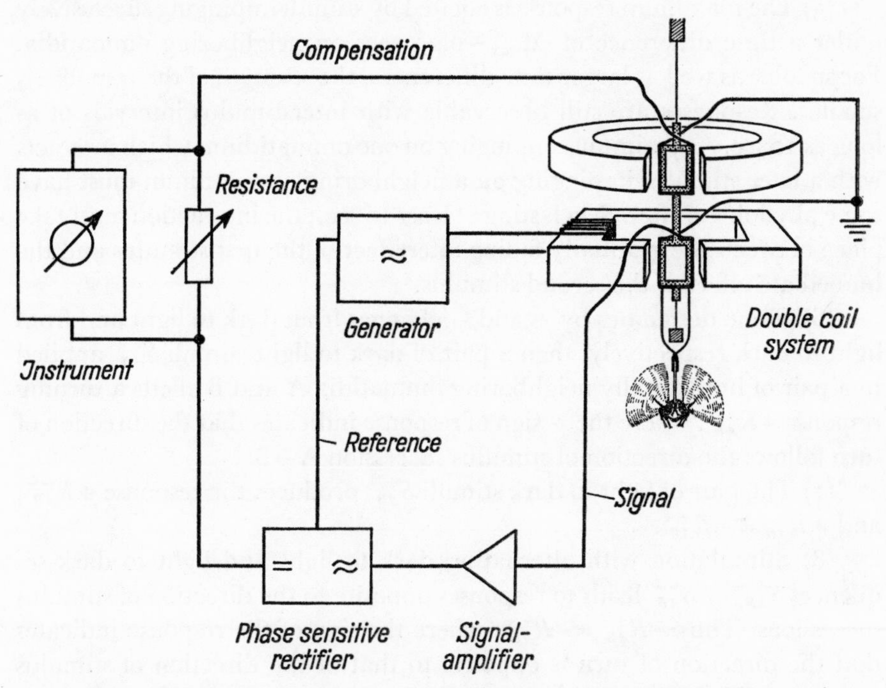

FIGURE 3. Schematic representation of the torque compensating apparatus for measurement of the optomotor response of fixed flying insects. Details described in the text.

(1) A sequence of two light stimuli impinging on adjacent facets, or ommatidia of the compound insect eye is the elementary event that evokes the optomotor response.

(2) Two stimuli which impinge on one and the same ommatidium do not release an optomotor response, even though every ommatidium contains more than one sensory (retinula) cell. The optical resolution of the compound eye, as found in optomotor response experiments, is determined by the angle sustained by the optical axes of adjacent ommatidia, and not by the angle sustained by optical axes of neighboring retinula cells (Hassenstein, 1959; Götz, 1965).

(3) In Chlorophanus, the stimulus received by one ommatidium can interact only with the stimulus received by the immediately adjacent om-

matidia and those once removed. No interaction for movement perception exists between ommatidia separated by more than one unstimulated ommatidium. In Drosophila and Musca the size of the group of interacting neighboring ommatidia has not yet been determined.

(4) The visual fields of adjacent ommatidia do not overlap in Chlorophanus (Varjú, 1959). However, they do overlap to some extent in Drosophila and in Musca.

(5) The maximum response is elicited by stimuli impinging successively under a time difference of $\Delta t_{max} \sim 0.25$ sec. on neighboring ommatidia. For smaller as well as larger time differences, the strength of the response is smaller. Responses are still observable with interstimulus intervals of as long as 10 sec. Any stimulus impinging on one ommatidium which interacts with a later stimulus impinging on a neighboring ommatidium must have some physiologic aftereffect lasting at least 10 sec.; the interaction must take place between the gradually fading aftereffect of the first stimulus and the immediate effect of the second stimulus.

(6) If one designates by + and – changes from dark to light and from light to dark respectively, then a pair of dark to light stimuli S_{AB}^{++} applied to a pair of horizontally neighboring ommatidia A and B elicits a turning response $+R_{AB}^{++}$, where the + sign of response indicates that the direction of turn follows the direction of stimulus succession A→B.

(7) The pair of light to dark stimuli S_{AB}^{--} produces the response $+R_{AB}^{--}$, and $+R_{AB}^{--} = +R_{AB}^{++}$.

(8) Stimulation with alternating dark to light and light to dark sequences S_{AB}^{+-} or S_{AB}^{-+} leads to responses opposite to the direction of stimulus successions. Thus $-R_{AB}^{+-} = -R_{AB}^{-+}$, where the sign of the response indicates that the direction of turn is opposite to that of the direction of stimulus succession. Thus the relation between stimulus input and response output follows the rules of algebraic sign multiplication (Hassenstein and Reichardt, 1956):

	S_A^+	S_A^-
S_B^+	$+R$	$-R$
S_B^-	$-R$	$+R$

(9) The strength of the optomotor response does not depend only on the speed of stimulus succession but also on the changes of light intensity representing the stimuli. If ommatidium A, for instance, receives stimulus intensity x and ommatidium B receives stimulus intensity y, the response R is proportional to $x \cdot y$, i.e. to the product of the two stimuli.

(10) Successions of stimuli in ommatidia ABCD··· of a horizontal row produce the response $R_{ABCD}^{++++}\dots$. This response is equal to the sum of all the partial responses evoked. Thus $R_{ABCD}^{++++}\dots = R_{AB}^{++} + R_{BC}^{++} + R_{CD}^{++} + \cdots + R_{AC}^{++} + R_{BD}^{++} + R_{CE}^{++} + \cdots$. Analogous results hold for the sequences $S_{ABCD}^{----}\dots$, $S_{ABCD}^{+-+-}\dots$ and $S_{ABCD}^{-+-+}\dots$.

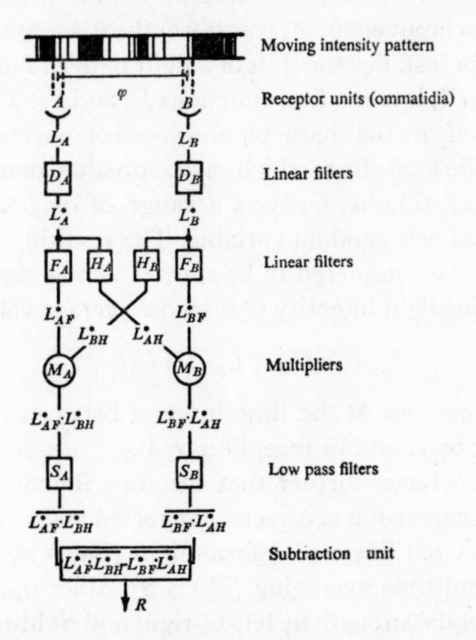

FIGURE 4. Minimal model for optomotor movement perceptions. The model consists of two direct and cross-connected information channels, originating at a pair of receptor units A and B, followed by three filter stages D, F, and H which feed into multiplier components M, followed by filters S with very long time constants whose outputs are finally subtracted in the last component, to produce the optomotor response R. The time functions representing the transformed stimulus light intensity sequence at various stages in the system are symbolized by L_A, L_{BF}^*, L_{AH}^*, etc. This information flow diagram is not intended to be a literal representation of actual neural pathways.

The results summarized here show that in the central nervous system of the insects studied there must exist physiological processes working in accordance with the mathematical operation of multiplication that links sensory input with motor output (Reichardt, 1957; Reichardt and Varjú, 1959).

C. THE MODEL OF THE MOVEMENT PERCEPTION SYSTEM

1. General Features

On the basis of the experimental conclusions concerning the properties of the optomotor response, a minimum mathematical model (Fig. 4) was designed that accounts for the functional properties of the physiological system. The model permits prediction of responses to previously untested stimulus patterns (Reichardt, 1957, 1961b; Reichardt and Varjú, 1959).

This model envisages two cross-connected information input channels from two ommatidia and a common output channel to the motor system. A and B represent two adjacent ommatidia, each of which acts as an input element that provides the minimum detector requirement for the optomotor response. The two input elements transform the space and time coordinates of the stimulus, for instance the shift of a light pattern L from ommatidium A to ommatidium B, into the time functions L_A and L_B. These functions are determined not only by the character and speed of movement of the pattern but also by the solid angle from which each individual ommatidium receives light. The original stimulus L covers a range of $0<L\leq L_{max}$ and may be either a structured or a random variable. The resulting time functions L_A and L_B may then be considered to be sums of an average light intensity C and of a fluctuating light intensity $G(t)$ whose average value is zero. That is

$$L_A=C+G(t);\ L_B=C+G(t-\Delta t) \tag{1}$$

where t is the time and Δt the time interval between the reception of a stimulus element by A and its reception by B.

The model envisages further that the time functions L_A and L_B are transformed by a succession of directly connected and crossconnected components that carry out linear transformations (filters D, F and H), multiplication (M), and time averaging (filters S). Since optomotor responses are evoked with equal strength by left-to-right and right-to-left movements, the channels must be symmetrical. Hence the transformation characteristics of the corresponding components in the two channels are equal: $D_A=D_B=D$; $F_A=F_B=F$ and $H_A=H_B=H$.

In particular, filters D transform L_A and L_B into L_A^* and L_B^*, which in turn are transformed into L_{AF}^* and L_{BF}^* and into L_{AH}^* and L_{BH}^* by filters F and H respectively. The multipliers M_A and M_B then multiply the signals of the two channels, $L_{AF}^*;L_{BH}^*$ and $L_{AH}^*;L_{BF}^*$, to yield the functions $L_{AF}^*\cdot L_{BH}^*$ and $L_{AH}^*\cdot L_{BF}^*$. The S components then process the transformed time functions $L_{AF}^*\cdot L_{BH}^*$ and $L_{AH}^*\cdot L_{BF}^*$ to the time averages $\overline{L_{AF}^*\cdot L_{BH}^*}$, and $\overline{L_{AH}^*\cdot L_{BF}^*}$. Since multiplication and time averaging of two time functions is called "first order correlation," the components M and S of the model are designated as "correlators." The correlator output of one channel is subtracted from the correlator output of the other channel in the last component, the subtraction unit, which controls the response R of the motor system.

Since the experimental data show that the transformations carried out by filters D, F and H are linear, the superposition principle holds and therefore their outputs for any arbitrary input may be expressed as the sum (integral) over elementary output signals triggered by a sequence of narrow input pulses into which the signal input may be decomposed. If a unit pulse (δ-function) with the mathematical properties

$$\delta(t-\zeta)=\begin{cases}+\infty & \text{for } t=\zeta \\ 0 & \phantom{\text{for }} t\neq\zeta\end{cases}\quad ; \quad \int_{-\infty}^{+\infty}\delta(t-\zeta)dt=1 \tag{2}$$

stimulates a filter input at time $t=\zeta$, the output responds with a time function $W(t-\zeta)$. W is called the weighting function of the filter. For an arbitrary input function $L(t)$, the output of the filter can be written as the superposition of the elementary responses $W(t-\zeta)\ F(\zeta)d\zeta$; with ζ covering the range $-\infty<\zeta\leq t$. This superposition process is expressed by the so-called convolution integral

$$L^*(t)=\int_{-\infty}^{t}W(t-\zeta)\ L(\zeta)d\zeta=\int_{0}^{+\infty}W(\zeta)\ L(t-\zeta)d\zeta. \tag{3}$$

If we designate with W_{DF} the weighting functions of the vertical and with W_{DH} the weighting functions of the crosschannels in Fig. 4, then we obtain at the inputs of the correlator M_A the time functions

$$L^*_{AF}=\int_{0}^{+\infty}W_{DF}(\zeta)\ L_A(t-\zeta)d\zeta$$

$$L^*_{BH}=\int_{0}^{+\infty}W_{DH}(\xi)\ L_B(t-\xi)d\xi \tag{4A}$$

and at the inputs of the correlator M_B the time functions

$$L^*_{AH}=\int_{0}^{+\infty}W_{DH}(\zeta)\ L_A(t-\zeta)d\zeta$$

$$L^*_{BF}=\int_{0}^{+\infty}W_{DF}(\xi)\ L_B(t-\xi)d\xi \tag{4B}$$

ξ and ζ in equation (4) being variables of integration.

The multiplier M_A multiplies the two time functions L^*_{AF} and L^*_{BH} to yield the expression

$$L^*_{AF}\cdot L^*_{BH}=\int_{0}^{+\infty}W_{DF}(\zeta)\ L_A(t-\zeta)d\zeta\cdot\int_{0}^{+\infty}W_{DH}(\xi)\ L_B(t-\xi)d\xi$$

$$=\int_{0}^{+\infty}W_{DF}(\zeta)\int_{0}^{+\infty}W_{DH}(\xi)\ L_A(t-\zeta)\ L_B(t-\xi)d\xi\ d\zeta \tag{5A}$$

and correspondingly the multiplier M_B multiplies the two time functions L^*_{AH} and L^*_{BF}, to result in the expression

$$L^*_{AH}\, L^*_{BF} = \int_0^{+\infty} W_{DH}(\zeta)\, L_A(t-\zeta)d\zeta \cdot \int_0^{+\infty} W_{DF}(\xi)\, L_B(t-\xi)d\xi$$

$$= \int_0^{+\infty} W_{DH}(\zeta) \int_0^{+\infty} W_{DF}(\xi)\, L_A(t-\zeta)\, L_B(t-\xi)d\xi\, d\zeta. \tag{5B}$$

The S components process the signals at the multiplier outputs to the time averages $\overline{L^*_{AF} \cdot L^*_{BH}}$, and $\overline{L^*_{AH} \cdot L^*_{BF}}$. If we express L_A and L_B in accordance with equation (1) by $C+G(t)$ and $C+G(t-\Delta t)$ and carry out the time averaging process under the integrals in equation (5) we obtain

$$\overline{L_A(t-\zeta)\, L_B(t-\xi)} = \overline{[C+G(t-\zeta)] \cdot [C+G(t-\Delta t-\xi)]}$$

$$= \frac{1}{2T}\int_{-T}^{+T} \Big[C+G(t-\zeta) \Big] \cdot \Big[C+G(t-\Delta t-\xi) \Big] dt$$

$$\lim T \to \infty$$

$$= C^2 + \Phi_{GG}\,(\zeta-\xi-\Delta t) \tag{6}$$

$$\text{with } \Phi_{GG}\,(\zeta-\xi-\Delta t) = \frac{1}{2T}\int_{-T}^{+T} G(t-\zeta)\, G(t-\Delta t-\xi)dt$$

$$\lim T \to \infty. \tag{7}$$

Φ_{GG} designates the autocorrelation function of the light intensity fluctuation $G(t)$. Substituting the two terms given above for $L_A(t-\zeta) \cdot L_B(t-\xi)$ in equations (5) and subtracting equation (5B) from (5A), the response R of the two ommatidia model to a moving contrasted light pattern is given by the expression

$$R = \overline{L^*_{AF}\, L^*_{BH}} - \overline{L^*_{AH}\, L^*_{BF}} = \int_0^{+\infty} W_{DF}\,(\zeta) \int_0^{+\infty} W_{DH}\,(\xi)\, \Phi_{GG}\,(\zeta-\xi-\Delta t)d\xi\, d\zeta$$

$$- \int_0^{+\infty} W_{DH}\,(\zeta) \int_0^{+\infty} W_{DF}\,(\xi)\, \Phi_{GG}\,(\zeta-\xi-\Delta t)d\xi\, d\zeta. \tag{8}$$

The response R depends on two "parameters" of the stimulus: the speed (proportional to $1/\Delta t$) and the structure of the moving pattern, which determines the autocorrelation function Φ_{GG}. The special character of the response also depends on the weighting functions of the filters D, F, and H which reflect functional properties of the optomotor response system.

2. Filter Transformations

The weighting functions which describe the properties of the filters D, F and H can be determined explicitly by analyzing relations between speed of pattern movement and optomotor responses. An appropriate experiment consists in the rotation of a drum carrying a sinusoidally changing contrast pattern. Under these conditions the nature of the relation between drum speed and response is such that the strength of response rises from zero to a maximum as speed increases and then falls again to zero at high speed. A plot of response against the log of speed yields a symmetrical curve. From the parameters of this curve the properties of filters D, F and H, expressed by their weighting functions can be determined by use of equation (8) (Hassenstein, 1959; Reichardt and Varjú, 1959; Varjú, 1959).

Taking the experimental data—especially from Chlorophanus—and solving equation (8) by Laplace-Transform technique one can show that the weighting functions of the filters F and H reflect first order ordinary differential equations describing the input-output relation of these filters whereas the weighting function of the D filters leads to a partial differential equation such as that describing one dimensional diffusion. The following time constants of the filters were derived for Chlorophanus:

$$\tau_F = 1.6\ \text{sec};\ \tau_H = 0.03\ \text{sec};\ \tau_D = \leq 10^{-4}\ \text{sec}. \tag{9}$$

The time constants of the filters describing the responses of Musca and Drosophila are of the same order of magnitude. A rigorous treatment of the determination of filter transformation can not be given here.

So far the experimental results have been used to infer the minimal properties of a quantitative mathematical model relating stimulus input to optomotor output. The model itself will now be tested for its predictive power.

D. TESTING THE MODEL FOR ITS PREDICTIVE POWER

1. Periodic Stimuli

If the fluctuating light intensity function $G(t)$ is determined for any arbitrary stimulus pattern, the optomotor response to this pattern moving at various speeds may be predicted from equation (8). Experiments can be performed to test these predictions and hence the validity of the model. An even more rigorous test of the model can be obtained by decomposing the one-dimensional light pattern and consequently the time function $G(t)$ into their Fourier components. Since the filters D, F and H are linear and hence the superposition principle holds, these Fourier components can interact with each other only in the correlators and not in the filters. However, the outputs of the correlators are *not effected* by *phase shifts* of sinusoidal functions in their inputs. Therefore all stimulus patterns differing from each other only in the phase relations of their Fourier components should produce the same optomotor responses over the whole range of effective pattern velocities (Reichardt and Varjú, 1959; Varjú, 1959).

This property of the correlation process can be derived by putting

$$G(t) = A \cos \left(2\pi \frac{t}{T} + \Phi\right) \tag{10}$$

where A designates the amplitude, T the period and Φ the phase angle of a periodic function. By application of equation (7) the autocorrelation function of $G(t)$ is then given by the expression

$$\Phi_{GG} = \frac{A^2}{2} \cos \left(2\pi \frac{\Delta t}{T}\right). \tag{11}$$

For the general case of an arbitrary periodic function

$$G(t) = \frac{a_0}{2} + \sum_{n=1}^{\infty} a_n \cos \left(n\, 2\pi \frac{t}{T} + \Phi_n\right) \tag{12}$$

where the a_n are the amplitudes, $\frac{n}{T}$ the frequencies and Φ_n the phases of the individual periodic components, the autocorrelation function becomes

$$\Phi_{GG} = \frac{a_0^2}{4} + \frac{1}{2} \sum_{n=1}^{+\infty} a_n^2 \cos n\, 2\pi \frac{\Delta t}{T}. \tag{13}$$

From these results one obtains the following general properties of the autocorrelation function of a periodic function:

(a) The autocorrelation function is periodic with the period of the given function.

(b) The autocorrelation function is a cosine series *dropping all phase angles* in the harmonics of the original function.

To test this predicted *invariance* property of the optomotor response, two superimposible stimulus patterns were chosen that consist of horizontal alternating black and white stripes with periods of 22.5° and 90° (Figs. 5a, 5b). These were put together in two ways: (1) contour edges of the 90° pattern were placed in the middle of two contour edges of the 22.5° pattern (Fig. 5c); or (2) contour edges were directly superimposed without any phase shift (Fig. 5d). These two patterns (Figs. 5c and 5d) therefore differ by 90° in the phase of the shorter period pattern. Experimental tests of the optomotor response to these two apparently very different light and dark patterns show that such phase difference has no effect on the response (Fig. 6). It follows therefore, that the optomotor response is invariant to

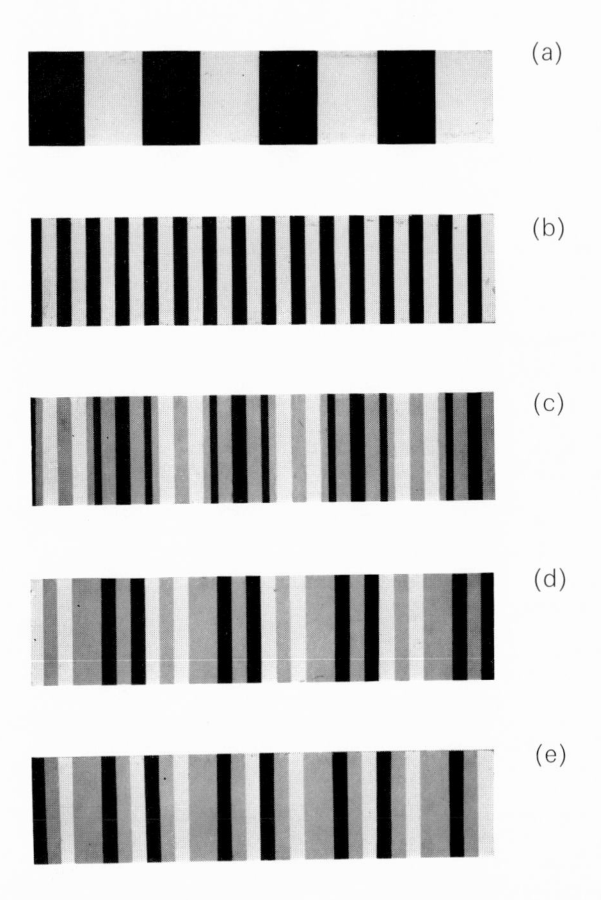

FIGURE 5. Periodic cylinder patterns to test effect of the phase of Fourier components on the optomotor response. Patterns *a* (90° period) and *b* (22.5° period) were superimposed so that vertical contours of *a* fall halfway between the vertical contours of *b* (Figure 5c), or so that the vertical contours of *a* and *b* coincide (Figure 5d). Figure 5e was obtained from *d* by interchanging two black and white stripes per period. Note that in superimposing black + black = black, white + white = white, and black + white = grey.

phase shifts of Fourier components in the visual surroundings of the insect.

The validity of this conclusion has been tested by interchanging a pair of black and white stripes in each period of Fig. 5d to obtain the pattern of Fig. 5e. This slight alteration of pattern markedly affects the response, because the interchange of stripes changes not only the phase relationships but also the amplitude distribution of the Fourier components. These responses to movement of patterns in Figs. 5c, d, and e agree well with the curves predicted from the mathematical model. It is concluded therefore that the model in Fig. 4 effectively describes movement perception through the insect's eye, at least in regard to its optomotor response to periodic light stimulus.

2. Random Stimuli

The stimuli described so far, being periodic, are quite different from the more random distributions of light points normally seen by insects in nature. Hence it may be asked whether optomotor responses require patterns of an ordered figural quality or whether such responses are evoked also by completely random patterns, or visual noise, containing contrast but having no order or figurative character (Reichardt, 1957; Reichardt and Varjú, 1959).

The experimental answer to this question is that an optomotor response *is* elicited by moving random patterns.

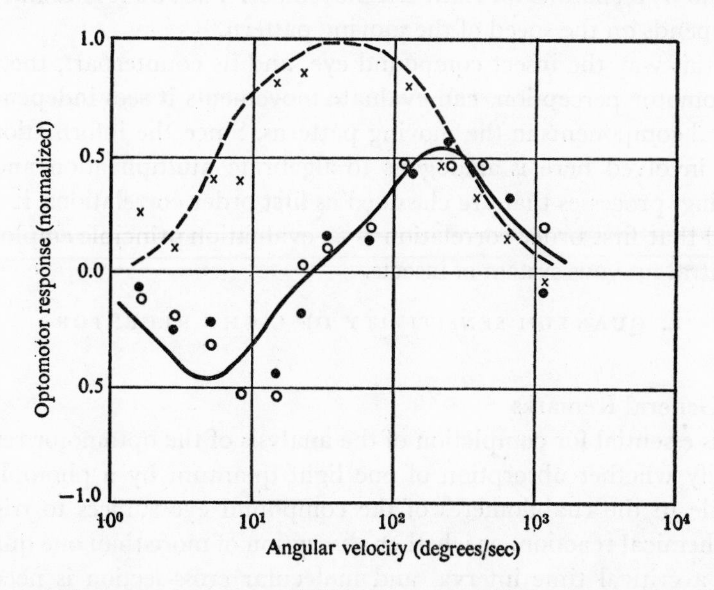

FIGURE 6. Optomotor responses of Chlorophanus to different angular velocities of the patterns in Figure 5c (open circles), Figure 5d (filled circles), and Figure 5e (x's). The two curves were predictions derived from the minimal model for optomotor perception.

The following qualitative considerations show that this is predicted also by the mathematical model. In these considerations the first crucial question obviously is whether or not the random signals received by receptors A and B are still random when they reach the inputs of the correlators. If the effects evoked in channel A by a particular light point of the stimulus pattern moving from A to B have been completely dissipated in filters D, F, and H by the time the same stimulus point reaches receptor B, then signals triggered at the inputs of the correlators by the four possible sequences dark-light, light-dark, dark-dark, and light-light are all equally probable and therefore elicit no response. But if the effects of the signal in channel A last longer than the time it takes a stimulus point to move from A to B, then the signals triggered by sequences light-light and dark-dark will constitute more than 50% of all signals entering the correlator M_A. However signals entering correlator M_B will still represent in equal frequency all four possible sequences since, as was shown in the preceding section, the time constant of the H filter is much smaller than that of the F filter. Since the strength of the signal effect in F, which determines the predominance of light-light and dark-dark coincidence at the correlator inputs, decreases with time, the quantitative excess of coincidence signals in the A channel is a function of the relative speed of the moving random pattern. These differences in correlator input in the two channels give rise in their correlator outputs to a net outflow of positive signals, in A channel for left-right movement and in B channel for right-left movement. The strength of this signal flow depends on the speed of the moving pattern.

In this way the insect compound eye, and its counterpart, the model for optomotor perception, can evaluate movements it sees independently of figural components in the moving patterns. Since the information processing involved here is analogous to algebraic multiplication and time averaging, processes that are classified as first order correlations, it can be asserted that first order correlation is an evaluation principle employed in the central nervous system of insects.

E. QUANTUM SENSITIVITY OF LIGHT RECEPTORS

1. General Remarks

It is essential for completion of the analysis of the optomotor response to clarify whether absorption of one light quantum by a photopigment molecule in the rhabdomeres of the compound eye suffices to trigger a photochemical reaction, or whether absorption of more than one quantum within a critical time interval and molecular cross-section is necessary. Furthermore, one would like to know especially whether a temporal and/or spatial coincidence of physiological events elicited by the photochemical reactions takes place *before* perception of motion is carried out by the central nervous system (Reichardt, 1965).

2. Theoretical Considerations

The following theoretical argument can be put forward: We consider an area F packed with N photopigment molecules, each molecule contributing to the area with a molecular cross-section q. It is assumed that in complete darkness all pigment molecules remain in their inactivated state. If a light beam j of n quanta per unit time t impinges on the area F, a pigment molecule may absorb one quantum. Let us first consider the case where the absorption of the quantum converts the inactivated molecule into an activated molecule and where the photochemically activated molecule sets in train sequential reactions which may culminate in a physiological event. In this case we may speak of a "one-quantum receptor," since absorption of one quantum sets off a physiological event with finite probability. We may next consider the case where absorption of a first quantum converts the pigment molecule from the inactivated state to a preactivated state and where absorption of a second quantum, during the life time τ of the preactivated state, is necessary for the conversion to the activated state. In this case, we may speak of a "two-quantum receptor," for here absorption of two quanta per molecular cross-section q and life time τ are required to elicit a physiological event with finite probability. This two-quantum reaction may be formulated as follows:

$$A_{inactiv.} \underset{\tau}{\overset{h\nu}{\rightleftharpoons}} A'_{preactiv.} \overset{h\nu}{\rightleftharpoons} A''_{activ.} \dashrightarrow B_{sequential\ prod.}$$

Based on these considerations, in the general case for the "z-quantum receptor" one finds that the average rate r of photochemical reactions per flux j is given by the expression

$$\frac{r}{j} = \frac{N \cdot q}{F}\left(j \cdot \frac{q \cdot \tau}{F}\right)^{z-1}. \qquad (6)$$

Two types of reactions may be considered:

(1) One-quantum reaction ($z=1$). Here we find $\frac{r}{j}=\frac{N \cdot q}{F}$. The quantum yield does not depend on the flux j. The molecular cross-section is a constant.

(2) Two-quantum reaction ($z=2$). Here we have $\frac{r}{j}=\frac{N \cdot Q}{F}$; where $Q= j \cdot \frac{q^2 \cdot \tau}{F}$. The quantum yield depends linearly on j, which amounts to saying that the effective molecular cross-section Q changes with the quantum flux j.

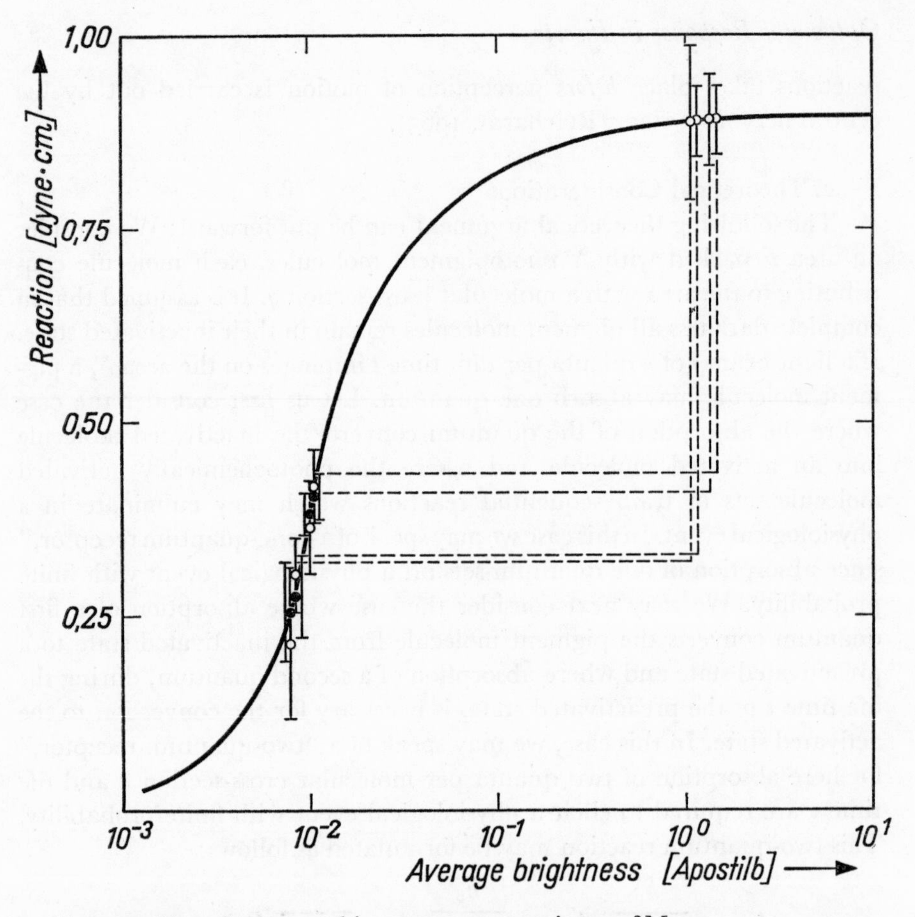

FIGURE 7. Test of one- or multi-quantum response in eye of Musca.

A) *Solid curve*: Optomotor reaction versus average brightness of patterned cylinder. Illumination with light program type 1 (constant intensity in time) provided by oscilloscope tube. Spatial wavelength of patterned cylinder $\lambda = 45°$; pattern contrast $m = 21\%$; cylinder speed $w = 49.25°$/sec. Response curve represents the averages obtained from five individual flies. These averages and standard errors are *not* shown in the figure. Average standard error amounts to ± 0.15 dyne cm.

B) *Full circles* and *open circles*: Full circles represent averages obtained from four individual flies with light program type 1. The standard error $\pm\sigma$ of each average, indicated by vertical lines, is calculated from five measurements taken from one individual fly. The open circles near the abscissa value 10^{-2} Apostilb represent response averages obtained from the same flies under light program type 2. Pairs of full and open circles located at the *same* abscissa value were obtained from one and the same individual fly. Light program type 2 consisted of pulsed light with the same average quantum flux as in program of type 1. Pulse frequency 500 c./sec; pulse length $1.5 \cdot {}^{-5}10$ sec; peak intensity of pulse to average intensity $I_2/I_1 = 133$. Open circles near abscissa value 10^0 Apostilb demonstrate optomotor response to be expected under light program type 2 for a two-quantum receptor if condition $\tau \leq 1.5 \cdot 10^{-5}$ sec. is fulfilled. The abscissa values of these open circles are derived from the abscissa values of open circles actually measured and multiplied by the factor $I_2/I_1 = 133$. This factor determines the increase in effectiveness during program type 2, since the effective molecular cross-section of the two quantum receptor changes by this factor. Horizontal dashed lines indicate these shifts of open circles, whereas vertical dashed lines give significant limits for one versus two quantum receptors. Experimental results plotted were obtained from oscilloscope illumination. Spatial wavelength of patterned cylinder $\lambda = 45°$; pattern contrast $m = 21\%$; cylinder speed $w = 49.25°$/sec.

Various programs of light stimulation are conceivable to test whether the response depends on a one- or on a multi-quantum reaction. In the experiments to be described now, the following two light programs were selected. Program type 1: the program consists of a light stimulus whose intensity is constant in time and equal to I_1. Program type 2: the program consists of a sequence of short light pulses of width Δt, peak amplitude I_2, pulse frequency $1/T$ and average intensity I. The average number of quanta applied per time unit evidently is the same in both programs. Programs type 1 and 2 applied to a one-quantum receptor should result in the same average rate of photochemical reactions. If programs are applied to a "multi-quantum receptor," program type 2 *may* trigger a greater photochemical reaction rate than program type 1. The maximum factor by which the reaction rates may differ is easily derived for the two-quantum case. Since I_2 is greater than I_1 by a factor $T/\Delta t$, the effective molecular cross-section Q is enlarged by this factor. Consequently under program type 2, the two-quantum receptor is $T/\Delta t$ times more efficient than under program type 1 and conversely the rate of photochemical reactions is enlarged by the same factor. Whether the effect predicted for the two-quantum receptor is actually found depends on the magnitudes of τ, Δt and T. One expects to find the predicted effect if $\tau \leq \Delta t$; if, however, $\Delta t < \tau < T$, the individual pulse becomes integrated over the life time τ of the preactivated state and the expected increase of the rate of photochemical reaction must be smaller than $T/\Delta t$. The increase factor declines to unity when $\tau \geq T$, since in this case two or more pulses are integrated in time, and the expected effect disappears. Under these conditions, multi-quantum receptors behave like a one-quantum receptor and do not respond differently to the two different light programs.

For any experimental test of the quantum efficiency, it is of greatest importance that one operates near the visual threshold. Only under these conditions is the frequency of quanta incident per receptor sufficiently low that the rates of photochemical reaction elicited do not produce significant changes in the concentration of inactivated receptor molecules. The kinetics of the photochemical reaction should then be determined entirely by the quantum flux and not by other processes.

The theoretical argument developed here is based on photochemical considerations but is valid also for spatial or temporal nervous coincidence processes, if the terminology used here is adopted to that domain.

3. EXPERIMENTAL RESULTS

A typical result of the critical test for deciding between one- and multi-quantum reactions is plotted in Fig. 7. In order to compare optomotor responses generated by the two different light programs, the average brightness at the surface of the rotating patterned cylinder has to be set to a very low quantum flux level. This condition is met if one operates near the threshold of reactions elicited by moving patterns of *high* contrast. In order to determine the associated average brightness range at the surface of the patterned cylinder, the strength of the optomotor response at various levels of brightness was measured with a group of dark adapted flies. All other parameters of the experiment (spatial wavelength of periodic pattern; pattern contrast; angular velocity of patterned cylinder) were kept unchanged. These measurements were carried out under constant illumination provided by the oscilloscope tube shown in Fig. 1. The result is given by the solid curve in Fig. 7, which represents averages obtained from various individual flies. The averages and the standard errors of these measurements are *not* shown in the figure.

The response curve (solid line) plotted in Fig. 7 shows that the response depends most critically on changes in brightness in the neighborhood of 10^{-2} Apostilb. This range was therefore selected for application of light programs of type 1 and type 2 in order to test whether the optomotor response reveals a requirement for absorption of two or more quanta. The average responses to program type 1 are represented by full circles and the average responses to program type 2 by open circles. Every pair of full and open circles, located at the *same* brightness level, are based on measurements made with one and the same individual fly. In Fig. 7 the light program type 2 consisted of pulses of length $1.5 \cdot 10^{-5}$ sec., a pulse frequency of 500c./sec. and an I_2 to I_1 ratio of 133. The near coincidence of the pairs of full and open circles at the same brightness levels near 10^{-2} Apostilb indicates that the reaction to the two different light programs do not differ significantly from each other. Hence in the perception of the special light program of type 2 applied here, the rhabdomers *behave* like one-quantum receptors.

One can easily estimate the strength of response which would be expected in the two-quantum receptor case if the coincidence time τ were of the same order of magnitudes as, or smaller than, the length of the individual pulses ($\tau \leq 1.5 \cdot 10^{-5}$ sec.). Under these circumstances the effective molecular cross-section of the photopigment system would have been increased by the factor $I_2/I_1 = T/\Delta t = 133$. Consequently the expected response can be estimated by shifting the open circles representing actually measured responses to light program of type 2 along the abscissa by the factor 133 and raising them to the response values on the solid line corresponding to the response at that average brightness under constant illumination. These shifts are indicated by the horizontal and vertical dashed lines. The vertical dashed lines then represent the differences in response strength between the one- and the two-quantum receptor case under type 2 programs.

The results presented in Fig. 7 tell us that the receptors of the fly Musca *behave* like one-quantum receptors. In reality they may be of the two-quanta type, of course, since no difference would have been detectable between light programs of type 1 and 2 employed here, if the coincidence time τ were equal to or greater than 1/500 of a second, the time interval between successive light pulses. Further experiments were carried out, therefore, in which the pulse frequency in light programs of type 2 was reduced to 200, 100, and 20 c./sec. Even at 20 c./sec. no significant differences between the responses to constant and pulsed light were found.

The method described in this paper is applicable down to light pulse frequencies of 20 c./sec. Below 20 c./sec., rotating stroboscopic patterns begin to influence the optomotor response. This influence can neither be compensated by lowering the velocity w of the rotating cylinder nor by changing the spatial wavelength of the pattern, since the ratio w/λ determines the strength of the optomotor response (other parameters such as brightness and contrast being kept constant). Therefore a new technique was recently developed which makes use of two types of quanta: one type determining the level of receptor adaptation and the other carrying motion information. The new technique enables one to apply frequencies of light pulses down to 1 c./sec. Even at this low pulse rate, however, no differences in response to the two light programs are observed.

The results of these experiments lead to the conclusion that coincidences of two or more quanta are not registered by the fly's eye—at least not for coincidence time intervals as long as 1 sec. The observations suggest that absorption of a single quantum elicits with finite probability a primary physiological event, which, in turn, with finite probability triggers a sequence of secondary events. Considering the low rate at which quanta impinge on the photoreceptors (at an intensity of 10^{-2} Apostilb, this rate amounts to less than 60 quanta per sec. per rhabdomere) the statement implies that in the optomotor response system of the fly Musca there are no temporal and/or spatial coincidences *before* perception of motion is carried out by the central nervous system.

REFERENCES

Fermi, G., and W. Reichardt. 1963. Optomotorische Reaktionen der Fliege *Musca domestica.* Kybernetik, 2: 15.

Götz, K. G. 1964. Optomotorische Undersuchung des visuellen Systems einiger Augenmutanten der Fruchtfliege Drosophila. Kybernetik, 2: 77.

———1965. Die optischen Übertragungseigenschaften der Komplexaugen von Drosophila. Kybernetik, 2: 215.

HASSENSTEIN, B. 1951. Ommatidienraster und afferente Bewegungsintegration. Z. vergl. Physiol., *33*: 301.

———1958a. Über die Wahrnehmung von Figuren und unregelmässigen Helligkeitsmustern. Z. vergl. Physiol., *40*: 556.

———1958b. Die Stärke von optokinetischen Reaktionen auf verschiedene Mustergeschwindigkeiten. Z. Naturforschg., *13b*: 1.

———1959. Optokinetische Wirksamkeit bewegter periodischer Muster. Z. Naturforschg., *14b*: 659.

HASSENSTEIN, B., and W. REICHARDT. 1956. Systemtheoretische Analyse der Zeit-, Reihenfolgen- und Vorzeichenauswertung bei der Bewegungsperzeption des Rüsselkäfers Chlorophanus. Z. Naturforschg., *11b*: 513.

KUNZE, P. 1961. Untersuchung des Bewegungssehens fixiert fliegender Bienen. Z. vergl. Physiol., *44*: 656.

MITTELSTAEDT, H. 1951. Zur Analyse physiologischer Regelungssysteme. Verh. dtsch. Zool. Ges. in Wilhelmshaven, *8*: 150.

REICHARDT, W. 1957. Autokorrelations-Auswertung als Funktionsprinzip des Zentralnervensystems. Z. Naturforschg., *12b*: 448.

———1961a. Nervous integration in the facet eye. Biophys. J., *2* (part 2): 121.

———1961b. Autocorrelation, a principle for the evaluation of sensory information by the central nervous system. *In* W. A. Rosenblith [Ed.] Sensory Communication. John Wiley, New York.

———1962. Nervöse Verarbeitung optischer Nachrichten im Facettenauge. Jb. der Max-Planck-Gesellschaft, 97.

———1965. Nervous processing of sensory information. *In* T. H. Waterman and H. J. Morowitz [Ed.] Theoretical and mathematical biology. Blaisdell Publishing Co., New York.

———1965. Quantum sensitivity of light receptors in the compound eye of the fly Musca. Cold Spring Harbor Symp. Quant. Biol., *30*: 505–515.

REICHARDT, W., and D. VARJÚ. 1959. Übertragungseigenschaften im Auswertesystem für das Bewegungssehen. Z. Naturforschg., *14b*: 674.

VARJÚ, D. 1959. Optomotorische Reaktionen auf die Bewegung periodischer Helligkeitsmuster. Z. Naturforschg., *14b*: 724.

GEORGE STREISINGER
Institute of Molecular Biology, University of Oregon, Eugene, Oregon

Terminal Redundancy, or All's Well That Ends Well

Having just gotten a degree for phage work with Salvadore Luria at the University of Illinois, I came to Max Delbrück's lab at Caltech in 1953 in order to develop a plaque assay technique for plant viruses. I no longer remember why I wanted to do this, but the choice of Caltech was obvious: everybody who was anybody in phage work spent at least *some* time at Caltech.

The work with plant viruses and plant tissue cultures was very fruitful at first. Harry Rubin and I shared a lab; Rubin worked on Rous-sarcoma virus and the by-products of his experiments were chickens (minus a wing, or leg perhaps, but still very good to eat), while I produced a vast excess of coconuts whose milk seemed to be essential for growing plant tissue cultures. Harry Rubin ate part of my coconuts, my family ate Rubin's partial chickens; an idyllic relationship which lasted for several months. After that, alas, my family got tired of eating chicken, I got tired of plant viruses not forming plaques and, besides, there seemed to be so many appealing experiments to be done with phage.

One of these phage experiments was a result of a series of six lectures that I delivered in Norman Horowitz's advanced class in genetics. Doermann, during the previous few years, had performed extensive mapping experiments with phage T4, as had A. D. Hershey with T2. It was clear from all of these experiments that there were three linkage groups in the phages T2 and T4. I had described these results to the class and, as a take-home exam question, asked them to outline an experiment that could test whether the three phage chromosomes remained associated with one another during recombination (in a meiosis-like process) or whether the chromosomes were dispersed in a pool, to be assembled randomly, during maturation of the progeny phage.

Much to my surprise, I was unable to answer the question myself before the class met again and, to my even greater surprise, none of the students arrived at a solution either. I settled down to work on this in earnest and, since I kept making mistakes in using the Visconti-Delbrück mating-theory equations, I got Victor Bruce (Delbrück's brother-in-law and an expert at handling the mathematics of that theory) to help. During the next week Victor Bruce and I thought of a number of four- and five-factor crosses, each more ingenious and complicated than the next, none however adequate to distinguish between the models. It finally occurred to us that the solution was exceedingly simple: we needed only to perform a cross using two linked markers on one chromosome and a third marker on another chromosome and to use a high multiplicity of infection of one parent and a low multiplicity of the other one. If we lysed the infected bacteria at very early stages of phage growth, and selected progeny that were recombinant for the linked markers, two possibilities existed for the assortment of the unlinked marker: (a) If meiosis occurred, the unlinked marker should assort randomly at each mating event and thus, at a very early time in phage development, nearly half of the selected recombinant progeny should carry the allele contributed by the minority parent. (b) If the chromosomes were randomly assorted in a pool, then the allele contributed by the minority parent should occur only rarely (with a frequency about equal to the ratio of the minority to the majority parent) among the selected progeny. We obtained the result expected for meiosis, and here matters would have ended, had we not performed several sets of crosses in parallel, in which the minority parent carried various genetic markers. To our great surprise the experiments showed that all markers, even those that had previously been assigned to different linkage groups, were linked on one chromosome (Streisinger and Bruce, 1960). At this point, Max Delbrück asked me to give a seminar. After the seminar, he took me by the arm to his office to tell me in confidence that it was the worst seminar that he had ever heard. It was not till years later, after I had met dozens of people each of whom had been told that *his* had been the worst seminar Delbrück had ever heard, that it dawned on me that Delbrück told this to (almost) everyone.

At about this time, ultracentrifuge analyses of the molecular weight of T4 DNA suggested that a particle of phage T4 contains about ten molecules of DNA, in apparent contradiction to the notion of a single genetic linkage group. (No one had yet realized that introducing DNA into an ultracentrifuge cell very efficiently sheared the DNA molecule.) Jean Weigle and I discussed various genetic experiments that might have tested for the presence of protein linkers welding the ten DNA molecules into a single chromosome, but luckily we decided that these experiments would be messy at best, and I was distracted by other experiments for several years.

I later moved to Cold Spring Harbor, and had been there for a while when in the spring of 1959 Charles A. Thomas came to give a seminar in which he described new experiments concerning the molecular weight of phage T4 DNA. Thomas then thought that there were two DNA molecules (not ten) in each phage particle and that there were no interruptions within the single polynucleotide strands of each double-stranded molecule of DNA. This did not make a very big impression on me until the beginning of the summer when Robert Edgar arrived at Cold Spring Harbor. Among the latest gossip Edgar brought us was the news that Doermann had found polar segregation from multiply-marked heterozygous phage particles and that this finding suggested that regions of heterozygosity involved interruptions within chromosomes. Having been a herpetologist during my high school years, I immediately thought of two snakes, each having hold of the tail of the other, as a model for phage heterozygotes. The snake-eat-snake model would account for two molecules of DNA per phage particle, as well as for interruptions of the DNA chain at regions of heterozygosity and for the lack of strand interruptions within the DNA molecules observed by Thomas. I thought it would be fun to do an experiment with Edgar during the summer and we discussed experiments that would test whether the genetic map was circular, as predicted by this model. We had a number of other ideas for other experiments that we could collaborate on, all of them pretty far-fetched and silly, but none sillier, it seemed to us, than the test for a circular map. But the summer was a busy one, and we did not get very far with any of our experiments.

Frank Stahl came to Cold Spring Harbor for the phage meetings late that summer and was very enthusiastic about the snake-eat-snake model; he almost persuaded me to take it seriously, and reminded me that some of his much earlier P^{32}-decay marker-inactivation experiments could best be explained if the T4 genetic map were assumed to be circular. He urged me to start to do experiments concerning this when I came to Oregon that fall (Frank Stahl and Aaron Novick had painted such a glowing picture of Oregon, the University, and the Institute of Molecular Biology there, that I had accepted a position at Oregon). The initial experiments that were needed were really quite simple: we had only to test for linkage between markers located at the two ends of the genetic map using the sensitive linkage tests developed with Victor Bruce earlier.

Two things persuaded me to start these experiments: First, I had brought a new assistant, Georgetta Harrar (now Georgetta Denhardt) to Eugene with me and this experiment would serve to train her in phage genetics so that she could then map phage lysozyme mutants, which was what I really wanted to do. Secondly, Frank Stahl pointed out that old, previously-published linkage experiments of Doermann suggested linkage between markers located at the two ends of the chromosome even though the statistical tests that Doermann employed led him to ignore this linkage.

It took a while to get a lab set up in Oregon and, besides, Mary Stahl convinced me that it was crucial to work on a congressional race in our district in Oregon that fall, so that we did not get to do any experiments for the first month or so. The very first experiments that Georgetta Harrar did showed quite clearly that the genetic map of T4 was circular. We were quite shocked by this result; I phoned Edgar and both he and I performed a number of new crosses and repeated some old ones to persuade us that the map really was circular.

Sadly we realized that this was just the beginning: Frank Stahl pointed out that genetic circularity was compatible with any of a number of different models other than the snake-eat-snake one. Around this time the results of Hershey, of Thomas, and of Davison and Levinthal showed that all the DNA of T4 was in a single molecule, and instead of the snake-eat-snake model we began to think of a hoop snake (*Farancia abacura abacura*) chewing on its own tail. Presently Matthew Meselson came to visit us at Eugene and we discussed circularity with him at length; a few days after he left, Meselson wrote us a post card implying that hoop snakes really do not roll downhill holding their tails in their mouths, and suggesting that a linear chromosome with a region of terminal redundancy of genetic markers was in every genetic respect equivalent to the hoop snake model; this had occurred to Frank Stahl and to me as well.

Nomura and Benzer had published results which suggested that two kinds of heterozygotes existed in phage T4 and we concluded that one of the classes of heterozygotes suggested by them was due to the region of terminal redundancy, the other to internal heterozygosity. It became clear to me almost immediately that the circularization of the genetic map of an infecting chromosome would come about through the replication of that chromosome and by recombination involving the region of terminal redundancy at the ends of the chromosome. This process would tend to lead to the formation of polymers of the vegetative genome, and the length of the DNA molecule of a mature phage T4 particle would thus have to be established by factors extrinsic to the chromosome itself, presumably the process of phage maturation. It seemed to me reasonable that during the assembly of phage-head-protein subunits a "headful" of DNA would get incorporated into a phage particle.

Janine Séchaud was working with me at that time on a very different problem, namely the synthesis of late proteins by replicas of the parental phage DNA in the intracellular pool of an infected bacterium and, one afternoon, we discussed the possibility of using fluorodeoxyuridine (FUDR), an inhibitor of DNA synthesis, for these experiments. It suddenly occurred to us that the use of FUDR could lead to a test of the terminal redundancy model: we realized that, while phage DNA synthesis is inhibited by FUDR, internal heterozygotes would be formed by recombination but would not be lost by replication. The frequency of internal heterozygotes should therefore increase as a function of the time that infected bacteria were incubated in the presence of FUDR. In contrast, the frequency of terminal redundancy heterozygotes, if these in fact existed, would not be expected to be changed in the presence of FUDR. We thought, on the basis of steric considerations, that deletion mutants would contribute to terminal redundancy heterozygotes but not to internal heterozygotes, whereas point mutants could contribute to both types of heterozygotes. We performed an experiment the next afternoon and found that the frequency of point mutant heterozygotes increased in the presence of FUDR, whereas the frequency of deletion mutant heterozygotes stayed constant. These results provided us with the first clear indication of the existence of terminal redundancy heterozygotes.

Some months later, we discussed deletion mutations in the b_2 region of the lambda phage chromosome with Maurice Fox. Both of us agreed that the molecular weight of the DNA of such mutant phage should be lower than that of normal lambda DNA. In the course of this discussion, it occurred to us that if the terminal redundancy model for the chromosome of T4 were correct, deletion mutation in T4 should increase the length of the region of terminal redundancy. It took several months to complete the first of these experiments: we found that strains of T4 carrying deletion mutations did in fact have longer regions of terminal redundancy.

All these results finally convinced me that the terminal redundancy model and its implications were correct. I had very mixed feelings about the terminal redundancy model; it seemed to me from the outset to be eminently reasonable and consistent, and yet it seemed so bizarre that the favorable outcome of the experiments that were performed to test the model came as a series of surprises.

A number of our own and Stahl's experiments concerning circularity and heterozygosis have already been published (Streisinger et al, 1964; Séchaud et al, 1965; Shalitin and Stahl, 1965; and Stahl et al, 1965); other experiments, all performed a few years ago, will soon (I hope) appear in print. I am amazed that the experimental results of my friends appear in print so soon after the experiments are performed. I find that it takes me a few months to believe that a result I have obtained is true, and several more months to believe that the result is interesting, though this second stage, unfortunately, is never reached for most of the experiments. By this time it is all too often clear that the experiments could have been performed more cleanly and elegantly—but is repeating them really worth while? About a year later I try to write the paper but it is so bad that I dare not look at it again for quite a long time. After I have finally rewritten the manuscript a few times I apologetically show it to some friends (nowadays usually Frank Stahl and Aaron Novick). Sometimes these friends take pity and rewrite the paper (Stahl claims that he wrote my 1960 paper with Bruce). Usually the friends tell me that the paper is really pretty dismal, or that the introduction and discussion miss the point entirely, and so I put the manuscript away for a while again . . .

Trying to delve into my subconscious I wonder whether these reasons for the sometimes less-than-prompt publication of my results are really the relevant ones. The real reason may be as follows: Max Delbrück often kidnaps people and locks them up in a room at the Caltech Marine Biological laboratory at Corona Del Mar; he gives them paper and a typewriter, and does not let them see the light of day till they have slipped a finished manuscript through the door. He has never done this with me . . . does he think any of my papers are worth writing? . . . perhaps not . . . in that case why write them?

Some readers, though not Max Delbrück, may wonder why this paper is the last one in the volume, rather than occupying a more reasonable place. The answer is simple: I had promised that my manuscript would be ready by September 1965. I did get some pages written then but there were lots of other things to do, and what I did write was hopelessly bad. After the papers of all the other contributors have been submitted and the volume has been set in type, I have finally managed to finish writing a draft. I submit it only because otherwise it *really* could never get into the volume, and also because I have been told by some of my co-contributors that Stent completely rewrites the manuscripts in any case; perhaps he will do the same with mine, if there is still time.

I have delayed also because I can still think of no way to describe the enormous influence Max Delbrück has had on me and on many others, and the very great love and admiration that so many of us feel toward him.

I find that I have neglected to mention Max Delbrück's reaction to our results concerning genetic circularity in T4 when I last discussed our experiments and models with him, several years ago: "I don't believe a word of it," he said.

REFERENCES

Séchaud, J., G. Streisinger, J. Emrich, J. Newton, H. Lanford, H. Reinhold, and M. M. Stahl. 1965. Chromosome structure in phage T4, II. Terminal redundancy and heterozygosis. Proc. Natl. Acad. Sci., *54*: 1333–1339.

Shalitin, C., and F. W. Stahl. 1965. Additional evidence for two kinds of heterozygotes in phage T4. Proc. Natl. Acad. Sci., *54*: 1340–1341.

Stahl, F. W., H. Modersohn, B. E. Terzaghi, and J. M. Crasemann. 1965. The genetic structure of complementation heterozygotes. Proc. Natl. Acad. Sci., *54*: 1342–1345.

Streisinger, G. and V. Bruce. 1960. Linkage of genetic markers in phages T2 and T4. Genetics, *45*: 1289–1296.

Streisinger, G., R. S. Edgar, and G. Harrar-Denhardt. 1964. Chromosome structure in phage T4, I. Circularity of the linkage map. Proc. Natl. Acad. Sci., *51*: 775–779.